ROCKET PROPULSION ESTABLISHMENT LIBRARY

Please return this publication, or request a renewal, by the date stamped below.

THE SPECTROSCOPY OF FLAMES

The Spectroscopy of Flames

A. G. GAYDON
D.Sc., F.R.S.

Warren Research Fellow
of the Royal Society
and
Emeritus Professor of Molecular Spectroscopy
Imperial College, London

SECOND EDITION

LONDON
CHAPMAN AND HALL

First published 1957
by Chapman and Hall Ltd
11 New Fetter Lane, London EC4 P 4EE
Second edition 1974
© 1957, 1974 A. G. Gaydon

Typeset by Preface Ltd, Salisbury, Wilts
and printed in Great Britain by
T & A Constable Ltd,
Edinburgh

ISBN 0 412 12870 5

Distributed in the U.S.A.
by Halsted Press, a Division
of John Wiley & Sons, Inc., New York

Preface to Second Edition

The spectrum of a candle flame was first mapped by Swan in 1857 and early observations on spectra of metals in flames were made by Bunsen and Kirchhoff around 1860. Since then investigations have continued at an increasing pace and knowledge obtained from these spectroscopic studies is accepted as making an important contribution to the understanding of flame processes. I have myself been working in this field for some 40 years now and have been associated with some of the recent advances in techniques.

My small monograph on *Spectroscopy and Combustion Theory* first appeared in 1942, and the revised edition in 1948. This summarized not only direct observations on the visible and ultra-violet emission spectra of flames, but also covered the infra-red region and absorption spectroscopy. Indirect applications, such as the use of spectroscopic data for determining dissociation equilibria and flame temperatures were also included. The spectroscopic observations were interpreted in relation to chemical processes of combustion, to the initial energy distribution resulting from the reactions, and to the state of equilibrium in the flame gases. This earlier monograph served to summarize work up to about 1947 and, I hope, to stimulate further research.

These research developments, and also the coverage of parts of the field in my other books, *Flames, their Structure, Radiation and Temperature*, jointly with H. G. Wolfhard, and my *Dissociation Energies*, led to a complete revision of the older monograph under the new title *Spectroscopy of Flames* in 1957. That edition covered many exciting new developments — the work on low-pressure flames, burners for flat diffusion and flat premixed flames, flames supported by free atoms, flash photolysis, early investigations in shock tubes, and the use of isotope tracer methods. The 1957 edition did indeed stand the test of time rather better than the older monograph, because the information about flame spectra and flame processes was by then much more definite.

During the 17 years since the previous edition was prepared there have been many refinements in techniques and some important related advances, such as the discovery of the laser and the use of atomic absorption and atomic fluorescence for analytical work. For this thorough revision the whole book has been completely re-worked, but the same plan has been retained so that it appears, deceptively, similar to the previous edition. It is hoped that the new book is now in a sufficiently definitive form to be of continuing value, because although further advances are to be expected and hoped for, the main facts, such as the identity of emitting and absorbing species and the reactions forming them in flames, are now established. Many of the older publications tend to get forgotten, but my long continuity of work in this field has enabled me to make use of old data where still relevant.

The book deals primarily with the use of spectroscopic observations to interpret combustion processes in flames and flame structure, including explosion flames and engines. This is not a book on flame photometry, but the understanding of flame equilibria, and departures therefrom, is relevant to this subject and in the final chapter I have commented on the use of flames for spectrochemical analysis including the recently developed atomic absorption and atomic fluorescence methods.

I have endeavoured to present the subject in such a way that the book may be of service to those engaged in research, but so that it may be read easily by the beginner and the non-specialist. The treatment is not mathematical, and where an understanding of the theory of molecular spectra is necessary I have endeavoured to give this theory in simple terms. I have included a fairly extensive Bibliography and an Appendix, giving tables of wavelengths, and in this edition wavenumbers as well, of all the main band spectra frequently found in flames and combustion systems; these should make the book valuable for reference purposes.

My early work owes much to the initial stimulation of the late Professors Alfred Fowler and Sir Alfred Egerton, and also I have benefited from collaboration, at Imperial College, with my former colleagues Drs H. P. Broida, R. A. Durie, P. F. Jessen, N. P. W. Moore, W. M. Vaidya and H. G. Wolfhard, to whom I express my thanks. I am grateful to my wife and Miss D. Waller for some help with indexing and proof reading. I would also like to acknowledge support over many years from the Warren Research Fund of the Royal Society.

Department of Chemical
Engineering and Chemical Technology,   A.G.G.
Imperial College,
London, S.W.7.
April 1974.

Contents

PLATES

Plates 1 to 8 appear between pages 212—213

Chapter I

Flame Spectra

The Purpose of Studying Flame Spectra

The chief distinction between combustion and other chemical reactions is the occurrence of a visible flame, that is the emission of light. It is therefore natural to expect that investigations of the quality and quantity of light emitted by flames will form an important part of the study of combustion processes. Studies of the quality, that is the spectrum, of the light from simple flames were indeed among the first spectroscopic observations ever made. Each flame has its own characteristic type of spectrum; the blue flame of burning carbon monoxide shows mainly a continuous spectrum from the green to the near ultra-violet with a large number of faint bands superposed on the continuum; flames of hydrogen show very little light in the visible but give strong bands in the near ultra-violet with an outstanding head at 3064 Å; Bunsen-type flames of hydrocarbons have a blue-green inner cone which shows characteristic banded spectra in the green, blue and violet; cool flames of ether etc. are bluish and show a relatively simple group of diffuse bands in the blue and violet. There is also some variation with fuel/air ratio; fuel-rich hydrocarbon flames show the green bands most strongly, while weak mixtures show the violet bands more prominently. Thus, with experience, purely empirical studies of flame spectra can be of use in distinguishing between different flames and different burning conditions; as an example the exhaust flames of aero engines usually show the spectrum characteristic of burning carbon monoxide, but if there is inefficient combustion in the engine we may find the bands characteristic of hydrocarbon flames occurring in the exhaust glow, while in some cases secondary injection of fuel or oil may produce the spectrum associated with cool-flame conditions.

While such empirical observations have their uses, we cannot expect them greatly to advance our knowledge of fundamental combustion processes. The great advances in the knowledge of the structure of molecular spectra, mostly made in the decade between 1920 and 1930, enabled us to assign most of the bands found in flame spectra to definite emitting species, especially to free radicals such as OH, CH, C_2 and HCO whose presence in combustion processes, or even whose existence, had been unsuspected before these advances in molecular spectroscopy. The discovery of these free radicals in flames had a great influence on the development of chain reaction mechanisms and indeed on chemical kinetics generally. One of the main objects of the study of flame spectra, both in emission and in absorption, has therefore been to identify the intermediate chemical species taking part in the reaction. For fundamental combustion studies this search for intermediaries must obviously be continued, but it needs to be accompanied by quantitative measurements to determine the concentrations of the various species and also by observations of the populations in the various energy levels to obtain information about the processes of formation and excitation of the species. Some of the most obvious spectroscopic features like the bands of C_2 and CH, are due to radicals which are still not usually considered to fit into the main combustion mechanisms, although in recent years some of the reactions forming and consuming these radicals have been determined.

There is no "open sesame" to the complete detailed understanding of flame processes. Ordinary chemical sampling, accompanied by modern analytical techniques such as chromatography and mass spectrometry, has assisted in the study of cool flames and pre-flame reactions by detecting aldehydes, peroxides and other intermediates, and has thus filled in part of the story; it cannot, however, give information about short-lived free radicals or about the very rapid reactions occurring in the reaction zone of hot flames. Physical measurements, such as those on burning velocity, on temperature profile through the reaction zone and on limits of flammability and ignition temperature help with understanding the basic processes of flame propagation (see for example Gaydon and Wolfhard, 1970), but cannot fill in the detailed chemical processes. Mass spectrometric studies have made a valuable contribution to knowledge on flame ionization and soot formation processes, but although some free radicals, like CH_3 and HCO can readily be detected there are severe limitations due to interference by the sampling probe and lack of knowledge about the cracking pattern of the free radicals. Optical spectroscopy, too, has its limitations; the stable products of combustion, CO, CO_2 and H_2O do not normally emit or absorb in the visible or near ultraviolet and a few radicals such as HO_2 and CH_3 are not normally observable. It is, however, the best method for detecting many of the simpler free radicals and there is the great advantage that spectroscopic studies do not interfere with the flame processes. Also, in addition to detecting species, there is the special advantage that by comparing the relative strength of the bands and line-structure in the spectrum we can obtain information about the energy distribution among its various possible forms in the newly formed radicals and this in turn may tell us about the chemical processes involved in their formation. There is, however, the limitation that the spectroscopic observations tend to throw undue emphasis on emission from a relatively small number of electronically excited radicals. It is only by a combination of spectroscopy and other evidence that we may hope to unravel the complexities of combustion processes.

The spectroscopic attack is especially valuable for study of the more physical processes of energy release and rate of equipartition of energy following combustion. Energy may take various forms, such as translational (kinetic) energy, internal vibration and rotation of molecules, electronic excitation, ionization and chemical dissociation. The use of spectroscopic methods to measure flame temperature is discussed most fully in *Flames* (Gaydon and Wolfhard, 1970). Quantitative measurements of relative intensities of bands and of lines of the rotational fine structure within individual bands can give information about the initial vibrational and rotational energy distribution in the newly formed molecules and about subsequent processes of energy transfer; such measurements will be fully discussed here.

Molecular spectroscopy has played an important part in interpreting spectroscopic observations of flames. However, studies of flame spectra have made a major contribution in the reverse direction, to the development of knowledge of molecular spectra and structure. Flames were among the earliest and most prolific sources of band spectra. Ever since Swan, in 1857, mapped the green bands in a candle flame — the bands we now know as the Swan system of C_2 — new band systems have frequently been discovered first in flame spectra. Bands of some radicals, such as HCO, ClO, IF, and NH_2 are obtained more readily from flames than from other sources. Thus flame spectroscopy has made and continues to make a useful contribution in the field of pure spectroscopy.

Flame spectra have quite a lot in common with the spectra encountered in astrophysics, and are especially similar to those of comets, which also show C_2, CH, CN, C_3 etc. Studies of flames may help to elucidate some astrophysical problems, such as the cause of the anomalous population of the Λ-doubling components of OH in interstellar space which results in strong 18 cm radio-wave emission and which may (Symonds, 1966) be related to the OH inverse predissociation in flames.

Studies of the composition of hot equilibrium flame gases containing metal or other additives are being used to an increasing extent to derive equilibrium constants, and thus the heats of dissociation of simple compounds. In simple flames, especially of oxygen/hydrogen/nitrogen mixtures with known amounts of additive, it is possible to study the influence of both temperature and composition on the concentrations of species present, often using spectroscopic methods to get these concentrations. Dissociation energies of some oxides, hydrides, hydroxides and halides have been determined in this way. In the non-equilibrium reaction zone there is some recent interest in developing chemical lasers, using chemiluminescent reactions to produce the population inversion necessary for laser action.

Chemical analysis by flame spectrophotometry has become of great importance for estimation of trace elements and for work on small quantities of sample. Developments in atomic absorption and fluorescence spectrometry have increased the importance of these methods. Knowledge about excitation processes and conditions in flames is essential for a full appreciation of the many factors influencing the accuracy and reliability of such measurements and for understanding mutual interference between elements. Although this subject cannot be dealt with at length here, more emphasis is given to it than in previous editions.

Types of Spectra

The three basic types of spectra are line spectra, band spectra and continuous spectra. Lines are emitted, or absorbed, by free atoms and each line corresponds to a transition from one stationary energy state of the electrons to another state, that is to an electronic transition. Banded spectra, or as they were formerly called fluted spectra, in the visible and ultra-violet regions correspond to similar electronic transitions for molecules, but the electronic change is then accompanied by simultaneous changes in the internal vibrational and rotational energy of the molecules; thus each electronic transition gives rise to a number of bands and each band has a fine structure of a large number of lines. The band spectra of diatomic molecules often show obvious regularity in both the gross band structure and the fine line structure of the individual bands, and the bands usually have sharp "heads" or edges which can be measured with precision. Band spectra emitted by polyatomic molecules may occasionally show sharp heads and some regularity, but more often present a much more complex appearance and individual bands are often headless and appear diffuse when examined with small spectrographs. Banded spectra in the infra-red are also due to molecules, but for these there is not usually any electronic transition, the vibrational and rotational energy alone changing. Continuous spectra are often associated with emission or absorption by solid particles or liquid droplets, but may also be due to gas-phase processes such as dissociation of molecules, ionization of atoms, or recombination processes. Chapter IV is devoted to a more detailed treatment of the theory of spectra of various types.

Band spectra are usually the most prominent features in flames, and have attracted the greatest interest. Free radicals like CH, OH, C_2, CN and NH are responsible for the strongest band systems, because for these free radicals the resonance transition from the ground electronic state to the lowest of the excited states requires comparatively low energy so that the bands lie in the visible or near ultra-violet and can be excited strongly at the limited temperature of flames. For the more permanent constituents of flame gases, such as O_2, N_2, CO and H_2O, the main resonance transitions correspond to band systems far down in the ultra-violet and although there is some weak O_2 and CO emission from very hot flames, the energy usually available in flames is inadequate for exciting these spectra. The well-known visible band systems of N_2 and CO, which occur so readily in discharges through these gases at low pressure, all correspond to transition between excited electronic states, and the energy available in flames is insufficient. Some weak band systems of polyatomic radicals such as HCO, NH_2 and C_3 are also emitted by flames. In the infra-red we observe the vibration-rotation spectra of the main products of combustion, H_2O and CO_2. Homonuclear diatomic molecules like O_2 and N_2 have no dipole moment and do not show vibration-rotation spectra.

Line spectra are not usually obtained from pure flame gases. In very hot flames, such as oxy-acetylene, the carbon line at 2478 Å has been observed, but the main resonance lines of C, O, N and H all lie in the extreme ultra-violet and cannot be observed in flames. Lines of metallic impurities, especially the D lines of sodium and the resonance line of Ca at 4226 Å are often observed. If suitable volatile salts are introduced into flames then relatively simple spectra, consisting mainly of the resonance lines, are excited for alkali and alkaline-earth metals. These simple spectra are especially suitable for spectrochemical analysis; in some cases they can also give information about departures from equilibrium in the flame gases.

The yellow "luminous" flames of hydrocarbons burning in air and the ordinary candle flame show strong continuous emission. This is due to the radiation from hot carbon particles. The intensity distribution on this type of continuum is related to the black-body radiation from a source at the flame temperature, but is slightly modified because the particles are usually small, often with diameters less than a wavelength of light, and so their emissivity tends to increase towards shorter wavelengths. Most clear flames also give some continuous emission. In hot flames recombination of ions with electrons, following ionization in the reaction zone, will contribute to this continuous spectrum. Recombination of atoms or free radicals to form molecules may also lead to some continuous emission, these recombination processes often being responsible for continuous spectra which are limited to less-wide regions of the spectrum. The blue colour of carbon monoxide flames is largely due to a continuum resulting from recombination of carbon monoxide and atomic oxygen, $CO + O = CO_2 + h\nu$, while the yellow-green continuum obtained from flames containing oxides of nitrogen is due to recombination of nitric oxide and atomic oxygen. These continuous spectra are usually regarded as rather a nuisance because they tend to mask the more definite banded features. However, we should not forget that they correspond to processes actually occurring in the flame gases, not just to the presence of a species, and their study may be helpful in studying the kinetics of flame processes.

The Structure of Flames

Spectroscopic observations are most frequently made on stationary flames on burners. Such flames fall into two classes, the so-called *diffusion flames* of fuels burning in the surrounding air or supporting gas, and *premixed flames* of which the ordinary Bunsen-type flame is the most familiar example.

In the diffusion flames the fuel and air are initially separate and the gases mix slowly by diffusion, or in the case of fast gas flows partly by diffusion and partly by turbulent mixing processes. In the very early days, candle flames were the subject of some spectroscopic study, but diffusion flames of organic compounds give strong continuous emission from soot particles and this masks other features of the spectrum so that for a long time there was little spectroscopic interest in this type of flame. However the development of the flat diffusion flame by Wolfhard and Parker (1949) has enabled the detailed structure of these flames to be studied spectroscopically and the results are among the most important contributions of spectroscopy to combustion. There is some mixing of fuel and air near the burner rim and the propagation of the flame down into this small region of mixed gases holds the flame to the burner. Above this small base region, fuel and oxidant are largely separated by a wedge of combustion products such as CO and H_2, this wedge increasing in thickness as we go higher above the burner. The shape of diffusion flames is discussed in *Flames* (Gaydon and Wolfhard, 1970), and spectroscopic observations are dealt with here in Chapter VII.

In the premixed flames, the fuel is previously mixed with the air or other oxidant and we have a travelling explosion front which is held above the burner by the faster flow of gases up the burner tube. Again the flame shape and the conditions governing the stability of such flames are fully discussed in *Flames*. If the oxygen in the mixture is in excess of the stoichiometric amount for complete combustion of the fuel then we have a single thin flame front, which usually emits quite a lot of visible light, and above this flame front there is a region of hot combustion products which give relatively little visible radiation. If, however, as is normally the case with a Bunsen flame, the fuel is in excess, then the primary flame front leads to partial combustion in which all the oxygen is used up, and the products from this partial primary combustion, which contain

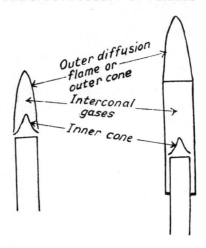

Fig. I.1.

hydrogen and carbon monoxide, will burn as a secondary diffusion flame in the surrounding air. This gives us at once the basic structure of the Bunsen flame. The bluish-green inner cone is the primary reaction zone which forms a roughly conical surface and is only a few tenths of a mm thick; for premixed flames with oxygen, such as the oxy-acetylene welding flame, the reaction zone may be as little as 1/50 mm thick. The outer pale blue or blue-violet sheath of the Bunsen flame, sometimes referred to as the outer cone, is the secondary diffusion flame of the excess carbon monoxide and hydrogen burning in air. By using a double tube (see Fig. I.1) it is possible to separate the inner cone on to the inner tube, leaving the outer diffusion flame burning on the outer tube. This is the well-known Smithells' separator, and with it, it can be seen that the 'interconal gases' between the inner cone and outer diffusion flame are almost non-luminous. In studying flame spectra we require knowledge of the separate spectra of the primary reaction zone (inner cone), the interconal gases and the outer diffusion flame. We also need information about the variation in spectrum through each zone, especially for the inner cone. This inner cone may be complex and a few special flames exist in which combustion takes place in two stages so that the inner cone becomes double.

Besides the steady diffusion flames and premixed Bunsen-type flames, we may also study the spectra of explosion flames travelling through premixed gases. In explosions at constant pressure we shall not expect any great difference between these travelling explosion flames and the steady flames. For explosions in closed vessels, however, there is a real and important difference because of the much higher temperature reached in the later stages of the explosion. This is because the final temperature is now determined by the specific heat at the constant volume, C_v, which is lower than that at constant pressure, C_p; also the higher final pressure tends to suppress dissociation of CO_2 and H_2O and so raises the temperature. The spectra of these hotter flames are thus likely to differ at least quantitatively from those of flames at constant pressure, and are of interest in connection with internal combustion engines, in which, indeed, they are best studied.

In explosions in long tubes (usually closed tubes) the flame may accelerate to produce the very violent type known as detonation, in which the flame front travels at several times the normal speed of sound. The method of flame propagation is then quite different, the detonation being propagated by a strong shock wave which heats and ignites the gas mixture. The processes of initiation of detonations and the structure of detonation waves and shock waves are outlined in the book by Gaydon and Hurle (1963). Knowledge of the spectra of detonations· is still limited because of the difficulty of handling this violent type of explosion, but some of the recent studies, especially of controlled detonations in shock tubes, are included here.

The Spectrum of the Bunsen Flame

While a full account of the spectra of various flames is reserved for later chapters, a brief description here of the main features of a Bunsen flame, burning either natural gas or manufactured gas, will assist with the continuous development of our subject. With a flame at maximum air-entrainment the mixture is still fuel-rich and shows both inner and outer cones, most of the light coming from the blue-green inner cone whose spectrum shows several characteristic band systems.

The regular system of bands which was first mapped by Swan is centred in the green. Strong groups of bands lie in the blue-green, green and yellow-green with similar weaker groups in the blue and orange. Each group consists of several heads shaded to the violet. The emitter of these Swan bands was for long a subject of controversy until in 1928 developments in the theory of molecular spectra, combined with a detailed study of the fine structure of the bands, proved (see later) that they were emitted by diatomic carbon, C_2. The visible spectrum of the inner cone also shows a strong violet—degraded band in the blue with its head at 4315 Å; this band shows quite open rotational fine structure. Another band in the violet with head near 3889 Å is degraded the other way, to the red, and shows open and even more obviously regular fine structure. These bands are due to separate systems of the same emitter, the radical CH.

In the ultra-violet the strongest band has its main head at 3064 Å; this is shaded to the red and again has open but rather complex fine structure and usually shows three heads. There is a less strong band of similar appearance with first head at 2811 Å, and weaker bands at 2608 and 3428 Å appear on heavier exposures. These bands also occur in the spectrum of the pure oxy-hydrogen flame and the structure shows that they are due to a diatomic hydride, the hydroxyl radical, OH.

These C_2, CH and OH bands are the main features of the inner cone, but some weaker bands in the near ultra-violet first observed by Vaidya in an ethylene flame and now usually known as the hydrocarbon flame bands, occur faintly in the Bunsen flame under conditions of high aeration. They consist of about a dozen rather diffuse red-degraded bands of roughly equal strength and a number of weaker bands, the system extending from 4200 to 2800 Å. They are known to be due to the triatomic formyl radical HCO.

The light from the outer cone is much weaker than that from the inner cone. It shows mainly the OH bands and some continuous emission, with weak narrow diffuse bands of CO_2 superposed; these are associated with the burning of carbon monoxide. This continuum is strongest in the blue, violet and near ultra-violet. The C_2 bands never appear in the true outer cone. Bands of CH and HCO do occasionally appear in the outer cone, especially if the flame gases are chilled by a cold surface or by turbulence.

Using a Smithells' separator we find that the interconal gases hardly emit at all in the visible, but that there is some OH radiation, although even this is much weaker than from the inner cone. We thus find that the main radiation comes from the regions of rapid chemical reaction and is due to simple intermediaries which are chemically unstable. The OH radical has, however, an appreciable equilibrium concentration in hot gases and so can radiate even from the interconal region.

Although the visible radiation from a flame is the most obvious, in actual amount the infra-red radiation is greater. The strongest band in the infra-red lies around $4.4\,\mu m$ and is due to CO_2; there is another strong band at $2.8\,\mu m$ due partly to CO_2 and partly to water vapour, and there are a number of other weak banded features mostly due to CO_2 and H_2O. These will be discussed in Chapter IX. These infra-red emission bands are not so strongly localised to the inner or outer cones as the visible radiation, but are shown by the whole body of flame gases.

The gases of a Bunsen flame are remarkably transparent in the visible and near ultra-violet. The OH band at 3064 Å can be obtained in absorption provided the experimental conditions are adequate, but the C_2 and CH bands for long defied observation in absorption and it is only in the last few years that it has been possible, using special techniques, to observe them in flames. In the infra-red, absorption by CO_2 and H_2O may be observed quite readily.

Equilibria, Radiation and Collision Processes

The radiation from any system in complete thermodynamic equilibrium will show a continuous spectrum identical with that from a black body at the same temperature. We know, however, that most small flames show banded spectra. This is because they are optically thin so that the radiation density is very low for most wavelengths. The flame is losing energy to the surroundings and the molecules in the flame tend to lose much more energy by emission of radiation than they gain by absorption.

It is only because flames are optically thin that we are able to make use of the spectrum to tell us about what emitting species are present, but the basic lack of radiative equilibrium, which is then an inevitable result, may cause certain difficulties if we try to use the strength of the banded radiation to estimate the concentrations of the emitting species.

The strength of emission of a particular spectrum line depends on the concentration of emitters in the appropriate excited state and on the rate at which radiation occurs from that state, or more precisely, on the transition probability for the particular transition. The observed strength may, however, be reduced by self-absorption. We have already seen that for most small flames the absorption is very weak, and for bands of CH, C_2 and HCO we do not usually suffer appreciable loss of intensity by self-absorption. For bands in the infra-red and for the OH bands, however, there may be significant self-absorption and this will both reduce the intensity of the bands and modify their appearance. Let us consider the effect on a band, such as that of OH, as we increase the thickness of the flame. The weakest lines in the fine structure of the band will not be affected by the increasing self-absorption and their increase in intensity will be proportional to the increase in flame thickness. For the strong lines the increase in intensity will be restrained, especially in the centres of the lines, by self-absorption. The emissivity cannot exceed 1, and for thermal equilibrium in the flame gases the line intensities cannot come above that for a black body at the flame temperature. Thus the tendency, as the flame thickness is increased, is for all the strong lines to be limited to the same intensity. There will, however, be some slight increase in width of the spectrum lines, and therefore some increase still in their overall strength. For a full understanding of the effect of self-absorption we have to study the "curve of growth" of the lines and are led into detailed considerations of spectrum-line contour.

If self-absorption can be neglected or corrected for, then the strength of emission can be used to determine the concentration of electronically excited species. We need to measure the light intensity in some absolute units (e.g. number of light quanta per unit flame area per unit solid angle), the flame thickness and the transition probability. Determination of the transition probability for the particular line or band requires an independent measurement from some equilibrium source, such as a high temperature furnace or shock-heated gases, where the species has a known concentration and temperature. Until recent years there has been great difficulty in determining transition probabilities reliably, but values of reasonable accuracy are now available for many of the important flame species.

If the gases are in thermodynamic equilibrium we can relate the number of molecules in an excited state to those in the ground state by using the Maxwell-Boltzmann distribution law. If the number in the excited state of energy E is N', and the statistical weight of the excited state is g', and the number in the ground state is N'' and the statistical weight of this ground state is g'' then

$$N' = N'' \, (g'/g'') \, e^{-E/kT}$$

where k is the Boltzmann constant and T is the absolute temperature. The statistical weights come in because there may be degeneracy, that is there may be more than one physically distinct state of the molecule having the same energy.

This relationship only holds in equilibrium. In an optically thin flame, radiation losses which are not balanced by absorption of radiation must tend to reduce the population in the excited state. This will be referred to as 'radiation depletion' and will be discussed in later chapters. We shall also find that in many cases chemical processes lead to the formation of molecules in electronically excited states, i.e. we have chemiluminescence. Apart from the direct formation of electronically excited molecules, strong exothermic combustion reactions will lead to the formation of molecules with excess energy in other forms, such as that of internal vibration, and this will take time to relax. The meaning of temperature, T, in the Maxwell-Boltzmann relationship is thus not simple, and we may have to use an effective temperature. The study and interpretation of flame spectra is thus linked with detailed energy transfer processes and determinations of effective temperature and this subject will be discussed in Chapter X.

When a molecule is formed in an excited state by chemical processes, the subsequent history of the molecule is very important. For an allowed electronic transition the average time before the molecule emits a quantum of light and returns to the ground state (i.e. the radiative lifetime) is between 10^{-6} and 10^{-8} seconds. Now in ordinary flame gases a molecule makes about 10^9 collisions per second. Thus the excited molecule will make between 10 and 1000 collisions before it radiates. Quenching, energy transfer processes and chemical reactions of excited species are thus very important. Recent work, by flash photolysis and shock-tube techniques, on energy transfer processes is thus very relevant. The laser is also becoming an increasingly important tool for this type of study. These techniques will be reviewed and the results used as appropriate.

Thus the study of flame spectra is closely related to work in other fields and in this book it will be necessary to make use of full data and take care in the interpretation of the observational material.

Chapter II

Experimental Methods

The Recording of Spectra

Spectra may be examined visually, photographically, photo-electrically or with a heat sensitive element such as a thermopile, bolometer or Golay cell.

Visual observation with a pocket spectroscope or constant-deviation spectrometer is often useful for preliminary work, but many of the most important bands, such as those of OH and HCO, fall in the ultra-violet and cannot be seen by eye. Many flames are too faint for the spectrum to be seen easily, and it is not practical visually to make measurements of wavelength or intensity.

The photographic method is relatively simple and straightforward. For bright sources, like an oxy-acetylene flame, a record can be obtained in a short time, while for weak sources intensity can be built up by giving a very long exposure. Accurate wavelength measurements can easily be made by putting a known comparison spectrum beside the flame spectrum, using a Hartmann diaphragm. The photographic method has the considerable advantage that it is two dimensional, so that by using a focused image of the flame on the spectrograph slit one can study the changes in spectrum with height through the flame. The limitation of the photographic method is that quantitative measurements of intensity require a rather complicated plate calibration, as discussed later.

In selecting the photographic plate one has to consider sensitivity, range in wavelength, speed, contrast and grain size. Fine grain and high contrast are desirable to obtain maximum resolution, but it is often necessary to sacrifice these for speed. When the spectrum is sufficiently bright a plate of fine grain and high contrast such as a Process or Rapid Process Panchromatic may be recommended. When high-speed plates are necessary, the author has used Ilford XK for the blue and near ultra-violet, HP3 and Astra III for the green and red, RL (ruby laser) for the far red, Q II for the ultra-violet below 2300 Å, and for the infra-red a range of special Kodak plates. The relative speed of a plate depends to some extent on the exposure time; thus for short exposures, especially for flash-tube work, the HP3 is much faster than the Astra III which is intended for long exposures on stellar spectra, but for weak flames the Astra III may be better. Many of the panchromatic and infra-red plates have a very uneven wavelength sensitivity; a lack of sensitivity in the blue-green is a common feature. This uneven sensitivity can easily distort the appearance of a band spectrum and with small dispersion may even cause a continuous spectrum to appear banded to the unwary. The author has found a developer of high contrast, such as caustic hydroquinone, helpful in displaying band structure and in getting the most out of weakly exposed plates, although it must be admitted that plates so treated are often dirty, grainy and less pleasing to look at than the cleaner ones obtained with a soft developer; for fine grain Microphen developer is good.

Photographic plates, especially on thin glass, are becoming more difficult to obtain and more expensive. Commercially, film is more used and cheaper. Film was long suspect for spectrographic work because it was feared that distortion of the film might lead to inaccurate wavelength measurements. However, provided the film is firmly held it is usually all right. For very long exposures slow buckling can cause a little trouble. For the visible region, fast films are Kodak 2475 and Ilford HP4. "Polaroid" is quite fast, but it is difficult to make measurements from opaque positives. In all long exposures it is necessary to control the temperature of the instrument to prevent loss of definition caused by thermal changes of refractive index of the prism or grating spacing.

A photo-electric recording system, with its attendant amplifying circuits, recording drum and gearing, is rather less simple than the photographic method. It has, too, a number of limitations, but it has the great advantage that unless overloaded it has a linear response and thus gives a direct measurement of relative intensity. It can also cover a much greater range of intensity, by use of electrical shunts, than is possible with a photographic plate. It is rather difficult to compare the ultimate sensitivities of photographic and photoelectric methods because of the differing criteria for speed (see later); for use with a high-aperture spectrograph on a very weak source, the plate is probably still superior, but for obtaining maximum resolution with a fairly large dispersion instrument modern photomultipliers, such as those made by E.M.I. in England and by R.C.A. in America, are better than plates or can at least give equivalent information in a shorter time. For use in the ultra-violet, photomultipliers built with a quartz window must be used. Photomultipliers with mixed silver and caesium oxide coated cathodes can reach 1.2 μm in the infra-red.

In general, photomultipliers can only be used satisfactorily with spectrographs properly designed for their use. High-aperture prismatic instruments with lenses usually produce a spectrum which is in focus on a curved field; to move the photomultiplier smoothly on a curved path is seldom practical; the prism is usually set near minimum deviation, so rotating the prism does not help. Some modern medium quartz spectrographs do have a practically flat field, and for these it is possible to design a system for moving the photomultiplier and second slit smoothly along the spectrum. Grating monochromators are more suitable, and plane gratings between concave mirrors (or one mirror used twice) are often used; these may be referred to as Czerny-Turner, Ebert or Fastie mountings; the grating or one of the mirrors may then be rotated so that the spectrum moves past the second slit in front of the photomultiplier. Systems of concave gratings or prisms in Littrow mountings are also used as monochromators, but it is then usually necessary to introduce an additional cam system to keep the spectrum in focus on the second slit. The precise determination of wavelength is more difficult with photoelectric recording. It is usual to superpose on the record some periodic signal operated from some part of the gearing system; these signals are then calibrated against a known spectrum. However, small deviations due to change of temperature, or of atmospheric pressure (which affects the refractive index of air), or to lack of instrumental rigidity can be troublesome. Any known lines in the spectrum being studied may be useful as internal standards. The output from the photomultiplier is usually amplified electronically and used to operate a pen-and-ink recorder which is geared to the system by which the grating or mirror is rotated. Highest sensitivity is usually obtained using a light chopper (e.g. a rotating sector) and alternating-current amplification, the amplifier being tuned to the chopping frequency; a further refinement is to use a phase sensitive detector working on a reference signal from the rotating sector. The output from a photomultiplier is very sensitive to any variation in the high voltage applied to the plates; good voltage stabilisation is essential.

The greatest limitation of the photo-electric method is the need to keep the flame absolutely steady and constant while the spectrum is scanned through. Any flickering may produce spurious structure; any drift in mixture composition or change due to warming up of the burner may upset the relative intensities of bands. For use with large dispersion instruments the scanning time may be considerable; thus Broida and Shuler (1952) scanned at 5 Å/min.; to cover the whole spectrum from 2000 to 7000 Å would take 16 hours! Photoelectric methods are thus not very suitable for mapping the spectra of turbulent flames, explosions or engine flames. The photoelectric method is also one-dimensional, so that if the spectrum of a flame is required at different heights in the flame it is necessary to scan through separately at each height. The method, using a light chopper, does, however have a special advantage for absorption spectroscopy of flames (see page 26).

The time resolution of a photo-electric system can, however, be made good use of in some cases. If a monochromator is set on some selected spectrum line or band and the photomultiplier output is displayed on an oscilloscope, the variation of light intensity may be measured with a time resolution of better than 1 μs. This type of system is suitable for studying the time history of emission from free radicals in explosions, shock-tube experiments and flash photolysis. For phenomena of rather longer duration it is also possible to scan rapidly through a portion of the spectrum, usually using a rotating cam device, and to display the output on an oscilloscope; thus Bullock, Hornbeck and Silverman (1950) used PbSe and PbTe photocells in the near infra-red, with a scanning rate of 125 scans per sec. over a 2 μm range, to study the development of CO_2 emission during explosions.

Spectra may also be recorded with a television camera; a normal spectrograph is used with the camera replacing the photographic plate. The image may be scanned off electronically and displayed on an oscilloscope. An early paper by Gloersen (1958) describes the use of this type of system for studying the spectra of the bright but short-duration emission from shock-heated gases; the sensitivity was around 10^3 greater than that obtained photographically. Image converters, with electronic gating, are coming into increasing use for the study of transient spectra.

For the infra-red, photoconductive cells of PbS and PbSe are sensitive to about 3 μm and 6 μm respectively and InSb will also reach about 6 μm. Germanium doped with gold will reach 8 μm. For the far infra-red, detectors cooled in liquid nitrogen or even helium are necessary. Germanium doped with Zn, Cu or Sb will reach 17 μm or further; for details of infra-red detectors and techniques see the book by Houghton and Smith (1966).

Thermal detectors, such as Schwartz thermopiles and other thermocouples, bolometers (depending on thermal changes in electrical resistance) and Golay cells (gas expansion, with optical-lever amplification) have a sensitivity which is independent of wavelength (apart from any limitations set by window materials) but are less sensitive than photo-electric devices. Response times are relatively long (usually in the range 10 to 100 msec) and light chopper amplification devices have to have a correspondingly low frequency so that it becomes necessary to scan through a spectrum rather slowly.

Spectrographic equipment

Although some flames supported by oxygen or by fluorine are quite bright, the majority of flames, especially the more ordinary types with air, are relatively weak sources of light. The first consideration in choosing a spectrograph must therefore be its speed; small-dispersion large-aperture prism instruments are, for most purposes, of more use than large-dispersion grating instruments. Many of the interesting features of flame spectra lie in the near ultra-violet and to reach this region quartz optics are necessary. Quite a lot of good work has been done with commercial small and medium quartz spectrographs, but faster types of larger aperture, such as those used for Raman spectroscopy, are better for exploratory work, and especially for weak sources such as cool flames or for studying single explosions. The author has found the small Hilger Raman spectrograph E 518, with interchangeable quartz and glass optics, portable and convenient. Although quartz is necessary for the ultra-violet, the dispersion in the visible and infra-red is very small and instruments with glass prisms give higher resolving power than quartz.

For some purposes, such as making detailed rotational analyses of new spectra or for measuring effective rotational temperatures, it is necessary to use instruments of higher dispersion. Either Littrow-type prism instruments or grating spectrographs are then suitable. A discussion of the various types of mounting and their relative advantages is beyond the scope of this work; some of the older books on the subject are useful for this purpose (Harrison, Lord and Loofbourow, 1949; Sawyer 1945). One of the main disadvantages of most grating mountings for flame work is that they suffer from astigmatism, that is the spectrum line is blurred along its length, so that we cannot in a simple way study the variation of spectrum with height in the flame. A concave grating in a Wadsworth mounting is, however, practically free from this defect; in this mounting the grating is illuminated with parallel light from a concave mirror and the spectrum is observed normal to the grating; the mounting gives good light gathering power but only half the dispersion obtained from most other mountings and it takes up more space than a grating in an Eagle mounting. The author has found a plane reflection grating in a Littrow mounting convenient; it is free from astigmatism and fairly portable. Gratings with fine rulings, e.g. up to 2160 grooves /mm or 55000/in., are best as it is then possible to work in the first order with little trouble from overlapping orders. Otherwise it is necessary to use filters or a fore prism to cut out overlapping orders. Ideally a set of gratings blazed for various regions of the spectrum is desirable, but for most work a grating blazed for around 5000 Å is adequate.

For photographic work on line spectra the speed is determined primarily by the angular aperture of the camera lens. If the focal length of the collimator lens is changed, the size, but not the brightness, of the image of the slit is altered. thus in high-aperture instruments it is common practice to have only a high aperture camera lens, and to use a longer focal length for the collimator; the result is a rather narrow spectrum, but there is economy in the cost of the collimator lens and the spectrograph is easier to fill with light. For prism instruments lengthening the slit only produces a wider spectrum strip, without affecting the brightness; for gratings in mountings suffering from astigmatism there will be some loss in brightness if too short a slit is used. For very narrow slits the brightness does, of course, vary with slit width, but for line spectra as soon as the slit width exceeds the resolving power (which usually happens around 20 μm) further widening only produces a wider, but not a brighter, spectrum line. For continuous spectra or wide unresolved bands, however, widening the slit does increase the brightness. For these types of spectra the brightness of the image also depends on the dispersion; lines merely become further apart with increasing dispersion, but continuous spectra are drawn out and weakened. For this reason, large dispersion tends to favour the observation of lines and of bands with resolved fine structure, such as those of OH and CH, while any diffuse banded structure, such as that of the CO flame or the hydrocarbon flame bands, is displayed more prominently with small dispersion.

The criteria for speed in a photoelectric instrument are quite different. Here the second slit covering the photomultiplier is also adjustable in both length and width and the optimum adjustment for this second slit is so that it just takes in the image of the first slit i.e. normally both slits should have the same length and width. The light intensity is then determined by the amount of light reaching the photomultiplier surface, that is by the product of the brightness, slit-width and slit-length. The aperture of the collimator lens is also important in this case, and speed can be gained by using as long a slit as possible, the limit being set by the photomultiplier surface. For large dispersion work with a photomultiplier, a grating spectrometer is probably most suitable. Gratings give straight spectrum lines, whereas in a prism instrument the image of the slit is slightly curved, so that for maximum resolution it is necessary to limit the length of slit used or else to use a special curved second slit. The astigmatism of the grating is no disadvantage when a photomultiplier is used as it is in any case not possible to study variation of spectrum with flame height in a single traverse of the spectrum.

In early work on flame spectra in the infra-red, prism mono-chromators with appropriate prism materials (quartz, LiF, KCl, NaCl, KBr) were used, but with improved detection equipment the recent trend has been to use grating monochromators. Order separation can be done with a selection of filters or with a fore prism of suitable material.

Flame spectra are seldom studied in the far ultra-violet because of absorption by oxygen in the flame and surrounding air, but Kydd and Foss (1967) have made a valuable extension of the spectrum to 1400 Å by working with a flame at reduced pressure and using a LiF window and evacuated grating spectrograph.

Recent developments towards faster instruments include the use of grill spectrometers and Fourier-transform spectroscopy. In the grill spectrometer (Girard, 1963) the two slits of a monochromator are replaced by a pair of "grills"; The first "grill" takes the form of a pattern of clear and opaque regions (often in the form of hyperbolic lines) of equal area; the second "grill" may be either identical to the first, or complementary, with the clear and opaque areas reversed. The grills transmit much more light than a slit system and, when used with photoelectric detection and suitable electronic amplification devices, can lead to an improvement in speed by a factor of around 100. In Fourier transform spectroscopy, the complex light pattern obtained from an interferometer with variable path length is analysed with a computer to derive the spectrum. These instruments have so far been used mainly in the infra-red, but may well prove useful for study of weak flame spectra in the visible.

Optical Systems

A condensing lens is usually employed to throw an image of the flame on to the slit of the spectrograph. For many purposes a simple spherical or plano-convex lens of fused quartz is adequate, but when investigating details of flame structure or making intensity measurements of spectrum features over an appreciable wavelength range, it is essential to use a good achromatic lens; for the ultra-violet this usually means a quartz-fluorite achromat; the author has found a quartz-lithium fluoride triplet very useful.

With a lens of focal length f, the flame must be at least $4f$ away from the slit for an image to be formed. If it is at a greater distance than $4f$ then there are two possible positions in which the lens may be placed, one giving an enlarged and the other a reduced image of the flame on the slit. Provided the aperture of the condensing lens is sufficient to fill the spectrograph the brightness of the spectrum will be the same for the two images, although the reduced image appears brighter on the slit because of the wider cone of light producing it. However, the light in the outer part of this wider cone is wasted because it only overfills the spectrograph; indeed it can be a nuisance by producing a background of scattered light within the spectrograph. Owing to certain differences between coherent and non-coherent light the spectrum produced by the larger image may give a rather better resolved spectrum (Harrison, Lord and Loofbourow, 1949). The choice of which image to use will, however, depend mainly on the size of the flame; for small flames, like that on an oxy-acetylene welding torch, it is preferable to have an enlarged image on the slit, whereas for very big flames it is obviously necessary to have a reduced image to get it on the slit. Also for deep flames, such as those on rectangular burners viewed end-on, the horizontal length of the image is less with the smaller image on the slit than with the larger image; it is convenient to remember that the horizontal magnification of a lens set-up (along the direction of travel of the light) is roughly the square of the vertical magnification.

The flame and condensing lens must, of course, be carefully lined up on the axis of the spectrograph. For weak flames and other difficult sources this can sometimes be done most easily by placing a torch in the position of the spectrum and following the light back through the optical system; small gas lasers are also very useful for this purpose. It is also very necessary to see that the condensing lens is of adequate aperture to fill the collimator lens of the spectrograph. If this is not so both speed and resolving power are lost. In checking this it is necessary to allow for the finite size of the image on the slit. Thus in Figure II.1, light from the tip of the inner cone of the flame falls on the centre of the slit and light will fill the collimator lens of the spectrograph from a condensing lens of the aperture shown; however, light from the top of the flame comes lower on the slit and then spreads out so that much of it comes below the collimator lens, which is no longer filled with light. To fill the spectrograph it is necessary to use a very much larger condensing lens as indicated in Figure II.2.

The author has always found relatively large condensing lenses of fairly long focal length most satisfactory. It is often impracticable to place a lens near a very hot flame or an enclosed flame, and the relatively large size of some flames also causes aperture difficulties, as indicated above, if small lenses of short focal length are used. The type of lenses supplied by some spectrograph manufacturers for use with arc and spark sources are unsuitable for flame work.

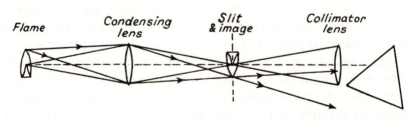

Fig. II.1

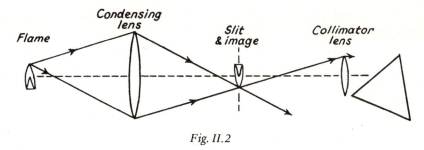

Fig. II.2

For an extended source, such as a large flame, it is not necessary to use a condensing lens. If the slit is placed near the flame so that the extended flame at least fills the aperture of the spectrograph then there is no loss of light. The author has often found this simple system suitable for studying the diffuse glow of a cool flame or for viewing an engine flame through a small window in the combustion chamber.

There are some optical systems which can give a modest increase in the brightness of the flame image on the slit. In the simplest system a concave mirror (preferably with aluminised front surface) is placed behind the flame with its centre of curvature in the flame so as to form an inverted image of the flame back in the flame; the light intensity is thus nearly doubled, but it is of course no longer possible, with the superposed inverted image, to study details of variation of the spectrum through the flame. An erect image can be obtained by placing a convex lens behind the flame and then a concave mirror at the place where an image is formed by this lens so that light is reflected back through the lens again and forms an erect image on top of the flame; allowing for reflection losses in the lens, a gain of around 70% may be expected.

Plyler and Kostkowski (1952) described an ingenious system of plane and concave mirrors which doubled light emerging in several directions back through the flame; allowing for losses at reflecting surfaces a gain in brightness by a factor of about three could be expected. Toishi and Muira (1955) have described a reflection system with specially cut prisms which should with 4% reflection loss give a 12-fold gain in light intensity and in practice gave about a 7-fold gain. However the best way of using an optical system to increase the light intensity from a small flame is probably to place it within a multiple-reflection mirror system of the type designed by White or by Jessen and Gaydon and described later (page 29) for use in absorption studies.

The gains possible with these various optical systems are, however, not very great, and the more usual way of gaining light is by design of burner to give a greater thickness of flame. This is dealt with later (page 41). It is also possible, of course, to use several flames in a row. Such gains in flame thickness are, however, accompanied by problems of how to retain sufficient angular aperture to fill the spectrograph. All these methods, giving either real or optically increased thickness of flame gases, may increase errors due to self-absorption of light in the flame.

The Study of Absorption Spectra

Absorption spectra may be regarded as falling into three groups (i) those with bands showing open discrete rotational structure, (ii) those with bands whose rotational structure is so close that it is unresolvable, at least near the band head, and (iii) those showing continuous absorption or diffuse bands.

The first group, which includes most of the diatomic molecules like OH, CH, O_2 and C_2, is the most difficult to observe experimentally. To detect a single spectrum line in absorption the resolving power of the spectrograph must be at least comparable with the width of the spectrum line itself. If this is not the case, the light of very slightly longer and shorter wavelengths present in the background continuum used to observe the absorption spectrum will spread across the narrow absorption line and prevent its observation. Normally the width of a line in the rotational fine structure of a band spectrum is very small; it depends on the Doppler broadening (which is temperature dependant) and on pressure broadening, this latter effect being, however, less important for band spectra than for atomic lines. The line width is usually of the order 0.2 cm^{-1} in flame gases. Thus it is necessary to use an instrument of high resolving power, such as a large grating spectrograph, to be sure of observing absorption bands with resolvable fine structure.

Spectroscopic methods are more suitable for detecting the presence of a molecule when its spectrum is of the second type, the rotational fine structure being unresolvable. This happens when the lines of the fine structure are so close that the spacing between them is less than the actual width of the lines themselves, or at least when the line width and line spacing are comparable. This most frequently happens at the head of the band or in piled-up Q branches of weakly degraded bands like the 3143 Å head of one of the CH bands or the Q head of NH at 3360 Å. Many polyatomic molecules, for which the rotational fine structure tends to be close and complex, show strong bands in absorption, good examples being the bands of benzene and formaldehyde. In these cases the absorption may easily be observed with small-dispersion instruments such as medium quartz spectrographs. Even in this type, quantitative measurements may be difficult, though. In a band where the separation between individual lines is comparable with the line width, some wavelengths will be almost completely absorbed and others almost completely transmitted, so that with small instrumental resolution the average effect is of moderate absorption. If the concentration or path length of absorbing molecules is increased the still greater absorption in the already extinguished parts produces relatively little change in the average absorption so that Lambert's and Beer's laws relating absorption to path length and concentration are not obeyed. For quantitative measurements it is therefore necessary to calibrate empirically. The appearance of a polyatomic spectrum often changes with temperature, high temperature causing the structure to become more complex and diffuse until the characteristic band features merge into continuous absorption.

Absorption spectra of the third group are shown by organic acids, peroxides and higher paraffins, which absorb the far ultra-violet, and by ketones which show diffuse absorption around 2800 Å. The difficulty with this type of spectrum is not so much in the observation as in the identification of the absorbing species. A diffuse band around 2600 Å obtained during the slow combustion of higher paraffins and the "pyrolysis continuum" found under incipient soot-forming conditions are the subject of some discussion later.

For the study of absorption spectra, a background source giving a bright continuous spectrum is required. Early work was mostly done with tungsten strip-filament lamps for the visible region and hydrogen discharge lamps for the ultra-violet. These sources are not, however, adequate for the photographic study of the absorption spectra of hot flames, because the brightness temperature of the background source must exceed the flame temperature, or more correctly the effective excitation temperature of the band system in the flame*. The pole of a carbon arc, which has a brightness temperature of about 3800 K, has also been used, but emission bands of CN, C_2 and CO from the arc spoil the continuum. The most favoured background sources now are the high-pressure xenon arc lamps and flash tubes. Xenon lamps are commercially available, and the hottest part of the arc may have a brightness temperature exceeding 5000 K, especially at short wavelengths. The continuum is fairly good, although there are a few broad Xe lines superposed on it. The Xe arc is a rather small source and its brightness varies rapidly across the arc; this causes difficulty in making quantitative temperature measurements, but for observing absorption spectra it is usually satisfactory. For photographic recording, alternating-current xenon lamps may be used, but for work with a photomultiplier it is necessary to use a direct-current lamp with a stable power supply free from any voltage ripple. Flash tubes have a very high brightness temperature but a short duration; the type developed by Garton and Wheaton (Wheaton, 1964) has a brightness temperature around 25000 K and a duration of about 5 μs. For use with instruments of moderate dispersion a single flash from this type of lamp gives an adequately exposed spectrum. For use with larger dispersion it is possible to use multiple flashes, a synchronised rotating sector being placed in front of the spectrograph slit to prevent excessive build-up of the emission from the flame between flashes (Gaydon, Spokes and van Suchtelen, 1960).

For studying steady flames, a photomultiplier and light-chopper system is by far the most sensitive and best. If the light chopper is placed between the background source and the flame (Fig. II. 3) the tuned recording system responds only to the modulated signal from the background and not to the steady light emission from the flame; it is thus possible to measure flame absorption even in the presence of flame emission. In early work (Silverman, 1949) the light chopper was run at a steady frequency and a simple electrical amplification system was tuned to this frequency. Modern systems (Bleekrode and Nieuwpoort, 1965; Jessen and Gaydon, 1969) use a tuned phase-sensitive detector controlled by a small monitoring light behind the light chopper. This method may be highly sensitive and avoids the need for the background to be brighter than the flame, but care is

---

*In the reaction zone chemiluminescence may lead to very high excitation temperatures, as discussed in Chapter VIII.

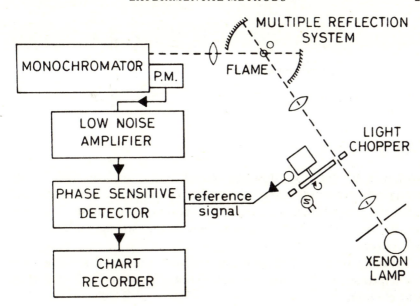

Fig. II. 3. Light-chopper and phase-sensitive detector system for studying absorption spectra.

needed to keep the signal-to-noise ratio high by having a very steady background source and preventing noise from fluctuating flame emission. It is not suitable for flames which flicker, turbulent flames or explosion flames.

The burnt product gases above the inner cone of a flame readily show absorption bands of OH in the ultra-violet and of H_2O, CO_2 and CO in the infra-red, but the more interesting species which are intermediates in the combustion are to be sought in the reaction zone or preheating region. In premixed flames, and in diffusion flames supported by air, the reaction zone is very thin (tenths of a mm or less) and as the concentration of these species is always low it is not possible to detect absorption in a single traverse of the reaction zone. A somewhat increased path-length may be obtained by having a flame on a rectangular slot burner and passing the light in an end-on direction through the nearly flat reaction zone. However, as will be seen from Figure II.4, if a light beam of reasonable aperture is used most of the light still does not pass through the reaction zone for any appreciable distance. If the light beam is restricted to a narrow pencil of very small aperture then there will be difficulty in filling the spectrograph with light and, much more serious, the light

Fig. II. 4

beam will be deflected downwards out of the reaction zone by the strong refractive-index gradient in the reaction zone and preheating zone; the temperature gradient in the reaction zone of a flame may be of the order 10 000 K/mm, producing strong schlieren-type deflections.

Special burners which give flat reaction zones and in some cases thicker zones, such as low-pressure flames, are described fully in the next chapter.

To increase the effective optical path through the flame gases a multiple-reflection system of mirrors is frequently used. The simple system devised by White (1942) using three or four concave mirrors (see Fig. II. 5) to reflect a narrow pencil of light backwards and forwards through the flame has often been used. With front-surface aluminised mirrors up to about 20 traversals of the flame may be obtained, although the light intensity is reduced and in some cases

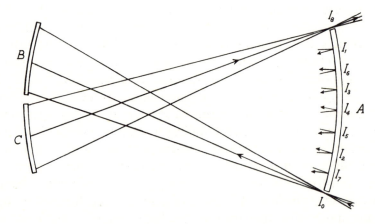

Fig. II. 5. White's multiple reflection system showing its use for 16 traversals and position of 7 images on mirror A.

the effective brightness temperature of the background may then fall below the flame temperature. Using dielectric coatings on the mirrors gives a much higher reflection, for the appropriate wavelength range for the coating, and up to 50 traversals have been used.

White's system is suitable for low-pressure flames, which have a thick reaction zone and for which schlieren-type deflections are unimportant because of the low gas density, but it is unsuitable for flames at atmospheric pressure because the beam tends to be much wider than the reaction zone and the nearly parallel light suffers from schlieren deflections which upset the focus. Jessen and Gaydon (1967) have designed a modified system (Fig. II. 6) which focuses the light into the flame, giving better spatial resolution and less trouble from schlieren deflections; even though the light beam is deflected it still comes from the same point in the flame and can be collected by a lens of sufficient aperture and brought to the spectrograph slit. Two flames, or two edges of the same flame, must be used, and there are some complications because the image of one flame on the other is inverted. However the system has proved useful and was used in the first observation of the main CH bands in absorption in a flame at atmospheric pressure.

Absorption spectra have usually been studied in the visible and ultra-violet to about 2300 Å. Below this there are troubles from decreasing plate sensitivity, limited transparency of optical materials and absorption by oxygen. The first difficulties can be overcome by

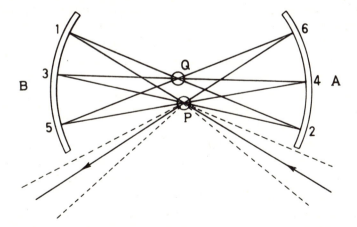

Fig. II. 6. Jessen–Gaydon multiple-reflection system, showing seven traversals though the two flame positions.

using suitable plates (e.g. Ilford QII) and concave grating spectro-graphs with mirrors, but the absorption by oxygen, which begins around 1900 Å in cold gas but at much longer wavelengths (2600 Å) in hot gas, is more serious. Garton and Broida (1953) made some preliminary observations on a flat diffusion flame surrounded by N_2 using LiF windows and a vacuum grating spectrograph and found continuous absorption, probably due to H_2O, below 1925 Å, and a few unassigned features. It seems that more work on the reaction zone of low-pressure flames might be worth while as this is a region where many organic compounds show strong absorption; Gaydon, Spokes and van Suchtelen (1960) found bands attributed to CH_3 and C_2H_5 below 2300 Å.

Wavelength Determination

Many commercial spectrographs have built-in wavelength scales which are very convenient for quick identification of the main features of a spectrum. Such scales are, however, of limited accuracy and a proper comparison spectrum should always be taken beside a flame spectrum showing unknown bands. For prism spectrographs this is done most conveniently with a Hartmann diaphragm over the slit, and for grating instruments with a shutter in front of the plate; the position of the diaphragm or shutter should be changed carefully, and the plate-holder should never be closed or disturbed in any way between the two exposures on the flame and comparison spectra. For photo-electric recording it is necessary to rely on calibration; monochromators usually have a wavelength calibration on the drum and in recording instruments a periodic signal is impressed on the pen-and-ink record. Internal standards in the flame spectrum should be used to check the calibration, in case of any error or drift due to temperature changes; often metallic lines, sometimes from impurities like Na and Ca will suffice, or known features like the OH or CH band heads may be used.

Ultimate standards of wavelength are determined interferometric-ally, and lines of iron, neon and other rare gases are used as secondary standards. Flame spectra are usually examined with rather small dispersion and high accuracy is not required. The iron arc is by far the best comparison for work with medium dispersion; for the ultra-violet only the central part of the arc, avoiding the poles, should be used as lines from near the poles are broadened and displaced by

Stark effect. The copper spark is best below 2200 Å. The copper arc is sometimes used with small dispersion, but the author has not found it very satisfactory as the spectrum varies with arc conditions and bands of NO often obscure parts of the ultraviolet. The spectrum of a mercury lamp is always easy to recognise and has its use for small dispersion but has too few lines for use in actual measurements. A neon discharge lamp is very good for the red region, where the iron arc spectrum is spoilt by oxide bands; the small indicator lamps, showing lines of Ne and often He and Hg, are useful for small-dispersion fast instruments. For the infra-red various rare-gas discharge lamps are useful. Hollow-cathode iron lamps are often recommended; the lines are fine, but the author has found these lamps rather too faint for use with large dispersion.

For accurate work the comparison source must be properly collimated like the flame; if the spectrograph is not filled with light, or not evenly filled, there may be some loss of resolving power and slight asymmetrical broadening of the lines. When the flame spectrum requires a long exposure it is best to take two comparison spectra, one before and one after the main exposure; this enables any temperature shift to be detected. For use with very fast spectrographs bright comparison sources like the iron arc are often difficult to expose properly; a photographic compur shutter built into the spectrograph or placed in front of the slit is the best solution. Sometimes it is sufficient to use scattered light from a sheet of white paper held in front of the slit, the iillumination of this being controlled by placing the arc at a convenient distance.

When trying to identify a complex band spectrum or diffuse structures the best way is by juxtaposition of plates, or by comparing enlargements made to the same scale. This subject is dealt with in *Identification of Molecular Spectra* (Pearse and Gaydon, 1965) in which other aspects of wavelength determination and identification are also discussed.

Intensity Measurements

Measurements of intensity, relative or absolute, of flame spectra can usually be made most accurately by photo-electric methods. Some limitations of photomultipliers, such as the need for a non-fluctuating light source, have already been mentioned. The great advantage of a photomultiplier is that it has a linear response, unless overloaded, and can measure intensities varying over several orders of magnitude. The equipment tends to be more elaborate and expensive than that used for photographic work, and it is usually necessary to have a constant-voltage supply available; however, the saving in microphoto-meter equipment for measuring plates must be remembered when making any cost comparison.

Photographic measurements of intensity are basically difficult because there is no simple relation between the photographic blackening or density and the intensity of light, and because a photographic plate does not integrate light; a short exposure to a bright light will give appreciably more blackening than, say, an exposure 100 times longer to a light of 1/100 of the intensity*. The book on *Practical Spectroscopy* (Harrison et al, 1949) discusses photographic and photo-electric methods of measuring intensity in some detail and here it must suffice to stress some of the particular problems encountered in the study of flames.

For photographic work with a spectrograph which does not suffer from astigmatism, calibration marks may be put on the plate with a neutral step-filter placed immediately in front of the slit. A rotating step-sector or a sector cut logarithmically is also frequently used; if the sector rotates fast enough the error due to reciprocity failure is small. For this work the slit, covered by the step-filter or sector, must be illuminated very uniformly. This can be done by omitting any condensing lens and placing instead a lens immediately over the slit; this lens is used to focus an image of a small source on to the collimator lens of the spectrograph; careful checking of this uniform illumination is essential, and the slit must be of good quality and wide enough to avoid errors due to its width varying with position. The calibration marks for intensity are best made with a source having a spectrum of similar character to that of the flame to avoid troubles from Eberhard effect[†], which will have more influence on a continuum or a broad line than on a fine line.

---

*Gerbach (1966) made measurements in which exposure time was varied between 3 and 100 s with the integrated intensity x time held constant and found that the optical density of the film decreased by an amount equivalent to a reduction of intensity by a factor of 4.

[†] The Eberhard effect is due to the developer being weakened locally, during development, near any heavily exposed regions, such as strong spectrum lines, so that neighbouring areas are starved and do not attain their full density. For all intensity work plates should be brushed constantly during development.

The characteristic curve for the plate may be drawn from measurements of the optical density of the calibration marks, and then this characteristic curve may be used to compare intensities of features in the flame spectrum A photographic plate has a threshold sensitivity, and light below a certain intensity does not produce any blackening. For high intensity, too, the curve flattens out so that there is little change in optical density with further increase of light intensity. The range of intensity which can be measured with reasonable accuracy on a single exposure seldom reaches 10 to 1. A special difficulty with flame spectra is that due to the continuous background; if this is strong enough to come above the plate threshold it can be measured near the feature of interest and its intensity can be deducted, but if it lies below or only just above threshold it is very difficult ot make an accurate correction; sometimes a supplementary exposure with a wider slit or longer exposure time will bring the continuum above background so that its intensity can be estimated and used to apply a correction. Under ideal conditions it should be possible to measure intensities to within about 2 per cent, but under practical conditions on photographic flame spectra the author has often found even 10 per cent difficult to achieve, especially when comparing intensities at widely different wavelengths.

The sensitivity of both photographic materials and of photo-multipliers varies with wavelength. The speed of the spectrograph or monochromator also changes with wavelength because of changes in dispersion, reflection losses, absorption in lenses and prisms, change of reflection by blazed gratings and sometimes because of change in aperture. To compare the intensities of light of different wavelengths it is necessary to use some standard source of known brightness at each wavelength. The ultimate standard is, of course, a black-body enclosure at known temperature (usually the gold point), but in practice it is more convenient to use calibrated strip-filament tungsten lamps. The calibration is made in terms of brightness temperature in the red (e.g. at 6600 Å) and the light intensity at other wavelengths can then be calculated from the Planck radiation law and the emissivity of tungsten (see *Flames*, Gaydon and Wolfhard 1970).

Intensity measurements on flame spectra may be classified into four groups. For studies of rotational intensity distribution and measurements of effective rotational temperature the strength of individual fine lines in the structure must be compared. For studies of vibrational intensity distribution it is necessary to compare the integrated intensities of individual bands of one molecule. For studies of the effect of some variable such as mixture strength or flame temperature on the flame spectrum we need to compare the relative strength of whole band systems of C_2, CH etc. in the various flames. For absolute quantitative measurements it is necessary to integrate the intensity over whole band systems and express it either in number of light quanta emitted or in energy flux, and then perhaps to relate this to rate of fuel consumption or flame volume.

For the comparison of individual lines of the rotational fine structure of a band, we need large dispersion adequately to resolve the structure. The lines are, however, usually sufficiently close for the change of photomultiplier or plate sensitivity with wavelength to be negligible, or at least small so that a relatively simple correction can be made. With a fine slit the lines will have an intensity contour caused by instrumental factors and by the natural line shape (Doppler broadening, pressure broadening and perhaps self-absorption). It is usually better not to attempt to measure the central intensity of the lines, but to widen the slit so that the intensity in the central portion of the wide slit image is uniform and proportional to the total strength of the line; particularly in photographic work the measurement with a microphotometer is then much better and less affected by graininess of the plate and curvature of the spectrum line. The main practical difficulty with this type of work is to get enough light intensity to enable adequate dispersion to be employed to give the necessary high resolution. If a considerable depth of flame is used, to gain light, then the risk of errors due to self-absorption will increase; self-absorption tends to occur most in strong lines and so to weaken these selectively. Another practical source of error is due to the continuous background, which is often invisible, below plate threshold, but still strong enough to increase the apparent intensity of weak lines; with photoelectric recording this difficulty is eliminated, as the background can usually be measured directly.

For studies of vibrational intensity distribution we have to compare bands of the same system which may be at appreciably different wavelengths so that calibration of the photomultiplier or photographic plate for varying wavelength sensitivity becomes necessary; in photographic work the contrast as well as the sensitivity changes with wavelength. To obtain the intensity, integrated over the whole of the rotational structure, for the various bands due to different vibrational transitions, it is usual to employ relatively small dispersion and a wide slit so that the rotational structure is blurred out; the total intensity may then be obtained by integrating the area from a record of the spectrum. The chief difficulty is due to overlapping of the various bands of the system; thus the (1,1) band frequently falls on the "tail" of the rotational structure of the (0,0) band. An approximate correction for the influence of the tail of one band on the head of another and vice versa is often made by extrapolating the intensity of the tail of the front band. With photoelectric recording it is possible to use higher dispersion and a fine slit and trace the individual lines of the tail through the next head, but this is seldom possible photographically because it requires a fine slit and measurements then become difficult, as indicated in the previous paragraph. Often the strength of the head of a band, or a region of maximum intensity under small dispersion, has been taken to represent the strength of the whole band. However, the rotational intensity distribution will also change with flame temperature or excitation conditions* and so the intensity at the head or maximum is not directly related to the integrated intensity.

In comparing the intensities of band systems due to different electronic transitions or due to different molecules we again encounter the complications of heterochromatic work, but troubles due to overlapping of bands may not occur. If, however we wish to compare the relative intensity of two band systems, such as those of C_2 and OH, in different flames, then we usually meet special difficulties because the flames will have different shapes, have reaction zones of different thickness and perhaps even be on burners of different sizes. These troubles are encountered in studying the influence on the spectrum of varying the mixture strength, dilution or addition of inhibitors; these special effects of flame shape are discussed in the final section of this chapter.

For absolute intensity measurements it is necessary to compare the spectrum of the flame with that of a standard, such as a calibrated tungsten strip-filament lamp. The energy radiated in unit

*This type of change in rotational intensity distribution is particularly strong for OH; in thermal excitation lines near the band head are strongest, but in chemiluminescent excitation in hydrocarbon flames the effective rotational temperature is abnormally high and the lines from high rotational levels are much stronger and the tail of the band is consequently much extended.

time by a surface of area A at absolute temperature T and emissivity E_λ at wavelength λ into a small solid angle $d\omega$ normal to the surface is

$$2Ahc^2 E_\lambda \cdot \lambda^{-5} \, d\lambda . \, d\omega /(e^{c_2/\lambda T} - 1)$$

where h is Planck's constant, c is the velocity of light and c_2 is the second radiation constant. In S.I. units $2hc^2 = c_1/\pi = 1.191 \times 10^{-16}$ watts m^2 where c_1 is the first radiation constant, and $c_2 = 1.4388 \times 10^{-2}$ m deg K. If we have a flame whose brightness is uniform over a reasonable area, we can make the comparison by forming an image of the flame on the slit, using a condensing lens which completely fills the monochromator or spectrograph, and recording the light intensity at the relevant wavelengths, and then making a similar measurement on the standard lamp The result gives the rate of energy emission per unit area of flame surface in the direction of the spectrograph.

However, many flames, such as the inner cones of premixed flames, are of uneven brightness, and we are more interested in the total radiation than in the emission per unit area. For this purpose the normal condensing lens is omitted* and the flame is set on the axis of the spectrograph or monochromator far enough away so that light from all parts of the flame comes within the cone accepted by the collimator. The intensity of the spectrum from the whole flame is then compared with that from a known area (selected with a stop) from the strip lamp; if the lamp and flame are at different distances from the slit the inverse square law may be used to relate the intensities. Gaydon and Wolfhard (1950) used a method of this type to compare the C_2 light yield in various low-pressure flames. The spectrum of the whole flame was compared with that of a known area of the filament of the strip lamp and the emission from the flame was plotted in watts/angstrom, and, after deducting the contribution from the continuous background spectrum in the flame, the intensity in all the C_2 bands was integrated graphically, giving a total C_2 emission in watts. This was then converted to number of photons emitted per second using the relation that the energy associated with each photon is $E = h\nu$ where h is Planck's constant and ν is the frequency. From the measured gas flow and Loschmidt's

---

*A lens may be used very close to the slit, focusing an image of the flame on the collimator lens of the spectrograph.

number the rate of emission of photons was then related to the number of fuel or oxygen molecules reacting per second.

Effects of Flame Shape

Comparison of the intensity of a band system in different flames is best done by comparing the light yield, either in absolute terms or relatively, as indicated at the end of the last section. The difficulty in making a simple comparison of the brightness of the bands in different flames has been noted, and is here explained in more detail.

For premixed flames of organic compounds the main spectroscopic interest is in the inner cone. This is a very thin luminous layer of roughly conical form, and if an image of such a flame is thrown on to the slit, this image will be brightest near the edges, where the light emerges tangentially from the cone. The maximum brightness at the edge depends on the radius of curvature of the section of the cone and should be greater near the wide base of the flame than near the tip. The light intensity is thus a rather complex function of the position of the image on the slit. When comparing different flames these will probably have different burning velocities and consequently inner cones of different height and form. It is obvious from Figure II. 7 that with a long inner cone (*a*) the light tends to come from a thinner layer of reaction zone than in case (*b*) where the cone is flatter and the light traverses it at an acute angle.

When comparing flames over a wide range of dilution or composition it is usually necessary to use burners of different sizes, with fast burning mixtures on small burners and slow burning ones on large burners.

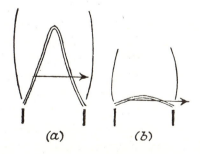

(*a*)   (*b*)

Fig. II. 7.

For band systems like C_2 and CH which come only from the inner cone, the intensity will depend on the area, thickness and character of the reaction zone, but the thickness is very small and difficult to measure, especially because it is often distorted optically by refractive index gradients in the flame, so it is usually best to relate the emission to the rate at which fuel is burnt. For comparing different fuels it may serve to relate the emission to the rate of oxygen consumption.

For band systems which are emitted by gases in thermal equilibrium, such as the various rotation-vibration bands of CO_2 and H_2O in the infra-red, the ultra-violet Schumann-Runge bands of O_2, and also for OH (for which emission may be both thermal and chemiluminescent), the emission depends on the volume and temperature of the interconal or burnt gases; while the size of the flame and the temperature distribution vary with flow rates, they also depend on heat losses to walls and on intermixing with surrounding cool air. A volume of gas will continue to emit, if kept hot, without the need for further combustion processes. For this type of emission it would be unreal to relate it to the fuel consumption as it is such a complex function of the burning conditions and gas flow. The radiation is better expressed as emission per unit volume of radiating gas, and to obtain this quantity it is necessary to measure the effective thickness of the flame, as well as its brightness.

An interesting example of the effect of flame shape on flame radiation is that of preheating on infra-red radiation. It was reported by von Helmholtz in 1890 that if a gas mixture was preheated the amount of radiation was reduced, whereas it would be expected that preheating would raise the final flame temperature and thus increase the radiation; the observation was confirmed by Haslam, Lovell and Hunneman (1925). It has now been shown (Guénebaut and Gaydon, 1957) that the effects are due to change in flame shape; without preheating the gases travel relatively slowly up the burner tube and are accelerated outwards as they pass through the rather long conical reaction zone (see Figure II.8a; with preheating (b) the thermal expansion accelerates the flame gases as they pass up the tube so that they issue from it at much higher velocity as a faster jet; also the burning velocity is higher so that the inner cone is shorter and at a less acute angle to the flow direction. Thus, as indicated in the figure,

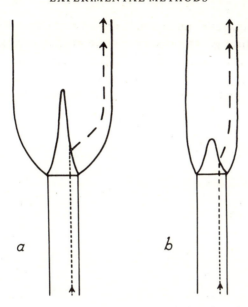

Fig. II. 8. Flow lines of flame without and with preheating

the flame of preheated gases is much thinner and rather longer. Because of the reduction in thickness the total radiation from the lower part of the flame is then reduced, despite the slightly higher flame temperature.

In some measurements of intensity, such as studies of the rotational intensity distribution of OH, self-absorption of the radiation may be important. For a conical Bunsen-type flame, radiation from the base of the inner cone passes through only a thin layer of interconal gases, but radiation from the tip of the inner cone traverses a much thicker layer and therefore suffers more absorption.

The brightness of a flame can obviously be increased by increasing the flame thickness; special flat flames are dealt with in the next chapter. For Bunsen-type flames rectangular slot burners are often used, the inner cone being examined end-on along the long direction. This type of burner does produce an appreciable increase in brightness and does weight the inner cone more highly compared with the interconal gases. The gain which can be obtained is, however, limited by aperture considerations as indicated in Fig. II.4 and discussed on page 27. If an image of the flame is formed on the slit by a condensing lens placed at some distance from the flame, so that the flame image is much smaller than the flame itself, then the cone of light through the reaction zone is narrower and a greater length can be used effectively, but the small image on the slit is difficult to focus accurately; a good achromatic condensing lens and a spectrograph which is free from astigmatism must be used.

Chapter III

Special Techniques

Direct observation of the spectrum of a simple flame gives only a limited amount of information and much of the progress in the spectroscopic study of combustion reactions has been due to the development of special techniques. Some of these aim at controlling the form of the flame so that its structure can be examined in greater detail, while others are directed towards fundamental studies of the spectra emitted or absorbed during controlled chemical reactions or well defined excitation conditions.

Flat Flames

Ordinary conical Bunsen-type flames on circular burner tubes are unsuitable for detailed examination of the structure of the flame front, because light from all parts of the reaction zone is superposed. Some improvement is obtained by working with a long rectangular-sectioned or slot burner. This gives a wedge-shaped flame, the sides of which are fairly flat and which can be viewed edge-on. There is some end effect, and there are the limitations which were discussed in the previous chapter, due to the need to use small apertures. However, a slot burner will give a useful increase in light intensity and for flames with a relatively thick reaction zone may be used to make some studies of the variation of spectrum through the reaction zone.

To obtain smooth steady flow on a slot burner, the slot must be accurately machined and the burner must be long enough to give established laminar flow. It has been pointed out in *Flames* (Gaydon and Wolfhard, 1970) that laminar flow conditions normally prevail only when the Reynolds number, R_e, is less than 2300 and that the length of the burner should be at least $0.05\,R_e l$ for the flow pattern to become established. Here $R_e = vl/\nu$ where v is the average gas velocity up the tube, v is the kinematic viscosity (= viscosity/density) and l is a characteristic length; for circular burners l is equal to the burner diameter; for a burner with rectangular sides of length a and b then $l = 2ab/(a + b)$, which reduces to $2b$ if $a \gg b$. Since v must appreciably exceed the burning velocity to prevent the flame striking back, there is a maximum burner diameter or slot width above which it is not possible to maintain laminer flow, and a little calculation shows that near this upper size limit we need a very long burner indeed to obtain smooth flow. With small burners, wall quenching becomes limiting.

Egerton and Powling (1948) designed a special burner (see diagram in *Flames*, p.68) which gives a large truly flat flame for gas mixtures whose burning velocity is below about 10 cm/sec. The flow pattern is rectified by two metal matrices in the burner tube and this tube is enclosed in an outer transparent tube with a flow of nitrogen between the tubes to isolate the reaction zone of the flame from the surrounding air. A metal gauze across the top of the outer tube holds the outer diffusion flame and prevents it burning back.

This type of burner with rectified flow pattern gives a large disc-shaped flame with a reaction zone several mm thick; it is only suitable for study of mixtures near the rich or weak limits. It was designed for measurements of burning velocity, but has proved useful for both emission and absorption studies of these slow-burning mixtures (e.g. Spokes, 1959). For faster burning mixtures a flat flame can be obtained by using a burner with a cooled sintered-metal plug (Botha and Spalding, 1954) but the reaction zone is thin and close to the metal plug and aperture considerations prevent this type of burner from being of much value for spectroscopic work.

For diffusion flames, a circular burner is again not very suitable for detailed examination of structure. The simplest flat diffusion flame is that obtained with a bat's-wing burner, but the special burner designed by Wolfhard and Parker (1949, 1952) is best and has led to really valuable information about the processes occurring in diffusion flames.

Figure III.1 shows diagrams of the burner and of the reaction zone of a typical hydrocarbon flame. The fuel gas and oxygen issue with equal velocities (of about 20 cm/sec) from parallel rectangular ducts which have one long side in common. The diffusion flame occurs at this common side. In Wolfhard and Parker's original design the rectangles were 5 cm x 0.7 cm. The flame is protected from

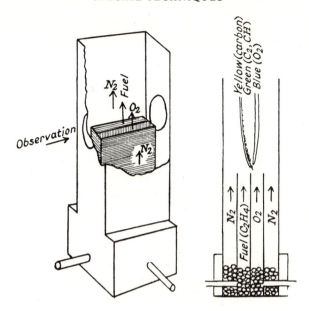

Fig. III.1. Wolfhard–Parker burner for diffusion flames.

surrounding air and draughts by an outer metal jacket, and a flow of
nitrogen at about the same speed is maintained in the space between
the burner and jacket. This jacket is fitted with quartz observation
windows and the flame is viewed along the direction of the long side
of the rectangular burner. Some heavy beads (lead shot or ball
bearings) in the bottom of the burner tubes were used to distribute
the gas flow; alternatively metal matrices may be used for this
purpose.

The thickness of the flame increases with height, and for flames
supported by oxygen is about 1 cm thick a few cm above the burner
rim. With flames supported by air it is much thinner; this contrasts
with effects in premixed flames for which air gives a thicker reaction
zone. For study of the spectrum across the reaction zone, either in
emission or absorption, it is best to turn the spectrograph on its side
so that its slit is horitontal; alternatively it is possible to use an
optical system to rotate the image through 90°. For studying this
type of flame a spectrograph which is free from astigmatism should
be used, and care must be taken to see that the aperture of the light
beam through the flame, or from it, is kept small.

For many fuels the diffusion flame, as viewed edge on, shows two
luminous zones with a darker central region. Thus hydrocarbons

show a bright yellow region due to soot formation on the fuel side, edged (especially at the base of the flame) by a greenish region showing C_2 and CH emission. The darker central region shows OH emission and absorption in the ultra-violet and vibration-rotation bands in the infra-red but little visible emission. Towards the oxygen side the flame appears blue and shows the spectrum of burning carbon monoxide and also bands of the Schumann–Runge system of O_2. For ammonia/oxygen flames the fuel side shows mainly NH_2 emission in the yellow-green and NH in the ultra-violet; the dark central region again shows OH and the blue zone on the oxygen side shows O_2 emission. Details of absorption and emission spectra will be discussed in the appropriate sections. Briefly it seems that complex fuels like hydrocarbons or ammonia are decomposed to simpler products and that the fuel and oxygen are separated by a wedge of combustion products (H_2O, CO, CO_2) which increases in thickness with height. Oxygen diffuses into this wedge from one side and the decomposition products (especially H_2) diffuse in from the fuel side.

Another useful type of flat diffusion flame is the counter-flow or opposed-jet diffusion flame (Pandya and Weinberg, 1963, 1964); a full description and diagram of the burner and discussion of the flame structure is given in *Flames*, p. 148. A mixture of fuel and diluent (N_2) flows up through one tube, and oxygen + diluent flows down through another tube held vertically above the first. A flat disc-shaped flame can be maintained between the ends of the two tubes. There are some important differences between this type of flame and that on the Wolfhard–Parker burner. Firstly the flame is out of contact of any walls and the slight premixing of fuel and oxygen which occurs near the base of flames on Wolfhard–Parker burners is eliminated. Secondly, the wedge of combustion products does not increase in thickness with distance, because the gas velocity also increases radially outwards. Thirdly, although the counter-flow burner produces a diffusion flame it is nevertheless possible to vary the ratio of fuel to oxygen and so obtain effects rather similar to those of changing the mixture strength. This type of flame is rather better for studying flames with air; in practice the nitrogen flow equivalent to that in air is usually divided between the fuel and oxygen side to maintain equal gas flows in the two tubes. Spectroscopic studies in counter-flow flames by Ibiricu and Gaydon (1964) and Laud and Gaydon (1971) are described later.

Low-pressure Flames

At reduced pressure the reaction zone or flame front for premixed gases becomes thicker, the thickness being roughly inversely proportional to the pressure. For most hydrocarbon/oxygen flames it is possible to reach about 1/100 atm. with burners and pumping systems of reasonable size, so that the reaction zone has a thickness of the order 10 mm instead of 1/10 mm. For some of the fastest burning mixtures, such as oxy-acetylene it is possible to reach even lower pressures, a little below 1 torr. Although a stoichiometric oxy-acetylene flame has a thickness of only about 1/50 mm at 1 atm, these low-pressure flames have a zone about 20 mm thick. With such a thick flame front it is possible to study its structure spectroscopically in some detail.

To obtain these low-pressure flames it is necessary to use very fast pumping and a burner of large diameter. The minimum diameter of burner on which a flame can be stabilized is again approximately inversely proportional to the pressure. The early work on this type of flame was carried out by Wolfhard (1939) and spectroscopic studies were made by Gaydon and Wolfhard. The technique is described in *Flames*, where a diagram of the simple pyrex burner used in early experiments is shown. Burner diameters up to 55 mm were used and the thick pyrex vessel itself had a volume of about 5 litres. In an improved later design, shown in Figure III.2, a large all-metal water-cooled burner was employed (Gaydon, Spokes and van Suchtelen, 1960); the height of the burner could be adjusted and a multiple-reflection mirror system was placed around the flame within the outer container. The vessel is pumped from the top through an adjustable valve, and a large (200 litre) reservoir between the valve and the pump is used to steady the rate of pumping. The flame is lit initially by a fairly powerful transformer discharge between electrodes sealed through the outer container.

Normally a thick dome-shaped flame is obtained; this corresponds to the inner cone of the ordinary Bunsen-type flame. The "interconal gases" above the reaction zone are practically non-luminous, and there is, of course, no outer cone because there is no air to maintain it. By adjustment of the gas flow and pumping rate it is possible to control the flame shape, and a flat disc-shaped flame can usually be obtained just above the burner rim. This is the most suitable form for detailed spectroscopic examination of the various zones.

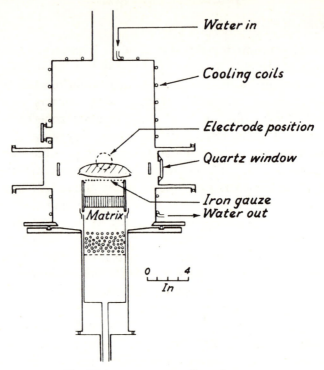

Fig. III.2. Low-pressure flame burner.

The spectra of these low-pressure flames are very similar to those of similar flames at atmospheric pressure, but there is some resolution of the reaction zone into regions in which the various band spectra occur with altered relative intensity. Thus hydrocarbon/oxygen flames show a relatively green base to the reaction zone, in which C_2 emission is strong, and a more blue-violet upper part from which CH and HCO emission is stronger. A colour photograph is reproduced in *Flames* and a spectrum, showing CH emission above that of C_2 in Plate 6*d*. OH bands are emitted strongly by all parts of the reaction zone, but only relatively very weakly by the hot products above it.

The partial resolution into zones of different colour and spectrum occurs with many flames and these are discussed in the appropriate sections. Methyl alcohol/oxygen flames have a bluish-white base which shows formaldehyde emission bands, the upper part of the reaction zone being blue-violet due to CH. Some flames, like that of iron carbonyl, show a clearly separated fore zone below the main luminous reaction zone.

Since the previous edition of this book was written, considerable progress has been made in studying the absorption spectrum of the reaction zone of low-pressure flames (Gaydon, Spokes and van Suchtelen 1960; Bleekrode and Nieuwpoort, 1965; Bulewicz, Padley and Smith, 1970). For these studies a simple multi-reflection system has been used to increase the effective optical path length. A number of measurements of effective excitation or rotational temperature through the reaction zone have also been made using low-pressure flames; the most detailed study is by Broida and Heath, (1957).

The basic chemical processes in flames at low pressure, even below 1/100 atm., do not seem to be very different from those at 1 atm. The facts that the thickness of the reaction zone is roughly proportional to the reciprocal of the pressure and that usually there is no marked change in burning velocity with pressure (see *Flames*) are consistent with a system of bimolecular reactions. The number of collisions a molecule makes in passing completely through the reaction zone is about the same at atmospheric and at reduced pressure. The low pressure will cause a slight reduction in final equilibrium flame temperature and some increase in dissociation.

Although the basic chemical reactions thus appear very similar, we may expect two differences with low-pressure flames. (1) Free atoms may be formed during the reactions and their removal normally requires three-body collisions; these will be much less frequent at low pressure and thus the removal of free atoms will be slower and afterburning effects may be increased. (2) The increase in the time between collisions will give a molecule which is formed in an excited electronic state a much better chance to radiate before it is deactivated by collision. Thus chemiluminescent radiation will be favoured at low pressure. An increase in the effective rotational temperature of OH in organic flames at low pressure is also attributed to the excited OH radicals suffering fewer collisions before they radiate. In making measurements of electronic excitation temperatures the reduced collision frequency at low pressure will tend to increase radiation-depletion effects (see page 13).

We may summarize this section by saying that these low-pressure flames are very suitable for detailed study of the structure of the flame front and that the more important reactions are similar in these and normal flames. Certain departures from chemical and thermal equilibrium will, however, tend to be greater at low pressure, and are indeed best studied under these conditions.

Temperature effects on Flames

Normally a flame of premixed gases attains nearly the adiabatic flame temperature for the particular fuel/oxidant composition. To distinguish between effects due to change of mixture strength and those due to temperature variation it is often desirable to control the temperature independently; this can be valuable for distinguishing between thermal and chemiluminescent excitation and for finding out whether processes require an activation energy.

The easiest way of reducing the flame temperature is with an inert diluent such as nitrogen or argon; a burner for stabilizing these very dilute flames is described later. Effects of dilution are somewhat akin to a reduction of pressure because the time interval between effective molecular collisions is increased and such flames tend to have a thicker reaction zone, but, unlike low-pressure flames, they have a lower burning velocity.

Flame temperatures can be raised slightly by preheating the gases, e.g. by an electric heating coil round the burner tube. For premixed flames the preheating can only be at most 300 or 400 degrees before the mixture ignites. For diffusion flames the gases may be heated separately to rather higher temperatures. For carbon monoxide flames Gaydon and Guedeney (1955) were able to preheat to 1000 K although this preheating by around 700 deg. only raised the final flame temperature of CO/air by about 250 deg. and of CO/O_2 by just over 100 deg. because the flame temperature is so strongly limited by increasing dissociation of CO_2; however it was shown that the banded CO_2 radiation was little affected by temperature because its emission is chemiluminescent, whereas the continuous CO/atomic oxygen radiation was strongly influenced by temperature.

Interest in flames at reduced temperature is stimulated by the relatively stronger emission from polyatomic molecules in these flames. These polyatomic emitters include Vaidya's hydrocarbon flame bands of HCO, the carbon-monoxide flame bands (CO_2), Emeléus's cool flame bands due to formaldehyde and the ammonia α band due to NH_2. The author (Gaydon, 1942a) found that the HCO bands were markedly enhanced when a flame was chilled by running it in a water-cooled burner tube and also the banded part of the carbon

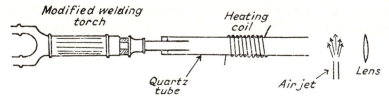

Fig. III.3. Burner etc. for highly diluted flames.

monoxide flame spectrum is relatively strengthened in cool or very dilute flames. Hornbeck and Herman (1951a) developed a simple burner for stabilising and studying flames of very weak or highly dilute mixtures. The flame was viewed end-on to obtain a greater path length, and the set-up is shown in Figure III.3. The gases were mixed in a commercial welding torch; the mixing baffle of this was removed and the end was cut off to permit high flow rates. The end of the torch was fitted into a quartz tube through a teflon plug. The centre of the quartz tube was initially heated with a winding of resistance wire, but after the mixture had been ignited and the tube had become hot enough to stabilize the flame, the external heating was cut off. A transverse jet of compressed air was used to divert the hot products so that the condensing lens and spectrograph were unharmed. This type of burner was very successfully used for studying the hydrocarbon flame bands of HCO (see Plate 4b) and for producing the CO_2 bands cleanly with adequate intensity.

For the study of the true cool flames given by various higher hydrocarbons, ethers and aldehydes at temperatures below the normal ignition temperature, Townend and colleagues (Topps and Townend, 1946) have used a conical pyrex tube in which to stabilize the flame (Fig. III.4). The mixture, consisting of oxygen saturated

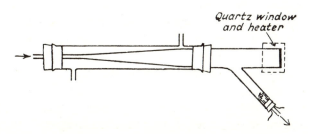

Fig. III.4. Tube for stabilizing cool flames.

with fuel at a selected temperature, entered at the narrow end of the tube and was sucked through the tube and over a platinum heating spiral by a water ejector-pump. The conical tube was surrounded with a water bath for temperature control, and the tube was viewed end-on through a quartz window, which was slightly heated to avoid condensation. Under suitable conditions of temperature and pressure a cool flame could be initiated at the platinum spiral, and this flame travelled back down the conical tube until it reached a point at which the gas flow was fast enough to arrest it. A steady flame was thus obtained and it was possible to give long exposures on the spectrum. In some cases Townend and colleagues succeeded in separating the first and second-stage cool flames.

Atomic Flames

Reactions of free atoms and free radicals are very important in the fast chemical processes, often involving chain mechanisms, which occur in flames and explosions. They are particularly important in the chemiluminescent processes which lead to light emission from the inner cones or organic flames. These free atoms and radicals can be produced in appreciable concentrations in electrical discharges through gases at low pressure, and when these gases are allowed to mix with organic vapours highly luminous glows, resembling flames, are observed. Spectroscopic studies of these "atomic flames" can give information about the specific reactions which occur in flames; in many cases these chemiluminescent reactions lead to the formation of molecules in electronically excited states or with excess vibrational or rotational energy and the low collision frequency in these atomic flames at reduced pressure facilitates observation of the abnormal energy distribution.

Early observations on the Lewis–Rayleigh afterglow of nitrogen and of the reactions of the products of a discharge through oxygen with organic and inorganic vapours were made early in the century, but the nature of active nitrogen was not properly understood and effects with oxygen were attributed to ozone formation (Strutt, 1911). Later, however, it was realised that free atoms, sometimes in concentrations up to 25%, could be produced in discharges and were responsible for the glows. Simple discharge tubes, running at gas pressures between 0.1 and 3 torr and maintained by high-voltage transformers using metal electrodes, were used for early work (Bonhoeffer and Harteck, 1928; Harteck and Kopsch, 1931; Bonhoeffer, 1936; Geib and Vaidya, 1941; Gaydon and Wolfhard, 1952). Atomic oxygen gives bright flames with most organic vapours, including methyl alcohol and some halides, such as chloroform, but there were no visible flames with methane or hydrogen. Spectra are generally similar to those of low pressure flames, but with some modification. Carbon monoxide and atomic oxygen give a pale blue glow which fills the reaction vessel but is not localised into a flame. There were many early reports of flames with atomic hydrogen, but it was usually necessary to condition the walls of the discharge tube with water or moist phosphoric acid, and it appears that most flames with atomic hydrogen and organic compounds are not real (Gaydon and Wolfhard, 1952), although luminous reactions do occur with some halides like CCl_4. Discharges through water vapour give products containing H, O and OH and these will react with many organic vapours to give bright flames. Active nitrogen, consisting mainly of free nitrogen atoms, but also vibrationally activated N_2 molecules, gives a characteristic biscuit-yellow afterglow, and when allowed to react with organic compounds gives very bright flames, usually blue or violet in colour, showing strong bands of the CN radical. This active nitrogen is not produced readily by a simple transformer discharge, but is obtained readily with a "condensed" discharge, i.e. with a capacitor and spark gap in series.

Recent work on atomic flames has been done with high-frequency electrodeless discharges; these avoid impurities from sputtering of the metallic electrodes. With radio-frequency discharges, and also the transformer discharges, it is often necessary to place an electrostatic screen of wire gauze round the tube to prevent stray discharges to the pumping system. Many recent workers have used microwave discharges; these are maintained by commercial microwave units, such as Raytheon giving 2450 MHz, and a short section of the discharge tube passes through a wave-guide tube. Discharges have been studied between gas pressures of 0.1 and 40 torr.

Estimations of the concentration of free atoms were initially made by measuring the heat release on catalytic surfaces, but are now made by gas titration methods (Kaufman and Kelso, 1957; Kaufman, 1958; Morgan, Elias and Schiff, 1961; Kistiakowski and Volpi, 1957). Atomic oxygen can be titrated by adding measured flows of

NO_2 ; with addition of a little NO_2 the rapid reaction

$$O + NO_2 = NO + O_2$$

produces nitric oxide which then reacts more slowly with atomic oxygen

$$NO + O = NO_2 + h\nu$$

giving the yellow-green "air" afterglow which spreads down the tube towards the pumps; with further addition of NO_2 a point is suddenly reached when all the atomic oxygen is used up and the afterglow suddenly ceases, this being the end-point of the titration.

Atomic nitrogen is titrated with NO. The rapid interchange

$$N + NO = N_2 + O$$

forms atomic oxygen and reduces the atomic nitrogen concentration so that the biscuit-yellow nitrogen afterglow is weakened, being partly replaced by blue and ultra-violet emission of NO bands. At the end point the afterglow is practically extinguished, and then suddenly with further addition of NO the air afterglow reaction $NO + O = NO_2 + h\nu$ starts, producing strong yellow-green luminosity filling the apparatus towards the pump.

If atomic oxygen is produced by a discharge through molecular oxygen the resulting atomic flames are initiated by free oxygen atoms but the undecomposed molecular oxygen then joins in the reactions so that the glow is partly a true atomic flame but partly a normal combustion reaction. However, it is possible to study flames produced by atomic oxygen which is free from molecular oxygen (Krishnamachari and Broida, 1961) if the atoms are produced by titration of atomic nitrogen with just sufficient NO. Figure III.5 shows the type of set-up, based on that used by Kiess and Broida (1959) and Krishnamachari and Broida. Ideally if the titration is perfect the gas should contain atomic oxygen in molecular nitrogen; in practice small residual amounts of either atomic nitrogen or nitric oxide may be expected, but Krishnamachari and Broida showed that the disturbance by these was small. Clough, Schwartz and Thrush (1970) have made further studies of flames of atomic oxygen by this technique and have passed the gases over glass wool coated with phosphoric acid to remove vibrationally excited nitrogen molecules, which are also known to occur in active nitrogen. Spectra of these atomic flames in which molecular oxygen is absent are different in

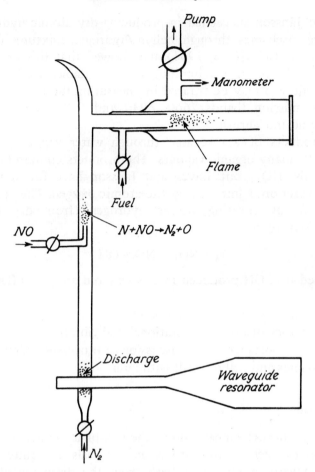

Fig. III.5. Apparatus for producing atomic flames from atomic oxygen which is
free from molecular oxygen.

many ways from the simple atomic flames with mixed O and O_2. A
simpler but less perfect way of obtaining atomic oxygen which
contains relatively little molecular oxygen is to use a microwave
discharge through an inert carrier gas (argon or helium) containing a
small percentage of oxygen (e.g. Bayes and Jansson, 1964).

Atomic hydrogen is readily produced in discharges through moist
molecular hydrogen, but not when dry hydrogen is used. The atomic
flames obtained with atomic hydrogen from the moist gas are
apparently due (Bayes and Jansson, 1964) to reactions of the
decomposition products of the water, mainly the atomic oxygen.

Bayes and Jansson succeeded in producing dry atomic hydrogen in microwave discharges through helium/hydrogen mixtures (see also Kaufman and Del Greco, 1961) and showed that this dry atomic hydrogen did not give a luminous atomic flame with acetylene. Atomic hydrogen may be titrated by measuring the intensity of the red HNO emission (bands 7660, 6930 and 6170 Å) when NO is added (Clyne and Thrush, 1961).

The products of a discharge through water vapour give good flames with many organic vapours. The products contain OH, O, H and possibly HO_2, but Bayes and Jansson have found that the luminous reaction is initiated by the atomic oxygen. They produced OH radicals by titrating atomic hydrogen (from dry $He + H_2$ discharges) with NO_2

$$H + NO_2 = NO + OH$$

and showed that OH produced in this way did not give a flame with acetylene.

The results of spectroscopic studies of these atomic flames, including studies of effective rotational and vibrational temperature, are of great importance in the discussion of processes which occur in ordinary flames and will be dealt with later.

Flash Photolysis

In ordinary photochemical studies the stationary concentrations of active species such as free atoms and radicals are quite low. If, however, a very intense light flash from the discharge of a large capacitor through a low-pressure gas discharge tube is used, a very high concentration of active species may momentarily occur. A similar flash tube which is discharged at a selected short time after the first tube may be used as background source to study the absorption spectrum of the free radicals produced by the first flash. Pioneer work on the development of this flash photolysis technique and its application to combustion problems was done by Norrish, Porter and colleagues at Cambridge; similar techniques have been developed and used by Herzberg and Ramsay in Ottawa.

Figure III.6 shows the principle of the optical arrangement. In the original work (Porter, 1950) the absorption tube A was 100 cm long and 2 cm diameter; it was made of quartz, with flat ends, and was filled at reduced pressure with the gas or gas mixture under

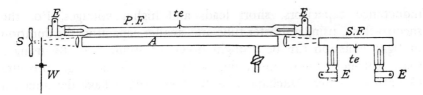

Fig. III.6. Early set-up for flash photolysis.

investigation. The photolysis flash tube P.F. was also 100 cm long and made of quartz. It was about 1 cm bore and was placed as close as practical to the absorption tube. These tubes were enclosed in a "reflector" (not shown) which was coated with magnesium oxide; this served to concentrate the light. The flash tube had stout tungsten electrodes E, and also a small trigger electrode *te*; it was filled with a rare gas (krypton or xenon) to a pressure of around 60 torr. In early work Porter used the discharge from a large bank of capacitors of from 48 to 480 μF charged to 4000 V. This voltage was insufficient to start the discharge and a trigger discharge from an induction coil was used to initiate and time the flash. The background source for the absorption spectrum, usually referred to as the spectro-flash tube SF was in Porter's experiments basically similar to the photolysis flash tube except that it was shorter with side electrodes 15 cm apart and it had a flat quartz end; this tube was filled with krypton to rather higher pressure, 100 torr.

In these early experiments by Porter a rotating wheel W with a speed up to 600 r.p.m. was used for timing the photolysis flash and spectro-flash. This wheel was placed in front of the spectrograph slit S and had a slot cut in it so that light from the spectro-flash could fall on the slit but scattered light from the earlier photolysis flash was intercepted. Moving contacts to the two trigger electrodes were attached to the wheel. By adjustment of these the time between the two flashes could be varied so that the absorption spectrum could be photographed at any desired time interval after the photolysis flash. In practice it is usual to take a series of exposures of the spectrum, side by side, with different delay times to obtain a time history of the variation of the absorption spectrum.

In this early work the duration of the photolysis flash was relatively long, lasting from ½ to 3 millisecs, but in later work (e.g. Calloman and Ramsay, 1957, Basco and Norrish 1961) this has been reduced to about 20 to 30 microseconds by using special low-inductance capacitors, short leads and higher voltages. For the spectro-flash, using a shorter tube and smaller capacitor, the duration of light emission can be brought down to about 3 μs. By using a spark through argon containing a trace of air at high pressure (3.7 atm) Fischer, Duchane and Büchler (1965) have developed a background source whose light emission has a half-life of only 28 nanosec. Even higher time resolution (18 ns) has been obtained by Porter and Topp (1968) using a Q-switched frequency-doubled argon laser. The light from the laser, at 3471 Å, is split into two beams; the first beam passes directly through the sample and serves as the photolysis flash; the second beam is delayed by an optical path and is then used to stimulate continuous-spectrum fluorescence from a dye, this serving as the spectro-flash.

Later developments are the use of electronic delay units instead of the rotating wheel to control the interval between the photolysis and spectro-flash, and tubes fitted with lithium fluoride windows so that the absorption spectra of the transient species can be studied in the far ultra-violet, below 2000 Å (Thrush, 1958).

This flash photolysis techinque was at first used for studies of photochemical decomposition and for detecting the free radicals so formed, and for studying their rate of recombination or reaction. Among the radicals which may be of special interest in combustion which can be studied by this technique we may note CH_3 and CH_2 (Herzberg 1961), HCO (Herzberg and Ramsay, 1955) and NH_2 (Herzberg and Ramsay, 1953). In many cases flash photolysis produces, either directly or as a result of subsequent reactions, molecules with high vibrational energy, and the subsequent energy transfer processes can be studied; thus Lipscomb, Norrish and Thrush (1956) found that in flash photolysis of NO_2 and ClO_2 oxygen molecules were formed with up to eight quanta of vibrational energy, probably by reactions $O + NO_2 = NO + O_2^*$ and $O + ClO_2 = ClO_2 + O_2^*$.

In the direct study of flash photolysis of molecules very large amounts of energy may be absorbed; thus the decomposition of NO_2 may be nearly complete and Porter states that for Cl_2 about 80% is dissociated; there is thus a very considerable energy input into the irradiated gas, raising it to a high temperature. This may produce chemical change and may also modify the absorption spectrum of any undecomposed molecules by giving them rotational and vibra-tional energy. If the flash photolysis is carried out in the presence of a large excess of inert diluent (e.g. argon or nitrogen) then the final gas temperature can be kept down. This is advantageous both because the spectra are simpler at low temperature and because it is easier to study the primary processes.

A photolysis flash may also be used to initiate explosive reactions in combustible mixtures. Reaction is initiated simultaneously throughout the whole mixture so that it is possible to study the spectrum of a long length (up to a metre) of reacting gas instead of that of a relatively thin flame front. For some fuels, such as acetone, diacetyl and ketene, the ultra-violet light from the flash is absorbed by the fuel molecules, which are decomposed and start the reaction, but most of the common fuels are transparent in the quartz ultra-violet, as also is oxygen, so that it is necessary to add a third gas as sensitizer. Nitrogen peroxide has been used successfully for this purpose and the absorption spectrum shows that it is almost completely decomposed to NO + O by the initial flash. Norrish and Porter (1952) used this method to study the hydrogen-oxygen reaction. The most important feature of the absorption is the OH band; its variation with time, mixture strength and pressure have given valuable information about the kinetics and reactions.

Acetylene-oxygen explosions initiated by the flash photolysis of nitrogen peroxide were studied by Norrish, Porter and Thrush (1953) and it was possible to observe absorption bands of C_2, CH, CN, NH and C_3 (4050 Å group) as well as OH. The C_3 occurred only in rich mixtures and was accompanied by regions of continuous absorption around 3800 Å and below 3000 Å. The OH absorption appeared first and in the weaker mixtures this continued to increase in strength over the period when the C_2, CH and CN were observed, this OH reaching its intensity maximum later than the other radicals. The time scale was fairly short the absorption passing its peak usually within 500 μsec.

The importance of the techinque is obvious. The long path length enables absorption by ground-state radicals like C_2 and CH to be studied readily and their concentrations measured as a function of time and other variables. In some ways, of course, the conditions in the reacting gases following photolysis differ from those in a flame front. While the NO_2 is largely decomposed the products certainly contain NO and possibly a residual trace of NO_2; it is known that only 0.06% of NO_2 in stoichiometric oxy-hydrogen will lower the ignition temperature by 200 K and it is difficult to feel quite convinced that the oxides of nitrogen are entirely inactive after the photolysis. For those fuels which absorb the ultra-violet light themselves, there must be a tendency for the fuel molecules to be decomposed to such an extent that normal reactions between oxygen and parent fuel molecules no longer occur as in the ordinary combustion. The photolysis process will also produce a very high concentration of radicals and electronically excited species, and even if these are undetected immediately after the flash (e.g. O atoms will not be detected) they may still influence the final concentration of other radicals such as OH which are in pseudo-equilibrium. The photolysis-initiated combustion also produces higher temperatures than in flames both because of the energy supplied by the flash and because it occurs under closed-vessel conditions for which the specific heat (at constant volume) is lower; these very high temperatures may modify the course of the reactions. However, while these limitations should not be forgotten, the technique is a very valuable one for the spectroscopic study of combustion processes.

Shock Tube Studies

The simple bursting-diaphragm shock tube is now an established tool for studying high temperature processes; the basic technique and its applications are reviewed in the book by Gaydon and Hurle (1963) and papers on the latest developments are collected in *Shock-Tube Research*, edited by Stollery, Gaydon and Owen (1971). In its simplest form a shock tube consists of a long tube, 6 ft or more in length and 2 inches or so diameter, divided into two sections separated by a thin metal or melinex diaphragm. One section is filled to high pressure (~ 10 atm.) with a low density driver gas, usually hydrogen or helium, while the other section is evacuated and filled at reduced pressure (~ 1/100 atm.) with the "experimental" gas or gas mixture to be heated. When the diaphragm is ruptured, either by piercing it or by raising the driver-gas pressure until it bursts spontaneously, the driver gas rushes out and drives a strong shock wave into the experimental gas. This shock wave compresses and heats the experimental gas. With the simplest design of tube a shock wave up to about a Mach number of 8 can readily be obtained, this raising molecular gases to over 3000 K and heavy monatomic gases to over 6000 K; much higher temperatures can be obtained if necessary by using combustion or detonating driver gas or using the region behind the reflected shock at the end of the tube.

This method of heating the gas has many advantages. The equilibrium temperature can readily be calculated from a knowledge of the initial conditions and a measurement of the shock speed, or it may be measured directly by a spectrum-line reversal method. The slug of shock-heated gas is very uniform, and the boundary layer at the wall is usually very thin. The heating process is almost instantaneous and initially takes the form of high translational energy, which is then converted by collisions to rotational, vibrational, electronic and chemical forms; in this respect the shock-tube is complementary to flash photolysis in which energy is supplied as light, producing electronically excited species and causing chemical dissociation before it is degraded to vibration, rotation and translational energy. The experimental gas can be heated to any selected temperature over a wide temperature range by using the appropriate pressure-ratio across the diaphragm, and it is possible to study the pyrolysis of pure fuels, combustible mixtures or gas mixtures which are outside the normal flammability limits.

Strong shocks produce a bright flash of light and the spectrum of this can be photographed, or alternatively a monochromator, photomultiplier and oscilloscope may be used to study the time history of a selected feature in the emission spectrum. The absorption spectrum of the reacting shock-heated gas can also be photographed at any selected time by arranging for a flash tube to be triggered, after a suitable delay time, by the shock wave; this technique is similar to flash photolysis with shock heating replacing the photolysis flash.

The method is particularly suitable for studying detonations. In a natural detonation the combustion wave is propagated by a shock which heats and ignites the mixture, the heat release and gas expansion from the combustion processes then maintaining the shock. Shocks from bursting diaphragms can be used either to initiate normal detonation or to force a detonation-type wave through a gas which would not normally support a detonation. As examples of early studies of the absorption spectra of detonations in shock tubes we may cite the work by Schott (1960) on OH concentration maxima in detonating oxy-hydrogen and by Lyon and Kydd (1961) who studied C_2, CH and OH absorption as a function of time in acetylene detonations. Our knowledge of the spectra of detonations is still somewhat meagre and further work using shock tubes would be helpful; in ordinary flames the propagation involves diffusion of free radicals and atoms which initiate the chemical chain reactions, but in detonations these chain reactions must be initiated thermally by gas-phase reactions which may involve peroxides or other intermediates in the early stages, so that some differences between flame and detonation spectra may be expected.

The shock tube has also proved useful in recent studies of activation energies and kinetics of pyrolysis processes and for detecting intermediates in soot formation.

Fluorescence and Laser—Raman Scattering

When a beam of monochromatic light passes through a flame it may either be absorbed, if its wavelength coincides with an absorption line of a species in the flame, or it may pass through unattenuated apart from very slight scattering.

If the light is absorbed it may lead to electronic excitation of the absorbing atom or molecule and may be followed by resonance fluorescence. The excited species may however be quenched by collisions with other flame molecules so that the strength of the fluorescence will depend on the initial light intensity, the concentration of absorbing species and also on the relative probabilities of radiative emission and collision quenching. Collisions may quench the electronic excitation or leave a molecule in an excited electronic state but with different rotational or vibrational energy, in which case the subsequent light emission will be at a different wavelength. The use of light-chopper techniques enables the fluorescent emission to be separated from the steady flame emission. Quantitative measurements of fluorescent intensity in various flames can give valuable background information about the basic collision processes actually occurring under flame conditions.

Atomic fluorescence from metals is now an established technique in spectrochemical analysis. Quenching by collision processes is very important in flames at atmospheric pressure, and in a study of sodium fluorescence Boers, Alkemade and Smit (1956) measured the ratio of the collision to radiative lifetimes; they showed that in normal flame gases collision processes maintained quite good equilibrium and that radiation depletion would only lead to a maximum error of 8K in temperature measurement by the sodium-line reversal method. Jenkins (1966) has made measurements of collision cross-sections for sodium for a variety of flame gases. In this type of work fluorescence is usually excited by the metallic line emission from a discharge lamp or hollow-cathode lamp. Measurements of OH fluorescence using OH emission from a microwave discharge through water vapour by Carrington (1959a) showed that even in low-pressure flames collision deactivation of electronically excited OH was dominant, but in later work in which the OH fluorescence was excited by a single line in the bismuth spectrum it was found that collisions produced a change in rotational energy without electronic deactivation about twice as frequently as they produced electronic deactivation; the total quenching cross-section of electronically excited OH was about twice the gas-kinetic value (Carrington, 1959b). This fluorescence technique is likely to become of increasing importance as recording methods improve. Thus as a recent example Sharma and Joshi (1972) have been able to measure the concentration of ground-state BO_2 molecules in atomic oxygen/triethyl boron flames by a fluorescence method.

When there is no absorption, and subsequent fluorescence, the scattering of light by a pure molecular gas is extremely weak; thus a beam of green light will travel about 100 miles through pure air before it is attenuated to one half its initial intensity. Most of the scattered light is then of the same wavelength as the original light (Rayleigh scattering) apart from small Doppler shifts due to random thermal motion of the scattering molecules. However a weak second-order effect, about 10^3 weaker than the Rayleigh scattering, is important; this is the Raman effect (see page 88). Changes in the vibrational or rotational energy of the scattering molecule result in the scattered radiation having its frequency altered to certain other discrete values.

The advent of the laser, an intense source of monochromatic light, has made possible the study of the Raman spectra of gases, and this could become important in flame studies (Widhopf and Lederman, 1971; Lapp, Goldman and Penney, 1972; Jessen, 1971; Vear, Hendra and Macfarlane, 1972) for determining concentrations and rotational and vibrational temperatures of ground-state species at well localised positions in the flame. Scattering is strongest for light of high frequency, and the best gas laser at present available is probably the argon ion laser giving a line in the blue-green at 4880 Å. If a vertical laser beam is brought to a focus in the flame the scattered light may be viewed horizontally, i.e. at right angles, and the image will then lie well on the vertical slit of the monochromator. It is usually necessary to use a good grating double-monochromator to reduce stray light. The light which suffers Rayleigh scattering is strongly polarized while the Raman scattering may or may not be polarized (this depends on the symmetry properties of the vibrational modes involved, see Herzberg, 1945) and it is often possible to filter out much of the Rayleigh scattering with suitably oriented polaroid. Lapp and colleagues have observed Raman scattering from nitrogen, oxygen and water molecules in hydrogen flames and shown that it is possible to determine the rotational and vibrational temperatures. Since the observations relate to the region of the focused laser beam, which is viewed at right angles, the spatial resolution is very good. Also the rotational temperature of the ground-state molecules is certain to be in almost complete equilibrium with the transitional temperature of the gas. Thus this seems an ideal method of measuring flame temperature, but the experimental equipment, involving laser, light chopper, good monochromator and very sensitive photomultipliers, will be expensive. These laser—Raman studies will probably be limited to the main flame gases and are unlikely to give information about radicals or other species present in small concentrations.

The laser also has new possibilities in flame studies for measurement of gas flow velocities by Doppler shift in the wavelength of the scattered light (Schwar and Weinberg, 1969a, 1969b) but this technique will not be discussed here.

Use of Isotope Shifts

In molecular spectra the vibrational frequencies depend on the masses of the vibrating atoms so the spectra of molecules composed of different isotopes show slight relative shifts. It is now possible to obtain separated stable isotopes for many elements and studies of the band spectra of isotopic molecules can be used in several ways.

The identification of an emitting species can sometimes by made or confirmed in this way. The best example is the Comet-head band group at 4050 Å; this was first observed in the spectra of comets and subsequently obtained in electrical discharges through flowing hydrocarbon vapours and in carbon arcs in a hydrogen atmosphere, and it also occurs in rich acetylene flames. The band was initially attributed to the radical CH_2, but attempts to observe an isotope shift when hydrogen was replaced by deuterium failed and clearly showed that the emitting molecule did not contain hydrogen (Monfils and Rosen, 1949); Douglas (1951) then made similar studies with the carbon isotopes C^{12} and C^{13} and various mixtures of these; the main band head was split into six components with the isotopic mixture and the emitter was certainly identified as the carbon radical C_3.

The origin of Vaidya's hydrocarbon flame bands was in doubt for many years, suggested emitters being HCO, CH_2 and HCOH; Spokes and Gaydon (1959) found that the spectra of flames of C_2H_2 and of C_2D_2 were different, proving that the emitter did contain hydrogen, but they also found that the spectrum of a flame of mixed normal and deutero-acetylene showed only a superposition of the spectra of flames of C_2H_2 and C_2D_2 without any new spectrum; if the emitter had been CH_2 then bands of three types due to CH_2, CD_2 and CHD should be obtained, and similarly formaldehyde or an isomer of it should show CH_2O, CD_2O and CHDO; thus the assignment to HCO was strongly supported.

Measurements of isotope shifts can also be useful when making the analysis of a difficult spectrum, being especially valuable for assignment of vibrational quantum numbers and location of the system origin. The analysis of the structure of the band system often gives valuable data about dissociation energies and thermodynamic quantities which are of indirect value in combustion studies.

Tracer techniques have also been used to elucidate the detailed chemical processes which may occur in flames. These are discussed in Chapter VIII, page 217. Most work has been done with flames containing hydrogen and deuterium to determine the processes by which excited CH and OH are formed. Similar work has also been done using flames supported by free atoms, these being more economical in the amount of isotope used than are ordinary flames at atmospheric pressure.

The formation of excited C_2 radicals in flames has been most difficult to understand. Ferguson (1955) used mixtures of acetylene containing the isotope C^{13} with ordinary acetylene and found that in the resulting C_2 emission the $C^{12}C^{13}$ ratio was almost completely randomised. This appears to disprove the theory of successive stripping of H from acetylene

$$C_2H_2 + H = C_2H + H_2 ; \text{ then } C_2H + H = C_2 + H_2$$

as this would leave the C^{13}/C^{12} ratio in the band spectrum as in the original acetylenes. Fairbairn (1962*a*) made some similar studies of C_2 emission from shock-heated acetylene/argon mixtures and again found more than 67% randomization; this shock-tube study is again relatively economical in isotope consumption.

So far this isotope technique has been applied mainly to emission spectra. Apart from a preliminary study of OH by Gaydon, Spokes and van Suchtelen, it has not been used for absorption, which might be more valuable as it gives information about ground-state molecules; now that improved techniques are available for detecting absorption bands in flames (see page 26) some isotopic studies of these might be helpful.

Chapter IV

Introduction to the Theory of Spectra

The object of this chapter is to provide a descriptive and entirely non-mathematical introduction to the general theory of spectra, especially of molecular spectra, to serve as a background for the reader who is unacquainted with spectroscopic theory. This monograph is not, of course, intended to serve as a textbook on spectra and it is necessary here to present the accepted facts and theory briefly as bald statements, without discussing the observations and theoretical developments which have led to their adoption. For a complete treatment of the theory of molecular spectra the reader is referred to the three books by Herzberg; these works on diatomic spectra (1950), infra-red and Raman spectra (1945) and electronic spectra of polyatomic molecules (1966) are excellent comprehensive treatises which are quite indispensable for anyone working in these fields, but are written at an advanced level and for less difficult introductions to the subject the books on the spectra of simple free radicals (Herzberg, 1971) and on flame spectroscopy by Mavrodineanu and Boiteux (1965) may be better. The book on Dissociation Energies (Gaydon, 1968) gives a rather similar treatment to that presented here but with some more detail.

As pointed out in Chapter 1, spectra fall into three main classes, line spectra due to free atoms, band spectra due to molecules, and continuous spectra due to processes such as ionization, dissociation or association and to emission from hot solids.

Line or Atomic Spectra

The spectra characteristic of many atoms, especially those in Groups I and II of the Periodic Table, appear fairly simple, but they nevertheless defied systematization for a long time. Balmer first arranged the lines of the very simple hydrogen spectrum into a series, the original arrangement in terms of the wavelengths of the lines being in a relatively complex form. The later realization that it was the reciprocal of the wavelength, that is the wave-number (proportional to the frequency) of the light which was, for theoretical purposes, the more fundamental, enabled the wave-numbers of the lines of the hydrogen spectrum, both of the visible series and of other series in the ultra-violet and infra-red, to be expressed in the simple form

$$\nu = R_H (1/n_2^2 - 1/n_1^2),$$

where n_1 and n_2 are integers such that $n_1 > n_2$ and n_2 takes the values 1, 2, 3 etc., and R_H is a constant known as the Rydberg constant for hydrogen. $m_2 = 1$ gives the far ultra-violet Lyman series; $n_2 = 2$ gives the visible Balmer series; $n_2 = 3$ gives the infra-red Paschen series, and higher values of n_2 give other series in the extreme infra-red.

Following this, it then became possible to arrange the spectrum lines of many other elements into series so that the wave-numbers could be expressed as the difference between two terms. The spectra of many elements can be represented by formulae of the type

$$\nu = R(1/(n_2 + a_2)^2 - 1/(n_1 + a_1)^2),$$

the Rydberg constant R having a value only very slightly different from that for hydrogen, and a_2 and a_1 being additional constants peculiar to the element whose spectrum is represented.

Bohr Theory

According to Bohr's theory, atoms and molecules can only exist in certain discrete energy states called stationary states which are selected by certain quantum conditions from the classically possible states. Thus for the hydrogen atom the electron may, classically, revolve around the nucleus in any form of ellipse, but on the Bohr theory the values of the major and minor half axes of the ellipse are restricted to certain values proportional to the squares of the natural numbers; the number n for the semi-major axis takes the values 1, 2, 3 etc. and is known as the principal quantum number. The number l for the value of the semi-minor axis takes the values 0, 1, 2, 3 etc. up to a maximum value of $n - 1$, and is known as the azimuthal quantum number.

According to Bohr, the emission or absorption of radiation occurs during the transition of the atom or molecule from one energy state to another, that is, for atomic spectra, when an electron changes its orbit. The relation between the frequency $\bar{\nu}$ of the radiation emitted to the energies E_1 and E_2 of the two states of the atom or molecule is given by the well-known expression

$$E_1 - E_2 = h\bar{\nu} = hc\nu,$$

where h is a universal constant known as Planck's constant, c is the velocity of light and ν is the wavenumber (in cm^{-1})*. The energies E_1 and E_2 correspond to the terms used in expressing the series in line spectra. The Rydberg constant used in these expressions can be derived in terms of Planck's constant and the charge e on the electron and has the value

$$R = 2\pi^2 e^4 / ch^3 \times mM/(m + M),$$

where m and M are the masses of the electron and the atom respectively. For $M \gg m$ this takes the value $109737.3 \ \mathrm{cm}^{-1}$.

All possible transitions between the energy states of the atom do not occur, and this is expressed by a number of *selection rules* regulating the possible changes in the various quantum numbers; originally these rules were largely empirical or were derived from the correspondence principle relating the results for high values of the quantum numbers to classically expected changes, but they can now be derived from wave mechanics. The line spectra, characteristic of free atoms, are thus due to electronic changes or *transitions*, and the first stage in the interpretation of a spectrum is to build up an energy-level scheme for the atom which will give the observed spectrum lines.

The simple Bohr model of the atom is not, of course, to be accepted too literally. According to more modern views the position of a moving electron cannot be specified exactly because of its own wave nature, so the Bohr orbits must be replaced by probablility

---

*The wave-number when expressed in cm^{-1}, i.e. the number of waves in a cm., is a convenient-sized unit for use with visible and near ultra-violet spectra, yellow light being $17\ 000 \ \mathrm{cm}^{-1}$; the S.I. unit, in waves per metre, involves seven significant figures and is inconvenient. Wavelengths are usually measured in air, but wavenumbers are always corrected to vacuum, so that they are proportional to frequency.

distributions or electron clouds. The symmetry properties of these electron clouds then replace the eccentricities of the Bohr orbits.

The electrons surrounding the nucleus tend to form closed shells of electrons, two in the innermost or K shell with $n = 1$, eight in the next L shell with $n = 2$, eighteen in the M shell with $n = 3$ and so on. It is the electrons which are left over outside these filled shells which are important for optical spectra. The movement of one of these electrons from one orbit to another is responsible for spectrum lines in the visible and near ultra-violet.

Alkali-metal atoms all have one electron outside the closed shells. The energy level depends mainly on the value of the principal quantum number n for this electron, but also to some extent on the azimuthal quantum number l, because the electron when in a more eccentric orbit penetrates at times through the screening inner electrons and comes nearer the highly charged nucleus. The constants a_1 and a_2 in the series formulae (page 66) depend on this penetration and thus on the quantum number l. The regular series in the simpler line spectra were for historical reasons known as the sharp, principal, diffuse and fundamental series and the initial letters s, p, d, f, are now used to denote orbits with azimuthal quantum numbers $l = 0$, 1, 2, and 3 respectively. There is no selection rule for n, which may change by any amount during a transition, but l may change only by ± 1 so that transitions $s \to p$, $p \to s$, $p \to d$ etc are allowed, but transitions $s \to s$, $s \to d$, $f \to p$ etc. are forbidden.

The electron has also an angular momentum of its own, usually called its *spin**, and the energy of the system is slightly modified according to whether the spin is with or opposed to the orbital motion. Thus all energy levels for alkali-metal atoms are doubled by the electron spin; this is the cause of the well-known doublets in these spectra, such as the yellow "D-lines" doublet of sodium.

For the alkaline-earth-metal atoms there are two outer electrons. The state of the atom is determined by the symmetry properties of the whole electron cloud. The small letters s, p, d, f, g etc are used to denote the azimuthal quantum numbers of individual electrons and the capitals S, P, D, F, G etc are used to denote the resultant orbital angular momentum of the electrons for the whole system; the quantum number for this resultant is L. Both the electrons have their

---

*The spin is denoted by a quantum number s which takes half integral values and for a single electron is $\pm\frac{1}{2}$.

own spin and according to whether the two spins are opposed or in the same direction, so the energy terms are single or split into triplets. For strong transitions the selection rules require that the multiplicity does not change, that is singlet-singlet and triplet-triplet transitions are allowed and singlet-triplet are forbidden.

Similar principles apply to the spectra of atoms with a greater number of outer electrons. Three electrons lead to doublet and quartet terms, four electrons to singlet, triplet and quintet terms. The full designation of an atomic energy level needs to give information about the orbits of individual electrons and about the resultant configuration. Usually it is sufficient to specify only the outer electrons, which are denoted by the value of the principal quantum number n and a letter s, p, or d etc which indicates the value of l. When there is more than one electron in a sub-shell the number of electrons is denoted by a superscript after the letter s, p etc, e.g. $3p^4$ to indicate four electrons with $n = 3$ and $l = 1$. The resultant orbital angular momentum of all the electrons is denoted by the capital letter S, P, D etc which indicates the value of L, and this is followed by a subscript which denotes the value of a quantum number J which is the resultant of the spin and the orbital angular momentum. Thus for example a level of Mg might be denoted

$$3s\ 3d\ ^3D_1$$

indicating two outer electrons, one with $n = 3$ $l = 0$ the other with $n = 3$ $l = 2$, the whole atom being in a triplet state with resultant orbital angular quantum number $L = 2$ and with final resultant $J = 1$. As a second example the ground state of Cr is

$$3d^5\ 4s\ ^7S_3$$

indicating five $3d$ electrons, one $4s$ electron and the complete configuration being a septet with $L = 0$ and $J = 3$; in this case although the multiplicity is seven the level S is still single; for P states the maximum splitting is into three, and for D terms is into five. The value of J is important in determining the number of levels into which a term is split by a magnetic field (Zeeman effect). The above discussion of the designation of atomic energy terms is that applicable to all but heavy atoms and applies in what is known as Russell–Saunders coupling. For some heavy atoms, or highly excited states of less heavy atoms, the orbital angular momenta and spins of the individual electrons are coupled to give resultants specified by quantum numbers j, and the final state of the atom is formed by a result of these and is designated by a value of J alone.

The Spectra of Diatomic Molecules: Vibrational Structure

Many band spectra which lie in the visible and near ultra-violet, such as, for example, the Swan bands which occur in hydrocarbon flames, show obvious regularity. A group of related bands of this type is known as a *system* and consists of a number of bands each with a sharp edge or *head* on one side and all shaded off or *degraded* the same way. When the spectrum is photographed under large dispersion each band is found to consist of a number of lines, also regularly arranged.

It is known that the position of the band system as a whole is determined by the electronic energy change, that is band spectra in the visible and ultra-violet are due to electronic transitions, as in atoms. The electronic energy levels of molecules are, however, usually much fewer in number than those of atoms, so that the number of band systems observed is much less than the number of lines observed in the spectrum of an atom. In only a few cases has it been possible to arrange band systems into series corresponding to the Rydberg series of line spectra.

The gross structure of a band system, that is the splitting of the system into a number of bands, is due to the contribution to the energy of the molecule made by the vibration of the atoms along the internuclear axis. Treating the molecule as a harmonic oscillator, it is found that the vibrational energy is quantized and takes values given by the product of the characteristic vibrational frequency, Planck's constant and the vibrational quantum number. This quantum number, denoted u, takes half integral values ½, 1½, 2½, 3½ etc., but it is usually convenient in practice to suppress this fact by using a quantum number v which takes integral values* and inserting the actual energy difference between the vibrational levels of lowest energy for the two electronic states instead of the change in electronic energy itself. With this simplification, each electronic energy level may be considered as split into a number of vibrational

---

*Thus $u = v + ½$. It is important to use u rather than v when calculating the isotopic displacement of vibrational energy levels.

levels given by

$$E = E_0 + v \cdot \omega$$

v being the vibrational quantum number taking values 0, 1, 2, 3 etc., ω a constant dependent on the vibrational frequency (usually referred to loosely as the vibrational frequency itself), and E_0 is the electronic contribution to the energy. Taking into account deviations from simple harmonic vibration

$$E = E_0 + v \cdot \omega - v^2 \cdot x\omega + v^3 \cdot y\omega + \text{etc.}$$

For electronic transitions, all changes of the vibrational quantum number are permitted so that the positions of the bands of the system may be expressed by the formula

$$\nu = E_0' + v' \cdot \omega' - v'^2 \cdot x'\omega' + \text{etc.}$$
$$-(E_0'' + v'' \cdot \omega - v''^2 \cdot x''\omega'' + \text{etc.})$$

where the primes $'$ and $''$ refer to the upper and lower electronic states respectively.

In making the vibrational analysis of a band system it is usual to try to arrange the bands into an array between the values of the initial and final vibrational quantum numbers v' and v''. Any band may be referred to by the values of the quantum numbers as the (v', v'') band. The wavelengths λ, wave-numbers ν and visual designations of intensity I for the strongest bands of the well-known Swan system of C_2 are given in Table IV.1 and the wave-numbers are then arranged in Table IV.2 in an array of the type mentioned. In this table the differences between the wave-numbers of the bands are printed in italics.

TABLE IV.1
Strong band heads of the Swan system of C_2

λ	ν	I	λ	ν	I	λ	ν	I
6191.2	16147	3	5501.9	18171	4	4697.6	21282	7
6122.1	16330	4	5165.2	19355	10	4684.8	21340	4
6059.7	16498	3	5129.3	19490	6	4382.5	22812	2
5635.5	17740	8	5097.7	19611	1	4371.4	22870	4
5585.5	17899	8	4737.1	21104	9	4365.2	22902	5
5540.7	18043	6	4715.2	21202	8			

TABLE IV. 2.

Vibrational array for the heads of the strong bands of the Swan system of C_2

v' \ v''	0		1		2		3		4
0	19355	1615	17740	1593	16147				
	1749		1750		1752				
1	21104	1614	19490	1591	17899	1569	16330		
	1708		1712		1712		1713		
2	22812	1610	21202	1591	19611	1568	18043	1545	16498
			1668		1671				1673
3			22870	1588	21282				18171
					1620				
4					22902	1562	21340		

It may be seen that the differences in any row, vertical or horizontal, are not far from constant, and that the wave-numbers of the bands may be expressed with reasonable accuracy by the formula

$$\nu = 19355 + 1770v' - 20v'^2 - (1625v'' - 11.5v''^2)$$

which is of the form given above, so that we see that the constants 1770 and 1625 correspond approximately to the vibrational frequencies ω' and ω'' of the molecule in the two electronic states. It is customary to use the symbols ω_0 and $x_0\omega_0$ for the frequencies defined in this way[*]; when insufficient data are available for calculating ω_0 or ω_e the energy separation $\omega_{1/2}$ between the two lowest levels is frequently given.

The bands which fall in diagonal rows often form close groups, and these are known as *sequences*; thus the (0,0), (1,1) and (2,2) bands at 19355, 19490 and 19611 cm^{-1} form such a sequence. Bands in the vertical and horizontal rows form more widely spaced series known as *progressions*; the (0,0), (0,1) and (0,2) bands at 19355, 17740 and 16147 form such a progression, all the bands

[*]The vibrational frequency constants for infinitesimal amplitude using the half-integral vibrational quantum number u are usually written ω_e and $x_e\omega_e$; in all these formulae $x\omega$ is treated as a single constant.

having $v' = 0$, and hence this is known as the $v' = 0$ progression.

If all bands of a system can be arranged into an array such as that of Table IV.2, then it is usually safe to assume that the emitter is a diatomic molecule. The value of the vibrational frequency depends on the binding force between the two atoms of the molecule and on their mass, and comparison of the vibrational frequency with those known for other molecules is often helpful in identifying the molecule responsible for the band system.

In making a vibrational analysis it is sometimes necessary to take into account certain refinements for which the reader is referred to the larger books on the subject. The array in Table IV.2 is drawn up from data for band heads so that the differences corresponding to the vibrational energy intervals are only approximate; correctly, data from band origins (see below, under rotational structure) should be used. The electronic energy levels are in some cases split into two or more components by the spin of the electrons; bands will then possess multiple heads, sometimes widely spaced, and this may complicate the appearance of the vibrational structure. It is a general but not quite invariable rule that the bands are degraded in the same direction as the vibrational structure; thus if $\omega' > \omega''$ the bands will be shaded to the violet, and if $\omega' < \omega''$ the bands will be degraded to the red; this is often of assistance in making the analysis. Systems for which ω' and ω'' are not very different form well-marked sequences, and when ω' and ω'' are very different the sequences are less obvious but the progressions are easier to pick out. When one or both of the atoms of the molecule possesses isotopes of comparable abundance, the bands of the spectrum are all split into components due to the various isotopes; the vibrational frequency is proportional to the square root of the force constant divided by the reduced mass $\mu = m_1 m_2 / (m_1 + m_2)$, and hence measurements of this isotope splitting are often useful in locating the (0,0) band, which has the smallest splitting, and in assigning the emitter of the system.

Rotational Structure of the Spectra of Diatomic Molecules

The fine, line, structure of individual bands, such as that which can readily be seen for the spectra of CH and OH in the plates, is due to the rotation of the molecule about its centre of gravity.

For the simplest type of electronic state, that known as $^1\Sigma$, the extra energy of the molecule due to its rotation may have the values given by the expression

$$E_r = B \cdot J(J + 1)$$

or with slightly greater accuracy by

$$E_r = B \cdot J(J + 1) + D \cdot J^2 (J + 1)^2 + \text{etc}$$

where J is the rotational quantum number, taking the integral values 0, 1, 2, 3 etc for a $^1\Sigma$ state, and B and D are constants. B depends on the moment of inertia, I, and has the value

$$B = h/8\pi^2 cI$$

where the moment of inertia is the product of the reduced mass and the square of the internuclear distance, $I = \mu r^2$. The constant B may be written as B_e, B_0 or B_v according to whether it refers to the completely vibrationless state, to the lowest vibrational level ($v = 0$) or to a vibrational level v.

It is found that for a transition between two such $^1\Sigma$ states the change ΔJ of the rotational quantum number is restricted to the values ± 1. Thus, using primes $'$ and $''$ to refer to the upper and lower electronic states respectively, the lines of any one band with origin v_0 are given by

$$v = v_0 + B' \cdot J (J' + 1) - B'' \cdot J'' (J'' + 1)$$

or since $J' = J'' \pm 1$
the lines fall into two series

$$v = v_0 + B'(J'' + 1)(J'' + 2) - B'' \cdot J''(J'' + 1)$$

and $$v = v_0 + B'(J'' - 1) J'' - B'' \cdot J''(J'' + 1)$$

These two line series are known as *branches* and are referred to respectively as the P branch and R branch. It can be shown by a little arithmetical manipulation of these formulae that if $B' = B''$ the two branches consist of lines separated by equal intervals of $B' + B''$ from each other, the branches forming a combination of each other apart from a single missing line at the origin v_0. If, however, $B' > B''$ the lines of the P branch start at the origin with the spacing $B' + B''$ but close up as we move away from the origin so that the separation between successive lines decreases each time by $2(B' - B'')$ until the

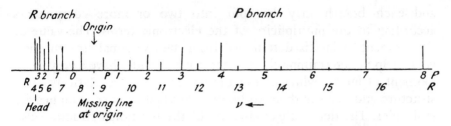

Fig. IV.1. The rotational structure of a $^1\Sigma$ to $^1\Sigma$ transition.

lines eventually reach a turning point, which corresponds to the head of the band; the lines of the R branch increase in spacing by steps of $2(B' - B'')$ each time. If, however, $B'' > B'$ the band is degraded to the red and the R branch forms the head; such a case is shown diagrammatically in Figure IV.1, the intensities of the lines being represented approximately by their lengths. The numbers indicate the values of J''.

For some other types of electronic transition the rotational structure may be less simple. In some cases ΔJ also takes the value 0, and the band then shows another branch between the others known as a Q branch; the lines of this start near the origin with a spacing which is initially zero or small and increases at each step, like the other branches, by $2(B' - B'')$ in the direction in which the band is degraded. A band of this type corresponding to what is known as a $^1\Pi$ to $^1\Sigma$ transition is sketched out in Figure IV.2. It will be seen that the Q branch does not form a true head, but the congestion of lines near the origin, especially when seen with small dispersion does suggest the appearance of a second head in this region.

The bands due to other types of transition have generally a greater number of branches; in some cases weak branches due to the rotational quantum number changing by ±2 or more are observed,

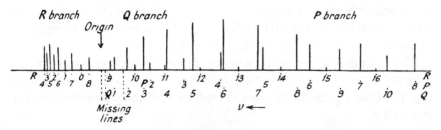

Fig. IV.2. The rotational structure of a $^1\Pi$ to $^1\Sigma$ transition.

and each branch may be split into two or more components according to the multiplicity of the electronic terms. The value of the moment of inertia determined from the rotational structure is a great help in determining the emitter; thus hydrides have a very small moment of inertia which causes the bands to have a relatively open structure and enables them to be distinguished at once from heavier molecules. The detailed examination of the rotational structure also enables the type of the electronic transition to be determined.

Electronic States of Diatomic Molecules

The electronic state of a molecule can usually be described by two quantum numbers. Λ is the resultant orbital angular momentum of the electrons along the internuclear axis, and corresponds to the quantum number L for the resultant orbital angular momentum in atoms; it takes the integral values 0, 1, 2, 3 etc, and according as this value is 0, 1, 2 etc the electronic state is referred to by Greek capital letters Σ, Π, Δ etc. The other quantum number of importance is the resultant spin of the electrons, S, which may be integral or half-integral according to whether there is an even or odd number of outer electrons; this determines the multiplicity $M = 2S + 1$ of the state. Thus any state is represented by the multiplicity, written as a superscript on the left, and its Λ value; thus a $^3\Pi$ state (read as triplet pi) has $M = 3$ (i.e. $S = 1$) and $\Lambda = 1$. In denoting a transition between two electronic states of a molecule it is customary to write the symbol for the state of higher energy first; this differs from practice in atomic spectra, for which the state of lower energy is given first. To distinguish between electronic states of similar character it is also the practice to "name" them by writing a letter, either small or capital, before the symbol. The ground state is usually designated by the letter x and successively higher states by A, B etc. Small letters are used when a molecule shows two multiplicities. Thus the electronic states of the NH radical, in order of increasing energy, are denoted x $^3\Sigma$, $a\,^1\Delta$, A $^3\Pi$, $b\,^1\Sigma$, $c\,^1\Pi$ and $d\,^1\Sigma$.

The coupling between the angular momenta due to rotation of the molecule, the electronic motion and spin may occur in several ways and the limiting methods are referred to as Hund's coupling cases a, b, c and d.

In Hund's case a, Λ and S combine to form a resultant Ω, and the rotational energy levels may be expressed in the form

$$E_r = B\,[J(J + 1) - \Omega^2\,]$$

where J can take the values Ω, $\Omega + 1$, $\Omega + 2$ etc. Thus, for example a $^2\Delta$ state, for which $\Lambda = 2$ and $S = \frac{1}{2}$, consists of two sub-levels with $\Omega = 1\frac{1}{2}$ and $2\frac{1}{2}$, the values of J being $1\frac{1}{2}$, $2\frac{1}{2}$, $3\frac{1}{2}$, etc for the first sub-level and $2\frac{1}{2}$, $3\frac{1}{2}$, $4\frac{1}{2}$ etc for the second. Thus the rotational levels corresponding to $J = \frac{1}{2}$ and $J = \frac{1}{2}$ and $1\frac{1}{2}$ are absent from the first and second sub-levels respectively; as a result, band spectra involving a $^2\Delta$ state have some additional missing lines near the origins of the bands.

This case a applies mostly to molecules containing at least one fairly heavy atom, when the spin splitting is relatively large (>100 cm^{-1}), and the results in the band being broken up into a multiplet of sub-bands each of which resembles a band of a singlet transition of similar type; thus a $^2\Pi \rightarrow {}^2\Pi$ tends to resemble two $^1\Pi \rightarrow {}^1\Pi$ bands. There are some minor differences in intensity distribution and number of missing lines near the origin.

In Hund's case b, which applies to all Σ states and to Π, Δ etc states of molecules composed of light atoms (roughly atoms of the first row of the Periodic Table) the effect of electron spin is much less important and shows only as a small splitting of the rotational energy levels. The electronic orbital angular momentum along the internuclear axis (represented by Λ) is coupled with the rotational angular momentum to give a resultant defined by a quantum number N which takes integral values of Λ, $\Lambda + 1$, $\Lambda + 2$ etc.*, and the rotational energy is given by

$$E_r = B \cdot N(N + 1) + \text{small spin terms}$$

These spin terms result in each main rotational energy level being split into $2S + 1$ sub-levels; the quantum number J for the final resultant including spin takes values from $N + S$ to $N - S$. Thus a $^3\Pi \rightarrow {}^3\Sigma$ band in case b (as for NH, Plate 3e) resembles a $^1\Pi \rightarrow {}^1\Sigma$

---

*There has been a change in notation here. This quantum number was previously (e.g. Herzberg, 1952) designated K and is referred to as K in the previous edition of this book. N is now internationally accepted.

band but with each line of the rotational fine structure split into .
three.

Hund's cases c and d are mostly encountered in spectra of very
heavy atoms or highly excited states and are of little importance in
the band systems encountered in flames. Many bands are, however,
in a coupling case intermediate between the limiting cases a and b
and then have a rather complex appearance.

There are some selection rules governing the possible electronic
transitions. Λ may change by 0 or ± 1 only. The multiplicity does not
change at all for strong band systems, but weak systems are known
corresponding to a change of ± 2 in M. The following are examples of
permitted transitions:

$$^1\Sigma \to {}^1\Sigma, \quad {}^2\Sigma \to {}^2\Sigma, \quad {}^1\Sigma \to {}^1\Pi, \quad {}^1\Pi \to {}^1\Sigma, \quad {}^2\Pi \to {}^2\Delta$$

and the following would not be allowed:

$$^1\Sigma \to {}^1\Delta, \quad {}^3\Sigma \to {}^3\Delta, \quad {}^1\Sigma \to {}^5\Sigma.$$

When Λ does not change (i.e. $\Sigma \to \Sigma$, $\Pi \to \Pi$, $\Delta \to \Delta$) the band
consists of strong P and R branches, and the Q branch is absent or
consists of only a few weak lines. If Λ changes by ± 1 (i.e. $\Sigma \to \Pi$
$\Pi \to \Sigma$, $\Pi \to \Delta$, $\Delta \to \Pi$) then the band shows a strong Q branch as
well, this being roughly twice as strong as the P and R branches. For
the higher multiplicities, especially when the coupling is intermediate
between a and b, additional weak O and S branches, for which
$\Delta N = \pm 2$, are also observed.

For states for which $\Lambda > 0$ (i.e. all except Σ states) there is a small
interaction between the rotation of the molecule and the resultant
orbital angular momentum of the electrons which causes a very small
splitting of all the rotational energy levels into two components; this
is known as Λ doubling. It is usually too small to detect but may be
observable in Π states; in some $\Pi \to \Pi$ and $\Pi \to \Delta$ transitions it causes
a small doubling in all lines, in addition to any structure due to spin
splitting.

For homonuclear molecules there is a degeneracy due to the
similarity of the two nuclei and this results in an alternation of
intensity of the lines of the rotational structure; the magnitude of
the effect is linked with the nuclear spin as well. For N_2 the lines in
the branches alternate with an intensity ratio $3:1$. For O_2 alternate
lines are entirely missing. This is also so for C_2, but for the $^3\Pi - {}^3\Pi$
Swan bands the effect is less easy to detect, as two Λ-doubling

components are present, and in one line the first of these components is missing and in the next line in the branch the second component is missing.

Σ states, which cannot show Λ doubling, are of two types, denoted Σ^+ and Σ^-, depending on certain symmetry properties of the molecule. Transitions only occur between Σ^+ and Σ^+ or between Σ^- and Σ^- and not between Σ^+ and Σ^-. In transitions between Σ and Π states the transition only occurs to one of the Λ-doubling components of the Π state, so that the rotational lines are not doubled. For homonuclear molecules there is also an additional symmetry restriction, states being designated u or g, usually written as a subscript after the main symbol (e.g. $^3\Pi_u$) and radiative transitions only occur between states of opposite symmetry $u \rightarrow g$ and $g \rightarrow u$.

Further discussion of electronic states is beyond the scope of this book, and the reader should refer to those by Herzberg (1950) and Kovacs (1969) for full details. It can be seen from what has been said that an examination of the rotational fine structure, including study of the number, type and intensity distribution of the branches and of missing lines near the band origin, usually enables the type of electronic transition to be determined unambiguously. The study of the rotational fine structure and assignment of the electronic transition is sometimes necessary in flame studies. Thus the multiplicity tells us at once whether there is an even or odd number of electrons in the molecule, and so assists in fixing the emitting species. The occurrence of alternating intensities or alternate missing lines, as in C_2, denotes a homonuclear molecule. Studies of the fine structure are also necessary when making measurements of the effective rotational temperature or in deciding the nature of a predissociation.

Vibrational Intensity Distribution: the Franck–Condon Principle

We have seen that the gross structure of a band system is produced by changes in the internal vibrational energy of the molecule, and it has been stated that for a diatomic molecule the change of vibrational quantum number is unrestricted during the electronic transition. It is now proposed to discuss the factors determining the relative intensities of the bands of a system. The intensity of any one band depends on the population of the initial state, that is the number of molecules with the appropriate initial value of the vibrational quantum number, v' for emission or v'' for absorption, and on the transition probability. The population depends on the energy of the particular vibrational level and on the temperature, or effective temperature; in thermal equilibrium this temperature dependence is given by the Maxwell–Boltzmann distribution law (see page 12). The transition probability depends, among other factors, on the relative position of the potential energy curves for the molecule in the two electronic states.

The potential energy curve of a diatomic molecule in a stable electronic state takes the form shown in Figure IV.3 (lower curve); the potential has a minimum at the equilibrium internuclear distance r_e, rises steeply as the atoms are brought closer together because of the strong repulsive force between the atoms when they are very close and rises to a limit, the *dissociation energy*, as the atoms are drawn apart. The molecule can only exist in a number of discrete vibrational levels, represented by the horizontal lines in the figure and marked by the vibrational quantum numbers v. A molecule in any state, say $v = 2$, may be visualised as executing an approximately simple harmonic vibration about the equilibrium position so that the internuclear distance and potential energy trace the path ABC, the fall in potential energy when at B being balanced by a gain in kinetic energy due to the rapid motion of the atoms through the equilibrium

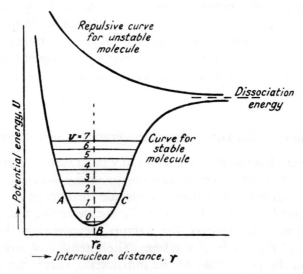

Fig. IV.3. Potential energy curves of a diatomic molecule.

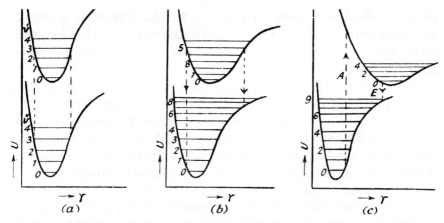

Fig. IV.4. Relative positions of potential energy curves.

configuration. The vibrational levels are approximately evenly spaced for low values of v, but close up (because of the $v^2 x\omega$ term) as the dissociation limit is approached. Potential energy curves of this type may be calculated from data obtained from the vibrational and rotational analysis of the spectrum, the value of B fixing the equilibrium internuclear distance, ω determining the shape round the minimum and $x\omega$ the approach to the dissociation limit. In Figure IV.3 the upper curve shown is the type which occurs when the two atoms do not form a stable molecule in the particular electronic configuration; it is sometimes known as a repulsive molecule state and an electronic transition to such a state results in dissociation and the spectrum is continuous instead of banded.

For an electronic transition, the distribution of intensity among the various bands of the system depends on the relative positions of the potential energy curves in the two electronic states. Figure IV.4 shows three such relative positions. If it is assumed that the electronic transition takes a time which is small compared with that for appreciable movement of the heavier atoms, then it follows that the internuclear distance r will not change appreciably during the transition. Classically it may also be seen that for a simple harmonic vibrator the atoms spend longest at either end of their path, so that the most probable starting values of r are the maximum and minimum ones for the vibrational level from which the transition is occurring. In Figure IV.4 (a), the potential curves are drawn roughly the same shape and with one immediately above the other. Considering a level of the upper state, say $v' = 3$, it will be seen that the ends of the level lie immediately above those for the same value of v'' in the lower state, that is $v'' = 3$. Thus the most probable transition from $v' = 3$ is to $v'' = 3$ and so on, so that the band system corresponding to two potential curves so arranged must show a strong sequence of bands (0,0), (1,1) etc.

With the curves slightly displaced, as in (b), it may be seen that the ends of, for example, the level $v' = 5$ come over the ends of the levels $v'' = 1$ and $v'' = 7$, so that the strongest bands with initial level $v' = 5$ will be the (5,1) and (5,7). Extending this, it can be shown that when the bands of a system are arranged into an array of the type shown in Table IV.2 the strongest bands will fall on a curve, in the array, of roughly parabolic form. This is usually referred to as a Franck–Condon parabola. When the relative position of the potential curves is similar to that of Figure IV.4 (a) the parabola on the array is narrow and lies about the (0,0) sequence which is strong. When the curves are relatively more displaced the Franck–Condon parabola is more open and the (0,0) band is weaker. When the potential curves are further displaced, as in (c), the parabola becomes very open and the strongest bands form long progressions, the $v' = 0$ progression often being strong in emission and the $v'' = 0$ in absorption; this is because in absorption the most highly populated level will be that with $v'' = 0$ and transitions from this will only occur to high values of v' or even to dissociation, while in emission most of the molecules will initially be in the $v' = 0$ level and transitions will occur to high values of v''. It will also be noticed that emission involves the small energy change E, so that emission bands tend to lie to long wavelengths, while the strongest absorption involves a big energy change A and lies at shorter wavelengths. This difference in the regions of the strongest emission and absorption bands is very noticeable in some cases, such as the Schumann–Runge bands of O_2 (see page 362) and is discussed again when considering the emitter of the carbon-monoxide flame bands (page 130).

The above simple treatment of the Franck–Condon principle, based on classical concepts, gives results which agree qualitatively with experiment, but is not quantitative and is incapable of explaining certain details of the vibrational intensity distribution. Thus weaker bands are frequently observed which lie right off the main parabolic locus and may indeed lie on subsidiary secondary or tertiary loci. To interpret these it is necessary to use wave-mechanics. For an emission band the transition probability can be shown to be proportional to

$$\nu^3 \left[\int_0^\infty \psi' M \psi'' \, dr\right]^2$$

where ν is the frequency, M is the electronic transition moment, ψ' and ψ'' are the vibrational wave functions for the upper and lower states, and r is the internuclear distance. ψ' and ψ'' are functions of v' and v'' and can be evaluated, usually with some simplifying assumptions, if the potential energy curves are known. Early calculations by Gaydon and Pearse (1939) for RbH assumed that the transition moment M was independent of r, and were able to explain well-marked secondary and tertiary intensity loci for this molecule, but did not give accurate quantitative agreement; more refined calculations (Jain and Sahni, 1966) have still not given perfect agreement. There is little theoretical guidance on the variation of the transition moment M with internuclear distance. The recent practice, following Nicholls and Jarmain (1956), has been to define a weighted mean value of r for each vibrational transition (the r-centroid) and to compare calculated Franck–Condon factors, $\left[\int \psi'v \cdot \psi''v \cdot dr\right]^2$, with observed band intensities at various values of the r-centroid and thus to deduce the variation of M with r. The positions of the effective potential energy curves also depend on the rotational energy, and this leads to the vibrational transition probabilities varying slightly according to the rotational energy; it has been shown that for OH especially this vibration-rotation interaction is quite important (Learner, 1962, Anketell and Learner, 1967). Vibrational transition probabilities have been calculated for most of the band systems encountered in flame spectra, e.g. OH (Anketell and Learner, 1967), C_2 (Tyte, Innanen & Nicholls, 1967), CH (Pillow, 1955; Halman & Laulicht, 1966), O_2 (Nicholls, 1964, Fraser et al, 1954; Hébert et al, 1967) and CN (Fraser et al, 1954; Brocklehurst et al, 1971). These calculated values are now reasonably accurate and serve to locate the Franck–Condon parabolae quite well and to show up any major anomalies in vibrational population, but should be used with caution when calculating precise vibrational temperatures.

Predissociation

Besides the normal radiative transitions between electronic states, in which a quantum of light is emitted or absorbed, radiationless transitions are sometimes possible. These only occur between levels of practically the same energy and when the potential energy curves of the two states cross or approach each other closely. There are also certain selection rules about changes of the various electronic and rotational quantum numbers. When radiationless transition is possible between two electronic states which both correspond to stable molecular states, then the result is some *perturbation* of the rotational energy levels in both states, and bands involving these states show some disturbances in both the positions and intensities of their lines. When, however, a radiationless transition is possible between levels of a stable electronic state and another state for which the molecule is unstable, having a repulsive potential curve of the type shown in Figure IV.3, then the results are more important In this case the transition will cause the molecule to dissociate into free atoms; this is known as predissociation.

When the predissociation is fully allowed by the selection rules, by energy considerations and by the relative position of the potential curves, then the rotational structure of the band disappears completely; in absorption the band becomes continuous in the region of the predissociation, while in emission it is usually absent. In many cases, however, the predissociation is much less strong; in absorption there is a slight widening of the affected spectrum lines, this being often slight and unobservable but sometimes causes an apparent strengthening of absorption when it is studied with instruments of limited resolving power; in emission the effect depends on the excitation conditions, but often depopulation of the excited state by radiationless transition into the unstable state followed by dissociation leads to a weakening in the emission lines affected. Frequently the observation is that the rotational fine structure of a band terminates abruptly above a certain energy. The phenomenon is important in determining dissociation energies of molecules and has been fully discussed in *Dissociation Energies* (Gaydon, 1968)

Predissociation has been detected in some molecules which are observed in flame spectra, and is particularly interesting because it may be used as a criterion as to whether or not the flame gases are in the thermal equilibrium. In complete equilibrium the radiationless transition followed by dissociation will be balanced by the reverse process, association of two atoms by collision to form a molecule temporarily in the unstable state which then undergoes the radiationless transition into the stable state. Thus in equilibrium there will be only a slight broadening of the lines produced by a weak predissociation, without any effect on the intensity. If, however, the population of electronically excited molecules is out of equilibrium then an intensity anomaly may occur. For CH (Durie, 1952a) lines affected by predissociation are normally absent the emission spectra of organic flames, showing that the concentration of excited CH radicals corresponds to a much higher temperature than does the population of free C and H atoms. For OH (Gaydon and Wolfhard, 1951b) there is a weakening of the affected lines in the emission from the inner cone of hydrocarbon flames, suggesting abnormally high electronic excitation of OH due to chemilumines-cence, but in hydrogen flames there is often a strengthening of affected bands due to an excess population of H and O atoms in the flame leading to inverse predissociation, sometimes referred to as pre-association.

Infra-red Spectra

Bands are observed in both emission and absorption in the near infra-red, between 2 and 20 μm, for many molecules. These are due to changes in the vibrational and rotational energy of the molecules only, the electronic state being unaltered during the transition. Such spectra are referred to as vibration-rotation spectra.

For diatomic molecules it is found that only heteronuclear molecules, like OH, CO and NO, give these infra-red spectra. For strong bands the vibrational quantum number changes by only one unit, $\Delta v = \pm 1$, although weaker bands are usually observable for other values of Δv. When the electronic ground state of the molecule is of Σ type the rotational quantum number changes by ± 1, so that the vibration-rotation bands show P and R branches, but the change in the rotational constant B between the two vibrational levels is small, so that the branches are only very weakly degraded to longer wavelengths and usually do not form a head. For molecules in a Π electronic ground state (e.g. OH) there is also a piled-up Q branch, $\Delta J = 0$. Homonuclear molecules (N_2, O_2, H_2 etc.) do not possess a dipole moment and are inactive in the infra-red and neither emit nor absorb radiation. Thus if, in combustion processes, homonuclear molecules acquire excess vibrational energy they cannot lose this by radiation but only by collisions and thus may serve as long-lived energy carriers; vibrational relaxation times for O_2 and N_2 are known to be relatively long.

For polyatomic molecules the spectrum is greatly complicated because in general there will be three moments of inertia, and for a molecule of N atoms there will be $3N-6$ fundamental vibrational frequencies. With many of the smaller molecules there is some simplification due to these molecules possessing some symmetry; thus methane has only four instead of nine vibrational frequencies. Linear molecules have only one moment of inertia, so that their rotational energy levels resemble those of a diatomic molecule, but they have in the general case $3N-5$ vibrational modes. For a linear triatomic molecule (e.g. CO_2) the perpendicular bending vibration ν_2 is degenerate, i.e. occurs twice, due to similar vibrations in two planes at right-angles to each other; however, due to coupling between the two degenerate modes it is still necessary to specify the vibrations by four quantum numbers, ν_1, ν_2, ν_3 and l. A useful simple account of the energy levels and spectra of polyatomic molecules is given by Mavrodineanu and Boiteux (1965); for a full treatment of polyatomic spectra it is necessary to study the symmetry properties of the molecule in some detail and to use Group Theory, and for this the best treatment is by Herzberg (1945).

If the vibrations of a molecule are considered as being simple harmonic, then, for the infra-red, changes of vibrational quantum numbers are restricted to one unit, and only one vibration at a time can alter. Thus only the fundamental frequencies should occur, without overtones or combinations of two or more frequencies. In practice the allowed fundamental frequencies give strong bands, but owing to the vibrations being appreciably anharmonic, weak overtone and combination bands do also occur. Thus the fairly strong band at $2.8\,\mu m$ in the spectrum of a carbon monoxide flame is due to a superposition of the two combination bands $\nu_1 + \nu_3$ and $\nu_3 + 2\nu_2$ of CO_2. Combination and overtone bands of the vibration-rotation spectrum of $\dot{H}_2O$ lead to weak emission from oxy-hydrogen flames even in the visible region of the spectrum. For molecules with a high degree of symmetry some of the fundamental vibrations may be absent from the infra-red spectrum, i.e. be "inactive", because emission or absorption only occurs when the vibration leads to a periodic change of the dipole moment of the molecule; for totally symmetric vibrations, such as the "breathing" vibrations of CO_2 and CH_4, there is no change of dipole moment and these bands are absent from the infra-red spectrum, although these frequencies can be studied from their Raman spectra.

For polyatomic molecules the rotational structure is, in the general case, very complicated, but is somewhat simplified if the molecule has an axis of symmetry. In these simpler cases we have two rotational quantum numbers J and K. If the oscillation of the electric moment is parallel to the axis of symmetry then we have what is known as a "parallel" band and the selection rules for changes of rotational quantum number are

$$\Delta J = 0, \pm 1 \quad \Delta K = 0$$

Since the moments of inertia of the molecule are very nearly the same in the unexcited and excited vibrational states, all the lines due to the K structure will be very nearly superposed. The J structure will then cause the band to have P and R-type branches, each line of which will be rather a broad cluster of fine lines due to the K structure; there will also be a strong central Q branch which will not usually be resolved. These infra-red bands are not strongly degraded, like electronic bands, because there is little change in the moments of inertia during the transition; they do not normally form heads, except perhaps for high overtone bands.

When the oscillation of the electric moment is perpendicular to the axis of symmetry, we have a "perpendicular band", for which

$$\Delta J = 0, \pm 1 \quad \text{and} \quad \Delta K = \pm 1.$$

This produces a relatively open K structure, spaced out like the P and R branches of a diatomic band, but with each line split into a whole sub-band with P, Q and R branches due to the change in J. The Q branches will normally consist of many unresolved lines. When the molecule lacks symmetry then for any value of the rotational quantum number J there will be $2J + 1$ sub-levels, these sub-levels being arranged without obvious regularity so that the band becomes very complex.

Infra-red spectra are of great value in determining the symmetries, structure and binding forces in polyatomic molecules. They are also very useful empirically in chemical analysis, especially of organic compounds. Infra-red gas analysers are in routine use for analysis of combustion products. Infra-red emission from flames is important for heat-transfer. It comes mainly from the stable combustion products CO_2 and H_2O and is discussed in the appropriate chapter.

Absorption spectra in the far infra-red have been observed for a few molecules. These rotation spectra, as they are called, involve changes in rotational energy only.

Raman Spectra

When a beam of monochromatic light (that is light of one wavelength) is scattered from a molecular gas it is found that the weak scattered light is mainly of the same frequency (Rayleigh scattering) but that a small fraction of the scattered light is of a different frequency corresponding to the original frequency less the vibrational or rotational frequency of the scattering molecule; in scattering from a hot gas we may also observe the anti-Stokes scattering in which the vibrational or rotational frequency is added to the exciting frequency*. This effect is known by the name of its discoverer, Raman. As with infra-red spectra, some frequencies are Raman active and others inactive, but the Raman effect depends not on variation of the electric dipole moment, but on the induced dipole or polarizability of the molecule. Frequencies which are infra-red inactive are nearly always Raman active, so that the Raman effect is a useful complement to infra-red studies for obtaining information about fundamental vibrational frequencies.

The Raman effect in gases is extremely weak, but the development of the laser, providing a very bright source of monochromatic light, has given an impetus to Raman studies. Raman scattering, like Rayleigh scattering, is much stronger for short wavelengths and so is better studied with blue or green light, rather than say a ruby laser. When viewed at right angles to the exciting beam, Rayleigh scattering is polarized, but Raman scattering is partly depolarized.

In the past, Raman scattering has only been of use for providing fundamental data about molecular structure, but recent developments (see page 61) promise that it may be valuable in flame studies for determining the concentrations and rotational and vibrational temperatures of just those molecules, especially N_2 and O_2, which cannot be studied either by their visible of infra-red spectra.

Electronic Spectra of Polyatomic Molecules

Band systems due to electronic transitions, lying in the visible and ultra-violet, are less commonly observed for polyatomic molecules

---

*The selection rule for rotational Raman effect is $\Delta J = \pm 2$, giving O and S branches with a spacing of $4B$ between lines. For the vibrational Raman effect $\Delta v = \pm 1$, $\Delta J = 0$ or ± 2, giving a piled up Q branch of usually unresolved lines and O and S branches of lines spaced $4B$ apart.

than for the diatomic ones. In hot sources such as arcs and powerful electric discharges these molecules tend to be decomposed. However, in flames, especially those that are not too hot, there are quite a few important emission band systems of simple polyatomic molecules; these include the carbon monoxide flame bands (CO_2), Vaidya's hydrocarbon flame bands (HCO), the Comet-head band (C_3), the cool flame bands (HCHO) and the ammonia α-band (NH_2). A few other systems (CH_3, NO_2, C_6H_6) have also been observed in absorption in flames, and the flash-photolysis technique has enabled a study of a number of polyatomic spectra.

For these electronic transitions we have up to $3N-6$ vibrational frequencies in both the upper and lower electronic state, so that instead of the single vibrational array of the type shown in Table IV.2, we have conceivably a very large number of such arrays, corresponding to simulataneous change in more than one vibrational quantum number. For diatomic molecules the change in vibrational quantum number is restricted only by the Franck–Condon principle, depending on the relative position of the upper and lower potential energy curves, but for polyatomic molecules there are certain additional restrictions. Most progress has been made in studying the vibrational structure of the spectra of molecules which have an axis of symmetry (for details see Herzberg 1966, 1971, W. L. Smith, 1966). If the electronic transition is fully allowed, then any change is allowed in the vibrational quantum number for totally symmetrical modes. For a mode of vibration which is antisymmetrical to any symmetry element, the change in vibrational quantum number is restricted to an even integer, with transitions in which the quantum number does not change at all being strongest. This may be summarized (for allowed electronic transitions)

$$\Delta v_{sym} = \text{any} \quad \Delta v_{antisym} = 0 \text{ (strong)}, \quad \pm 2 \text{ (weak)}, \quad \pm 4 \text{ (v. weak)}$$

Band systems which are forbidden by the normal electronic selection rules often occur with moderate strength in polyatomic molecules because the vibrations modify the symmetry properties. In these cases the vibrational quantum number for antisymmetrical modes changes only by ± 1. Thus (for forbidden transitions)

$$\Delta v_{sym} = \text{any} \quad \Delta v_{antisym} = \pm 1$$

In addition to these selection rules, the intensity distribution is still controlled by the Franck–Condon principle. However, for polyatomic molecules we no longer have simple two-dimensional potential energy curves. There will be several internuclear distances which may change during the transition from one electronic state to the other, and also the shape of the molecule may change. A particular example of importance is when a triatomic molecule changes from a bent (i.e. triangular) form to a linear form; this leads to strong excitation of the bending vibration ν_2 as a result of the electronic transition (e.g. for CO_2, page 130).

For linear polyatomic molecules the electronic structure may be designated in a way similar to that for diatomic molecules, and since these molecules have only one moment of inertia their rotational structure is also similar. Thus for linear C_3 individual bands resemble those of a normal $^1\Pi - ^1\Sigma$ diatomic transition, although the larger number of vibrational modes results in the occurrence of many more bands. In other cases, however, such as CO_2, degeneracy of the bending vibrational mode results in a coupling between the electronic and vibrational structure and the vibronic levels have a rotational structure which varies from level to level according to the value of l (page 86). For non-linear molecules the multiplicity, due to electron spin, is retained, but the designation of molecular electronic states is more complex (see Herzberg, 1966).

The rotational structure of bands involving electronic transitions is somewhat similar to that of infra-red bands, but the moments of inertia will in general be different in the two electronic states so that the branches may be strongly degraded and form heads, and the strong piled up Q maxima present in infra-red bands may be absent. For molecules with an axis of symmetry we again have bands of parallel and perpendicular types with J and K structure. It is possible for the J and K structure to be degraded in opposite directions, so that the fine structure of the band is degraded one way but the group of heads degrades in the opposite direction. The direction of degradation depends on the change in moment of inertia which in turn depends on both inter-nuclear distance and angular configuration. In some types of electronic transition the type of band (parallel or perpendicular) will depend on the vibrational transition, so that both types of band occur in the same system.

Continuous Spectra

While line and band spectra are due to transitions between quantized energy states, continuous spectra (in gases) are due to transitions in which at least one of the states involved is unquantized, possessing free kinetic energy. Continua therefore correspond to *processes*, such as ionization, dissociation and association. Thus while the study of band spectra tells us what molecules and radicals are present in flames, the study of the continua might be expected to give even more valuable information about the actual processes which are occurring. The analysis of a band spectrum gives values of molecular constants which usually enable the system to be assigned unambiguously to a particular molecule. With continua, the lack of characteristic features like band heads makes assignment of the particular process difficult, and it is necessary to make deductions from other evidence, such as observation of the chemical species present and knowledge of the amount of energy likely to be absorbed or emitted during a process. The origins of the various continuous spectra found in flames are thus often in doubt, although good progress has now been made in assigning the main sources of continuous emission.

Ionization Continua. The absorption spectra of atoms consist of series of lines (Rydberg series) which converge to a limit, and beyond this limit a region of continuous absorption is observed. The lines correspond to excitation of an electron to the various outer orbits, and the continuum to complete ejection of an electron, the electron and positive atomic ion separating with free kinetic energy. Similar absorption spectra due to ionization of molecules also occur, these usually lying well down in the vacuum ultra-violet.

The reverse process, the association of a positive ion and an electron to form a neutral atom or molecule, with emission of light is also possible. The chance of such a transition, emitting light and stabilizing the system, occurring during the short time of a collision is relatively small. However, it can be understood from the principle of micro-reversability that such processes must occur in an equilibrium situation. This association, an ion-electron recombination, is responsible for most of the continuous emission from hot plasma, such as arcs, and is responsible for some of the weak continuous background in those flames which are hot enough to show appreciable ionization. In addition to the association process, continuous emission may also result from the retardation of fast electrons in the field of positive ions (bremsstrahlung), but this process is not important in flames.

In absorption by cold gases, ionization continua commence near the series limit and extend to shorter wavelengths. In emission the ion-electron recombination continua also occur at longer wavelengths because the process may lead to formation of atoms or molecules in excited electronic states. Theory (e.g. Maecker and Peters, 1954) indicates that in plasma discharges the intensity of the continuum, per unit frequency interval, should be frequency-independent at low frequency; near the series limit and beyond this limit the intensity should fall off as $e^{-h\nu/kT}$. Parkinson and Reeves (1961) studied various flash tubes experimentally; for co-axial and Lyman-type tubes the intensity was indeed frequency independent over the range studied (4500—2700 Å) and for capillary-type flash tubes it was constant to about 3400 Å and then decreased towards the further ultra-violet. It seems that the intensity distribution in these ionization continua is very flat and not likely to give any information about the type of positive ion involved.

Dissociation Continua.   The commonest process responsible for regions of continuous absorption in molecules is dissociation. The absorption of light excites the molecule from its ground electronic state to an upper state with a repulsive energy curve (e.g. Fig. IV.3, p. 80) or to a part of a stable potential curve which is above the dissociation limit (Fig. IV.4c, p. 81) and the molecule dissociates to two fragments which fly apart with an amount of kinetic energy equal to the excess above that required just to reach the dissociation limit. The reverse association process is considered in the next sub-section. Strong predissociation (see p. 83) can also lead to regions of continuous absorption or diffuse bands in absorption, and these are particularly common for polyatomic molecules.

It is also possible for a dissociation process to produce an emission continuum. There is one well-known example of this, the hydrogen continuum; hydrogen molecules in a highly excited but stable $^3\Sigma_g^+$ state undergo transitions to a lower excited but unstable $^3\Sigma_u^+$ state, with consequent dissociation to two normal hydrogen atoms and the emission of strong ultra-violet light; this is the origin of the strong continuum from hydrogen or deuterium lamps which were formerly so useful for a background source in absorption spectroscopy. This type of continuum requires high excitation energy and neither the hydrogen continuum nor others of this type are likely to occur in normal flames.

Association Continua.   While direct bimolecular associations, unassisted by a third body, are rare, they can occur with appreciable probability under certain circumstances. The association is the

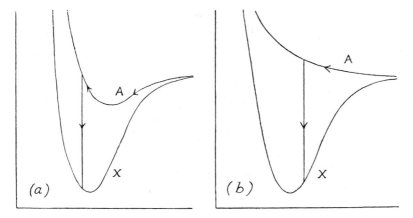

Fig. IV.5. Potential energy curves of a diatomic molecule illustrating the processes leading to association continua.

reverse of the photo-dissociation process. It occurs most frequently when the colliding particles, in their ground electronic states, can form a molecule in two different electronic states and when an electronic transition is possible between these two states. This is illustrated in Fig. IV.5, for the simple example of the association of two free atoms to form a diatomic molecule.

In Fig. IV.5a it is assumed that the ground states of the two colliding atoms can give rise to two stable electronic states of the molecule, an excited state with potential curve A, and the ground state with potential energy curve X. As the two colliding particles approach they attract each other and rush towards each other, their internuclear separation and potential energy tracing out part of the curve A. Normally the two particles will swing together, and then, being unable to get rid of the excess energy will swing apart again, i.e. there will be an elastic collision. There is, however, a possibility that an electronic transition to the ground state X may occur during the brief time of the collision. If this happens, then energy will be disposed of as light emission and the molecule will be stabilized.

If the particles come together on a repulsive potential curve, as shown in Fig. IV.5b, an association may still take place provided the colliding particles come together with sufficient kinetic energy to approach to a reasonable distance, i.e. if the gas is hot.

The duration of a collision between two atoms usually lasts a time of the order 10^{-12} or 10^{-13} s, and the electronic transition probability for a fully allowed transition in the visible region may be of the order 10^7 or 10^8 per sec. Thus there may be a chance as high as one in 10^5 that a collision between atoms will result in the emission of light and formation of a molecule. An association of the type illustrated in Fig. IV.5a may occur in cold gases, and in this case there will be a definite short-wave limit to the continuous spectrum emitted. The intensity distribution within the continuum will depend mainly on the relative positions of the potential curves and will not change rapidly with gas temperature.

An association continuum of the type illustrated in Fig. IV.5b will, however, only occur with hot gases and the strength of the continuum as a whole and the intensity distribution within it will depend on the temperature.

For simplicity we have dealt with collisions between two atoms forming a diatomic molecule. Similar considerations apply to collisions between more complex particles forming polyatomic molecules, but the situation will then be more complex as we shall have to consider potential energy surfaces instead of curves. It is then also possible for short-lived collision complexes to be formed (sticky collisions); in a system containing three or more atoms energy may be distributed among several vibrational modes and it is possible that during a collision this may occur, but because the system still retains sufficient energy for its own spontaneous decomposition it will eventually break up, but the complex will have a longer life than the duration of a simple elastic collision. If the collision complex has a long life the light emitted will tend to be composed of diffuse bands, corresponding to an inverse predissociation, but usually the spectrum will still be continuous but with some intensity fluctuations.

Flames contain many free radicals and free atoms, and weak continuous emission, due to association processes, occurs in many flames. Continua of particular importance which will be discussed in later chapters are those due to $CO + O = CO_2 + h\nu$ and $NO + O = NO_2 + h\nu$. The $CO + O$ continuum is mainly responsible for the strong blue colour of carbon monoxide flames; it seems to be of the type illustrated in Fig IV.5b and is strongly temperature dependent. The $NO + O$ continuum is responsible both for the yellow-green air afterglow in discharge tubes and for a pale yellowish-green emission from many flames containing oxides of nitrogen; it serves as a test for the presence of free oxygen atoms, but there has been some discussion whether it is a simple bimolecular recombination or whether it requires a third body; if the third body also quenches the excited NO_2 then the kinetics are similar. Strong continua are also emitted when alkali metals are added to flames; this was reported by Hartley as long ago as 1907 and the author at one time attributed these continua to ion recombination processes, but later evidence (James and Sugden, 1958) favours associations of the type

$$A + OH = AOH + h\nu$$

where A is an alkali-metal atom.

The Width and Shape of Spectrum Lines

In an optically thin gas, i.e. when self absorption is negligible, there are several factors which contribute to the broadening of spectrum lines, the most important in flames being Doppler broadening and collision or Lorentz broadening.

If an emitting or absorbing molecule is moving with respect to an observer with line-of-sight velocity u then the Doppler effect causes a spectrum line of frequency ν_0 (wavelength λ_0) to be displaced by an amount $\Delta\nu = \nu_0 u/c$ or $\Delta\lambda = \lambda_0 u/c$ where c is the velocity of light. The random thermal motion of gas molecules causes a broadening. We define the Doppler half-width $\Delta\nu_D$ or $\Delta\lambda_D$ as the width of the spectrum line measured between those points on the intensity curve at which the intensity is half the maximum. It can be shown that for a gas in equilibrium at temperature T

$$\Delta\nu_D = 2\nu_0(2kT\ln2/mc^2)^{\frac{1}{2}} = 0.716 \times 10^{-6}\nu_0(T/M)^{\frac{1}{2}}$$

or

$$\Delta\lambda_D = 0.716 \times 10^{-6}\lambda_0(T/M)^{\frac{1}{2}}$$

where M is the molecular weight (or atomic weight) on the scale $C^{12} = 12.00$. At flame temperatures, lines in the visible region have a Doppler half-width of around 0.15 cm^{-1} or 0.05 Å, and in the near ultra-violet around 0.3 cm^{-1} or 0.02 Å. Doppler broadened lines have a Gaussian distribution of intensity and have relatively weak wings to the lines as the intensity falls off exponentially as $e^{-B(\nu-\nu_0)^2}$ where $B = mc^2/2kT\nu_0$.

Spectrum lines have a "natural" breadth due to the finite lifetime in the excited state, which limits the length of the wave-train which can be emitted, and thus the monochromaticity of the light. This is related to the Heisenberg uncertainty principle and leads to a breadth

$$\Delta \nu_N = \frac{1}{2\pi} (1/\tau_1 + 1/\tau_2)$$

where τ_1 and τ_2 are the radiative lifetimes in the two electronic states. When only a single electronic transition from the upper state to the ground state is involved $\tau_2 = \infty$ and $\Delta \nu_N = 1/2\pi\tau_1 = A_{12}/2\pi$, where A_{12} is the transition probability. For fully allowed electronic transitions $\Delta \nu_N$ is only about 0.001 cm^{-1}, and for partly forbidden transitions of longer life the value of $\Delta \nu_N$ is still less.

The Lorentz-type collision broadening may be treated like the natural broadening, but with the radiative lifetime replaced by the collision lifetime $= 1/Z$, where Z is the collision frequency. This leads to $\Delta \nu_L = (Z_1 + Z_2)/2\pi$ where Z_1 and Z_2 are the collision quenching frequencies in the two electronic states; for a transition to the ground state this reduces to $\Delta \nu_L = Z/2\pi$. The collision frequency, and therefore the Lorentz broadening, is proportional to the gas density and $T^{1/2}$; at constant pressure this leads to a $T^{-1/2}$ dependence. There is some difficulty in defining a collision, as a slight perturbation of energy levels may occur when the colliding molecules are relatively far apart, so the effective collision cross section may exceed the gas kinetics value derived from viscosity measurements. For flame gases the Lorentz and Doppler broadening is of the same magnitude; thus collected data given by Mavrodineanu and Boiteux (1965) give $\Delta \lambda_L = 0.029$ Å and $\Delta \lambda_D = 0.044$ Å for Na 5890, and $\Delta \nu_L = 0.023$ and $\Delta_D = 0.017$ for Sr 4607 Å in an acetylene air flame at 2480 K. The Lorentz broadening tends to be greater for free atoms than for molecules and is especially large for transitions involving states of high principal and azimuthal quantum number, e.g. the very wide lines of the "diffuse" series of alkali and alkaline earth metals.

Collision broadened lines have a different contour from Doppler broadened ones. The intensity distribution in a Lorentz-broadened line is given by

$$I(\nu) = I_0/(1 + [2(\nu - \nu_0)/\Delta \nu_L]^2)$$

This leads to very much wider "wings" to the lines, as illustrated in Fig. IV.6. These stronger wings are particularly important at high optical density when they cause the line to be much wider than in the case of pure Doppler broadening.

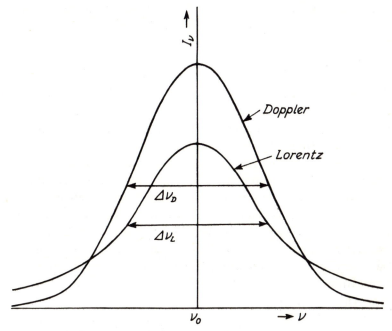

Fig. IV.6. Comparison of the intensity contours of lines having the same half-width and integrated total intensity when the broadening is due to Doppler effect alone and to Lorentz pressure broadening alone.

In practice lines are broadened by both Doppler effect and collision. The two effects are usually treated as independent* in first approximation, but this still leads to a more complex line profile, as discussed in the old book by Mitchell and Zemansky (1961) and by Penner (1959). The combined half-width is, of course, less than the simple sum $\Delta \nu_D + \Delta \nu_L$.

These are the main causes of line broadening in flames. In hotter ionized plasma (e.g. arcs and strong shock heating) other broadening effects occur due to random perturbations of the energy levels due to the electric fields of neighbouring ions and electrons; this is related to the Stark-effect splitting of levels and leads to considerable asymmetrical widening and displacement of lines. It is the main influence in broadening lines of diffuse and fundamental series in

*It could, however, be argued that a fast moving molecule with a large Doppler shift would, because of its high velocity, also have a higher probability of making a collision.

ionized plasmas, and the Holtsmark broadening of the hydrogen Balmer lines has been used to measure ion concentrations and temperature.

The line contours considered so far have been for optically thin gases. With increasing concentration of emitter and increasing path length, self absorption becomes important. This limits the intensity at any wavelength to that of a black body at the flame temperature. Thus with increasing concentration the centre of the line tends towards a limiting brightness but the weaker wings continue to strengthen, also towards this limit, so that the observed line contour is modified. The total intensity of an emission line increases at first linearly with concentration and path length, but then less rapidly. For pure Doppler broadening after the initial linear increase the line intensity only increases very slowly indeed. With Lorentz broadening the stronger wings to the lines cause the total intensity, after the initial linear phase, to increase roughly as the square root of the concentration (James and Sugden, 1955a; Penner and Kavanagh, 1953). These "curves of growth" or "working curves" for the increase in line intensity with concentration are particularly important in flame spectrophotometry (see for example Herrmann and Alkemade, 1963).

Chapter V

Hydrogen Flames

The OH Bands

The existence of strong short-wave radiation from hydrogen flames was noted by Stokes in 1852 and the banded ultra-violet spectrum was first mapped independently by Liveing and Dewar (1880) and Huggins (1880). The strongest band has its main head at 3063.6 Å, is degraded to the red and shows a very open but rather complicated rotational structure. This strong (0,0) band has two other strong heads at 3067.6 and 3090.4 Å with a fourth less definite head around 3078 Å. There are similar weaker bands with first heads at 3428 (0,1), 2811 (1,0) and 2608 (2,0). Wavelengths are given in the Appendix, page 364, and the structure of the (0,0) band is shown in Plate 3a. Other plates show the bands with smaller dispersion.

Detailed analysis of the rotational structure (Watson, 1924; Jack, 1927, 1928) shows that the bands are due to a $^2\Sigma - ^2\Pi$ transition in a diatomic molecule with a small moment of inertia. This at once limits us to a hydride with an odd number of electrons and the hydroxyl radical OH is the only possibility.

The analyses by Watson and Jack were made while the notation for molecular spectra was still developing. The most complete report on the OH bands is by Dieke and Crosswhite (1962) who give tables of wavelengths, wavenumbers, band and branch assignments and measurements of intensity in an oxy-hydrogen flame for all rotational lines between 2811 and 3546 Å. Bass and Broida (1953) have shown large-scale tracings using a photoelectric recording system for the region 2600–3520 Å, with rotational assignments.

The rotational and vibrational transition probabilities for OH are particularly important because they are needed for rotational and vibrational temperature measurements. The basic formulae for the rotational transition probabilities for lines of the main branches of a

TABLE V.1

Rotational intensity factors for the main branches of a $^2\Sigma-^2\Pi$ transition (e.g. OH)

$R_1 = \{(2J + 1)/(2J + 2)\}\ \{(2J + 1) + U[(2J + 1)^2 + 2(Y - 4)]\}$

$R_2 = \{(2J + 1)/(2J + 2)\}\ \{(2J + 1) + U[(2J + 1)^2 - 2Y]\}$

$Q_1 = \{(2J + 1)/(2J^2 + 2J)\}\ \{(2J + 1)^2 - 2 + U[(2J + 1)^3 - 8J + 2(Y - 4)]\}$

$Q_2 = \{(2J + 1)/(2J^2 + 2J)\}\ \{(2J + 1)^2 - 2 + U[(2J + 1)^3 - 8J - 2Y]\}$

$P_1 = \{(2J + 1)/2J\}\ \{(2J + 1) + U[(2J + 1)^2 - 2Y]\}$

$P_2 = \{(2J + 1)/2J\}\ \{(2J + 1) + U[(2J + 1)^2 + 2(Y - 4)]\}$

where $Y = A/B$ (A is the multiplet splitting of the Π state and B the rotational term constant) and $U = \{(2J + 1)^2 + Y(Y - 4)\}^{-\frac{1}{2}}$

There are also weak satellite branches with the U term replaced by $-U$.

$^2\Sigma-^2\Pi$ transition (see Earls, 1935, Dieke and Crosswhite, 1962) are given in Table V.1. Dieke and Crosswhite have listed numerical values of these rotational intensity factors for OH using $Y = A/B = -7.55$. Actual line intensities depend on the product of this factor, a ν^4 term and the vibrational transition probability, and also, of course, on the upper state population, which depends on concentration and temperature.

Calculations of the vibrational transition probabilities from the Franck—Condon principle are difficult for OH because of lack of theoretical knowledge of the variation of the electronic transition moment M with nuclear separation (see p. 83) and older values by Shuler, quoted in the previous edition, are now known to be inaccurate. Dieke and Crosswhite calculated the relative intensities of bands in a flame at 3000 K from experimental measurements on individual lines, and their values, based on a scale of 1000 for the (0,0) band are shown in Table V.2.

From these experimental values Anketell and Learner (1967) have derived the variation of M with internuclear distance and calculated vibrational transition probabilities for the whole system. For OH, especially for bands of the (0,1) sequence, the overlap integral of the vibrational wave functions is quite sensitive to the positions of the potential energy curves and since the effective potential energy curves, including rotational energy, are different from the curves without rotational energy there is considerable rotation-vibration interaction so that the vibrational transition probabilities depend also

TABLE V.2
Vibrational intensity distribution for OH in a flame at 3000 K
(after Dieke and Crosswhite)

v' \ v''	0	1	2	3
0	1000	4	2	
1	86	156	23	
2		35	8	5
3				

TABLE V.3
Vibrational transition probabilities for OH, showing effect of
vibration-rotation interaction. From Anketell and Learner.

v'	J''	v'' 0 P	Q	R	1 P	Q	R	2 P	Q	R
0	1½	1.000	0.996	0.990	0.017	0.016	0.015	0.002	0.002	0.002
	10½	0.941	0.917	0.890	0.018	0.013	0.009	0.002	0.002	0.002
	15½	0.864	0.830	0.794	0.016	0.010	0.006	0.002	0.002	0.001
	20½	0.761	0.720	0.678	0.013	0.007	0.003	0.002	0.002	0.001
	25½	0.640	0.595	0.550	0.010	0.004	0.001	0.002	0.001	0.0007
1	1½	0.449	0.450	0.453	0.780	0.775	0.765	0.025	0.024	0.022
	10½	0.430	0.439	0.447	0.725	0.688	0.647	0.026	0.019	0.013
	15½	0.413	0.423	0.433	0.645	0.593	0.539	0.022	0.014	0.007
	20½	0.390	0.400	0.407	0.537	0.476	0.415	0.017	0.008	0.003
2	1½	0.144	0.146	0.148	0.685	0.685	0.686	0.586	0.579	0.567
	10½	0.144	0.154	0.165	0.646	0.646	0.644	0.533	0.488	0.440
	15½	0.149	0.163	0.178	0.602	0.598	0.588	0.453	0.392	0.332
	20½	0.158	0.175	0.192	0.538	0.524	0.502	0.344	0.278	0.215
3	1½	0.043	0.043	0.045	0.339	0.342	0.346	0.751	0.749	0.745
	10½	0.046	0.051	0.058	0.335	0.352	0.369	0.692	0.674	0.651
	15½	0.053	0.061	0.071	0.338	0.359	0.378	0.614	0.582	0.542
	20½	0.065	0.077	0.091	0.339	0.356	0.368	0.500	0.449	0.390

N.B.   these vibrational transition probabilities have to be multiplied by ν^4 when calculating intensities in emission, or by ν in absorption; ν is the wavenumber of the line.

on the rotational quantum numbers in the upper and lower states, J' and J''. Anketell and Learner have given an extensive table of vibrational transition probabilities for OH for lines of the P, Q and R branches for J'' values of 1½, 10½, 15½, 20½, and 25½. Part of this table is extracted in Table V.3. It will be noted that even for the strong (0,0) band there is over 30% variation in vibrational transition probability between P lines of low J'' and R lines of high J''. This is an important correction when making measurements of rotational temperature. Dieke and Crosswhite's rotational transitional probabilities, based on the formulae of Table V.1, should be multiplied by the vibrational ones, listed in Table V.3, to get the full transition probabilities for measuring rotational temperatures.

Hydrogen/Air Flames

Flames of hydrogen with air, either premixed or of the diffusion type, are practically non-luminous in the visible region. Normally impurities such as sodium and calcium give flickers of orange coloration, and sulphur, even in very small amounts, causes the flame to show a blue-violet coloration; in the premixed flame this may appear as a bluish inner cone, the spectrum showing S_2 bands. The only radiation from pure hydrogen/air flames is, however, that due to the OH bands. This OH radiation is of the same order of strength as from the interconal gases of hydrocarbon/air flames and is very much weaker than the OH radiation from the inner cones of hydrocarbon flames. Hydrogen flames do not usually show any definite inner cone, and when the image of a hydrogen/air flame is thrown on the slit of a spectrograph the OH radiation is usually fairly uniform over the length of the flame. From this relative weakness of the OH radiation and its uniform distribution throughout the flame it seems that the excitation is mainly thermal. However, anomalies occur in some flames.

In relatively low-temperature flames, obtained from either hydrogen-rich mixtures or by dilution with extra nitrogen, the thermal radiation of OH in the ultra-violet is very weak, and it is then possible to see a region of relatively strong OH emission at the very base of the flame; Charton and Gaydon (1958) have shown photographs of the spectra and made quantitative measurements. The effect appears to be due to some chemi-excitation of OH by the reaction

$$H + OH + OH = H_2O + OH^*$$

This excitation will vary as the cube of the concentration of free radicals, and it dies out rapidly higher in the flame as the excess radicals recombine; we shall see later that the combustion reactions produce free atoms and radicals so that their concentration in the reaction zone may greatly exceed that for equilibrium at the flame temperature.

An anomalous vibrational intensity distribution is found in some hydrogen/air flames. In the (1,0) sequence starting at 2811 Å the bands (1,0), (2,1) and (3,2) usually become progressively weaker; this is so for oxy-hydrogen flames (Table V.2) and is expected, from the vibrational transition probabilities, for a flame at ordinary temperatures; for stoichiometric hydrogen/air at about 2100°C the expected intensity ratio is about 1 : 0.28 : 0.029. However, in some small hydrogen/air flames the intensity distribution is abnormal with the (1,0), (2,1) and (3,2) bands having about the same intensity; this is shown in Plate 2a. This observed distribution is more like that expected for infinite temperature and clearly indicates some lack of equilibrium. The exact conditions under which hydrogen/air flames show this abnormality most strongly have not been established; a small high-velocity jet of hydrogen burning as a diffusion flame in air shows the effect well and it also occurs in premixed flames cooled with extra nitrogen (Charton and Gaydon, 1958). The effect occurs in those vibrational levels, $v' = 2$ and 3, which are predissociated (see page 110) and is due to the inverse process, an association; an excess population of O and H atoms leads to formation of OH in an electronic state, now believed to be $^4\Sigma^-$, of low stability, which undergoes a radiationless transition to levels $v = 2$ and 3 of the $A\,^2\Sigma$ state from which radiation occurs. Even in stoichiometric hydrogen/air the equilibrium concentration of free radicals is quite low (OH 1.0%, H 0.2%, O 0.05%) and in the cooler flames will be even less, so that any persistence of free atoms and radicals produced in the reaction zone or diffusion of them out of the flame into the surrounding cool air may cause a significant excess and lead to this abnormal radiation showing up more strongly than the thermal radiation of OH. This effect should be proportional to the square of the free radical concentration, whereas the strengthening of the OH (0,0) band in the reaction zone depends on the cube.

Hydrogen/Oxygen Flames

Flames with oxygen, unlike those with air, do give moderately strong visible emission. A premixed flame, such as an oxy-hydrogen blowpipe, has a bluish white core surrounded by a less luminous outer zone of a more orange hue, while the tip shows a pale yellow or yellow-green coloration. The flat diffusion flame between hydrogen and oxygen on a Wolfhard—Parker burner is fairly bright bluish white; the tip is again yellowish.

A rather striking flame is that of oxygen burning at a quartz jet in an atmosphere of hydrogen. It has a blue base and is surrounded by a red mantle. This red mantle is particularly strong when the supply of hydrogen is only slightly in excess of the amount required for complete combustion of the oxygen; the lower intensity of the mantle in faster flows of hydrogen is probably due to the cooling effect of the hydrogen which has a high thermal conductivity. In the author's experiments the red mantle was observed to extend from 1 to 3 cm. above the blue part of the flame, the red dying out gradually without any sharp limit.

The strongest feature of all H_2/O_2 flames is, of course, the spectrum of OH; this usually has a normal appearance without the vibrational intensity anomaly found in some H_2/air flames. The spectrum of the yellow or greenish yellow tip to flames surrounded with air is continuous and is due to the NO + O reaction, discussed below; in hydrogen/air flames the formation of nitric oxide is very slow and equilibrium is not attained, but if a little air is entrained into the much hotter oxy-hydrogen flame then some NO is formed. The blue part of the flame shows numerous fine lines superposed on a continuous background. Wolfhard and Parker have shown that the rather complex structure of fine lines is due to the Schumann—Runge system of O_2 and recent work by Padley shows that the reaction $H + OH = H_2O + h\nu$ contributes to the continuous background. The orange outer region of premixed flames and the red mantle of the diffusion flame of O_2 burning in H_2 show a complex banded structure increasing in strength towards the infra-red; this structure is mainly an extension of the infra-red vibration-rotation spectrum of H_2O but also shows some of the vibration-rotation bands of OH as well.

The O_2 Schumann—Runge Bands. These are strongest in the 3000—4000 Å region, but extend considerably further into the ultra-violet and also some way into the visible. The bands are so strongly degraded to the red that they do not usually show clear heads but only a mass of open rotational structure due to overlapping bands. They are due to a $^3\Sigma_u^- - ^3\Sigma_g^-$ transition, but the spin splitting in the Σ states is very small so the usual appearance of the bands under moderate dispersion is of apparently single P and R branches. Alternate lines in the rotational structure are missing, due to the homonuclear nature of the emitter (see page 78) and the structure is therefore more open than would otherwise be expected for a molecule having the moment of inertia of O_2. The stronger OH bands overlap and confuse the appearance of the Schumann–Runge bands in most flames. Some of the O_2 structure is, however, visible in Plate 1b. The most easily located features are the heads or convergence points of the following bands: $\lambda\lambda 3370$ (0,14), 3517 (0,15), 3671 (0,16), 3743 (1,17), 3841 (0,17) and 4096 (1,19).

Wolfhard and Parker (1949, 1952) have shown that in flat diffusion flames the O_2 emission lies fairly well towards the O_2 side of the hottest region. These bands require the relatively high energy of 6.09 eV or 590 kJ/mole for their excitation. Feast (1950) studied O_2 excitation in high-tension arcs and suggested that the excitation mechanism was recombination of excited and normal O atoms,

$$O(^3P) + O(^1D) + M = O_2(^3\Sigma_u^-) + M$$

However, Wolfhard and Parker found that in the H_2/O_2 diffusion flame the O_2 reversal temperature is close to the adiabatic flame temperature, indicating thermal rather than chemiluminescent excitation. The above reaction may, of course, play a part in maintaining the thermal population of excited O_2 molecules, but there is no evidence from the O_2 emission of any disequilibrium in these flames. Quantitative measurements on premixed flames (Shuler, 1951) are also consistent with thermal excitation of O_2.

The Blue Continuous Spectrum. This continuum is strongest in the blue and near ultra-violet, having a flat maximum around 4500 Å. It is overlaid by other structure, especially the OH and O_2 bands, and also at the long-wave end by H_2O vibration-rotation bands, but it appears to extend from 2200 Å to at least 6000 Å. The hot oxy-hydrogen flame contains high concentrations of free radicals (OH 10%, H 8% and O 4%) and there are a number of exothermic association reactions which may be considered. Padley (1960) has used spectroscopic techniques with added metals to follow the concentrations of H, OH and O and has studied the dependence of the strength of the continuum on these concentrations. Of the six possible reactions considered, all except $H + OH = H_2O + h\nu$ and $H_2 + O = H_2O + h\nu$ can be eliminated, and the latter seems unlikely because the production of water in a singlet electronic state from $O(^3P) + H_2(^1\Sigma)$ would break the spin conservation rule. Padley showed that the strength of the continuum closely followed the product of the concentrations [H] [OH], a slight discrepancy being attributed to quenching by water molecules. This blue continuum can therefore be assigned almost certainly to

$$H + OH = H_2O + h\nu.$$

The Yellow-green (NO + O) Continuum. The coloration at the top of oxy-hydrogen flames in contact with air seems to be associated with nitric oxide formation. Many other flames containing oxides of nitrogen also show similar coloration. Early observations on such flames were made by Lord Rayleigh, and the author (Gaydon, 1944, 1946) extended these observations and identified the yellow-green continuum with the air afterglow obtained in discharge tubes.

An electrical discharge through air at reduced pressure gives a strong yellow-green afterglow. A similar afterglow is obtained from oxygen unless it is carefully purified, and it has often been referred to as the oxygen afterglow. Lord Rayleigh showed that oxides of nitrogen are essential for its production, however, and Spealman and Rodebush (1935) demonstrated that the afterglow is due to reaction between nitric oxide and free oxygen atoms. We may therefore assign it to the association

$$NO + O = NO_2 + h\nu$$

The continuum extends from around 4000 Å to the near infra-red but is strongest in the yellow and green and gives a whitish yellow green colour to afterglows and flames. The energy liberated is about 300 kJ/mole corresponding to 3960 Å, which agrees closely with the short-wave limit. Since the reaction occurs in afterglows in the cold, the upper electronic state is presumably stable, as in Fig. IV.5a (page 93), and may tentatively be identified with the excited state of NO_2, transitions to which cause the visible absorption bands and the brown colour of the gas. In afterglows there are some faint diffuse NO_2 emission bands superposed on the continuum. Some authors have suggested that the reaction might involve a three-body reaction

$$NO + O + M = NO_2^* + M$$

followed by quenching $NO_2^* + M = NO_2 + M$ competing with the emission of light $NO_2^* = NO_2 + h\nu$. This would give similar kinetics at high pressure. However, Reeves, Harteck and Chace (1964) have shown that the kinetics are second order down to a pressure of only $3\mu m$ of mercury and the reaction is almost certainly the simple bimolecular one. Kaufman (1958) found the intensity was proportional to [NO] [O] and was independent of inert diluents. The association reaction occurred once in every 10^7 collisions. He developed the method of using nitric oxide to titrate for atomic oxygen (see page 51).

The author has found it possible to test for the presence of free oxygen atoms in flames by adding a little nitric oxide to the flame gases before combustion and observing with a spectroscope whether the yellow-green continuum occurs. In some cases the effect of testing a flame with nitric oxide is very striking and makes a pretty demonstration experiment. Hydrogen flames contain a fair amount of atomic oxygen when oxygen is in excess, but enclosed flames of rich mixtures do not show much. Flames of carbon monoxide also contain a lot of atomic oxygen, but premixed hydrocarbon flames, studied with a Smithells' separator, only show atomic oxygen in the outer cone, not in the inner cone or interconal gases. James and Sugden (1955b) have used the test semi-quantitatively to show the presence of excess atomic oxygen near the reaction zone of weak oxy-hydrogen flames and have studied the decay of [O] above the reaction zone.

The Vibration-Rotation Bands of H_2O. These were first recorded by Kitagawa (1936, 1939) who studied the red mantle of the flame of O_2 in H_2. He measured bands between 5683 and 6922 Å and provisionally assigned them to H_2O, because of coincidence of four of the strongest bands with absorption bands of steam at 4 atm. pressure. Gaydon (1942b) extended the observations into the photographic infra-red and made a detailed study of the fine structure of one band, confirming the assignment to H_2O; the bands are thus an extension of the infra-red spectrum of H_2O which is discussed in a later chapter. The intensities of the bands tend to increase towards longer wavelengths. They have a complex but resolved rotational structure, as expected for a triatomic hydride, but some are sufficiently degraded to longer wavelengths to show sharp heads, those at 6165, 6457, 6919, 7164, 8097, 8916, 9277 and 9669 Å being relatively prominent; a spectrum is shown in Plate 1c and details are given in the Appendix.

Kitagawa pointed out that excitation of these vibration-rotation bands in the visible required quite high energy and discussed various exothermic reactions which might lead to formation of H_2O with high vibrational energy. However, the radiation does not appear to be strong in the reaction zone, and there is no evidence that it is other than thermal excitation.

Vibration-Rotation bands of OH. Oxy-hydrogen flames, and also the outer cones of hot hydrocarbon flames like C_2H_2/O_2 also emit OH bands in the photographic infra-red (Hornbeck and Herman, 1951; Herman and Hornbeck, 1953; Déjardin, Janin and Peyron, 1953). The strongest bands are the vibrational transitions $4 \to 0$, $5 \to 1$ and $6 \to 2$ at 7461, 7849 and 8278 Å. Déjardin, Janin and Peyron (1952) also reported weaker bands attributed to the $11 \to 5$, $10 \to 5$, $9 \to 4$ and $8 \to 4$ transitions. The main OH vibrational excitation is probably thermal, but it seems possible that bands arising from these high vibrational levels could result from the $H + O_3$ reaction which is believed to be responsible for the Meinel bands in the night-sky glow spectrum and which has been shown directly to excite bands to $v = 9$ (McKinley Garvin and Boudart, 1955).

Pressure Effects

Early observations by Liveing and Dewar (1891) showed that the visible radiation from oxy-hydrogen flames increased rapidly with pressure. This early work is still of value because of the considerable difficulty in handling flames at high pressure. They studied flames of O_2 in H_2 and of H_2 in O_2. In the flame of O_2 burning in H_2 the blue radiation increased as the square of the pressure in the range from 1 to 3 atm. For H_2 burning in O_2 the flame was brighter but its luminosity increased less rapidly with pressure. These results can perhaps be explained by assuming that the flame of O_2 in excess H_2 gave mainly the continuous spectrum and that this depends on the product of H and OH concentration and goes as the square of the pressure, while the flame of H_2 in excess of O_2 emits mainly Schumann–Runge bands of O_2 which are excited thermally and depend mainly on the O_2 density and therefore increase less rapidly with pressure*.

With flames at reduced pressure, the luminosity continues to fall with pressure, and the premixed low-pressure flames down to 5 torr studied by the author and Dr Wolfhard were quite invisible even in a completely darkened room. The only visible indication of the presence of a flame was a slight heating of the burner rim and some bright hot spots on the walls of the outer reaction vessel; these are probably due to recombination of excess free atoms on specks of dust. These flames were so non-luminous that it was necessary to mix a little hydrocarbon with the hydrogen when lighting, and to turn this off when the flame was seen to be stabilized. The absence of O_2 emission at low pressure may be due to radiation depletion, because the normal processes of thermal excitation involve collisions which are relatively inefficient at interconversion between translational energy and that of electronic excitation; the number of these collisions falls, of course, with pressure.

Kondratiev and Ziskin (1937a) first commented on the strength of the (2,1) band of OH in diffusion flames of H_2 burning in O_2 at 10 torr. This selective excitation of the (2,1) band, and also of the (2,0) band, has also been found (Gaydon and Wolfhard, 1951b) in the premixed flames down to 5 torr. In the premixed flame there is no well-defined reaction zone, as with hydrocarbon flames. The (0,0) band of OH is emitted from the whole body of burnt gas, as is the (1,0) band, but these bands fall off slightly in intensity with increasing height up the flame. The bands with $v' = 2$ increase in strength upwards, however. This is shown in Plate 2b, in which the (1,0) is rather stronger than the (2,1) at the base of the flame, but the (2,1) is relatively much stronger and definitely exceeds the (1,0) higher in the flame. The intensity distribution in this flame is quite different from the H_2/air flames previously discussed; in the low-pressure flame the (3,2) is very weak.

---

*This is supported by later work of Diederichsen and Wolfhard (1956) who studied the flat H_2/O_2 diffusion flame to 40 atm. and found that the O_2 emission increased by a factor of 400 and the continuum by a factor of more than 1000.

Kondratiev and Ziskin suggested that the OH radicals were produced in the excited state by chemiluminescence by the reaction

$$H + H_2 + O_2 = H_2O + OH + 417 \text{ kJ/mole}$$

However, this reaction would also be expected to excite bands with $v' = 0$ and 1 as well, and not to be selective in the excitation to the $v' = 2$ level which actually requires 453 kJ/mole; it does not explain the relative increase in strength towards the top of the flame where the concentrations of H_2 and O_2 would be expected to fall; also the OH radiation is quite weak and not as strong as would be expected for chemiluminescence.

Gaydon and Wolfhard (1951b) have therefore sought an explanation involving the predissociation of OH which is known to set in at $v' = 2$. This predissociation was first observed as a weakening of bands with $v' \geqslant 2$ and a weakening of very high rotational levels of bands with $v' = 0$ and 1 in the inner cone of hydrocarbon flames in which there is definite chemiluminescent excitation of OH. One spin component of the $A^2\Sigma^+$ state (the upper state of the main 3064 Å system) is predissociated more strongly than the other component. The initial explanation suggested that it was due to a radiationless transition between $A^2\Sigma^+$ and a $^2\Sigma^-$ state of low stability. This predissociation has since been discussed in many papers; Gaydon and Kopp (1971) favour a transition to a $^4\Sigma^-$ state. *See also* Palmer and Naegeli, *J. Chem. Phys.* 59 994 (1973).

In the H_2/air flame at 1 atm. we attributed the strength of bands with $v' = 2$ and 3 to pre-association due to an excess population of H and O atoms which formed OH in the low-stability $^4\Sigma^-$ state, from which transitions populated $v' = 2$ and 3 of $A^2\Sigma^+$. In the low-pressure H_2/O_2 flame the concentrations of H and O will be quite high in equilibrium (about 11% and 4% respectively) and while some excess of free atoms in the flame is not unlikely it could not, on energetic grounds alone, amount to a big proportional increase. On our plates the (2,1) band is from 5 to 10 times stronger compared with the (1,0) than it should be for equilibrium. We therefore attribute the vibrational intensity anomaly to lack of radiative equilibrium. In these low-pressure flames normal collision processes are unable to maintain the population in the $v' = 0$ and 1 levels at the Maxwell—Boltzmann equilibrium level because of radiative depopulation (the radiative lifetime for OH is around 2×10^{-6} sec) i.e. there is radiation depletion. For the $v' = 2$ level, however, in addition to normal collisions, the predissociation/preassociation equilibrium will also help to maintain the population at its equilibrium level and so bring the (2,1) and (2,0) bands up to nearer their equilibrium intensity and reduce the radiation depletion effect. This type of anomaly only occurs in low-pressure H_2/O_2 flames where the collision frequency is low and the concentration of free atoms is high. The (3,2) band should also be helped up to its thermal equilibrium strength, but this band will be much weaker than the (2,1) because of the smaller equilibrium population in the $v' = 3$ level. Radiation depletion has been discussed by Gaydon and Wolfhard (1970) in connection with spectrum-line reversal measurements of flame temperature; at atmospheric pressure the effect is usually small, but it becomes appreciable for low-pressure flames or inert gases (e.g. shock-heated argon).

Absorption Spectra

The OH bands are obtained fairly readily in absorption. Since they have an open rotational fine structure it is necessary for sensitive detection to use spectrographs of high resolving power (see page 24) but the (0,0) band is sufficiently strong in most flames for the main heads to be observable with a medium quartz spectrograph in good adjustment; a sufficiently bright background, such as a flash-tube, is of course necessary. The (1,0) and (2,0) bands at 2811 and 2608 Å can also be observed with a good spectrograph.

At shorter wavelengths, oxy-hydrogen flames also show quite strong absorption due to hot oxygen. In cold oxygen the main Schumann—Runge system gives bands in the vacuum ultra-violet up to about 1950 Å, but as the gas is heated new bands, arising from higher vibrational levels of the ground electronic state, appear, and with very hot flames such as the oxy-hydrogen flame the absorption extends to 2600 Å or even further. As with the emission bands of this system, which are usually observed at much longer wavelengths, the rotational structure of individual bands is fairly open and simple in appearance, but the bands do not show very obvious heads because they are so strongly degraded and there is much overlapping of bands so that the structure appears confused. Again, the bands are best seen with large dispersion but can be seen to some extent with a medium quartz spectrograph.

These are the only two absorption systems in the near ultra-violet whose existence has been definitely established. Hydrogen peroxide shows continuous absorption in the ultra-violet; this starts weakly around 3700 Å and extends with increasing strength to beyond 1900 Å. H_2O_2 is known to be formed in discharge tubes and in chilled low-pressure explosion flames but there is no conclusive evidence from absorption spectroscopy of its presence in flames; weak continuous absorption in H_2/O_2 explosions at reduced pressure could be due to H_2O_2 or hot O_2 (Broida, Everett and Minkoff, 1954). Similarly ozone, which is known to show strong absorption bands around 3100–3400 Å, has not been observed in flames. Kinetic studies of hydrogen explosion limits require the formation of the radical HO_2 and this would be expected to show some absorption in the near ultra-violet but no spectroscopic evidence for its existence has been obtained and it is also difficult to record with a mass spectrometer.

Garton and Broida (1953) made a preliminary study of the H_2/O_2 flat diffusion flame in the vacuum ultra-violet. On the O_2 side the Schumann–Runge bands caused complete absorption, but towards the H_2 side a continuous absorption below 1925 Å was provisionally attributed to hot H_2O, and six unknown weak bands were reported between 1940 and 2200 Å; attempts to study some of these under larger dispersion, by the author and F. Guedeney, were unsuccessful.

Kinetics of Reactions of OH Radicals. Early studies of OH absorption in low-pressure H_2/O_2 flames by Kondratiev and colleagues (e.g. Kondratiev and Ziskin, 1937b) indicated an OH concentration far in excess of the equilibrium value; their estimate of an excess concentration by a factor of 1000 has been criticized and is indeed probably too high, but there is no doubt that there is a major departure from equilibrium in the reaction zone.

Norrish and Porter (1952) studied low-pressure explosions of H_2/O_2 initiated by flash photolysis, using NO_2 as sensitizer. They rightly stressed the difficulties of making precise quantitative measurements from absorption spectra because of the effects of temperature on rotational and vibrational intensity distribution and on line width. Also their conditions, with rapid wall cooling and in the presence of oxides of nitrogen are not quite the same as for a steady flame. They found that the OH absorption reached maximum intensity very quickly, in less than a millisecond, and they concluded that three-body collisions could not be important in its formation and that the bimolecular processes must have a high efficiency, not less than 10^{-3}. With excess hydrogen present the OH decay was rapid but with excess O_2 it was much slower. Evidence for a departure from equilibrium was supported by experiments in which heavy water was added as diluent; the absorption bands of OD were less strong, compared with those of OH, than for equilibrium.

There have been a number of time-resolved studies of OH absorption during the shock heating and ignition of H_2/O_2 mixtures at various temperatures and these have given actual rates for the major reactions involving OH radicals. For this type of work a background discharge-tube source giving fine OH emission lines has often been used, but changes in absorption-line contour with temperature and pressure may limit the accuracy of the results. The main initiating reaction is

$$H_2 + O_2 \xrightarrow{k_0} 2 \, OH$$

with chain propagation

$$OH + H_2 \xrightarrow{k_1} H_2O + H$$

and chain branching

$$H + O_2 \xrightarrow{k_2} OH + O$$

and

$$O + H_2 \xrightarrow{k_3} OH + H.$$

Expressing the results in Arrhenius form, with the activation energy and the gas constant R in calories, Jachimowski and Houghton (1970, 1971) found

$$k_0 = 1.70 \times 10^{10} \exp(-48150/RT) \text{ litres mole}^{-1} \text{ sec}^{-1}$$

$$k_2 = 9.9 \times 10^{10} \exp(-15020/RT) \quad \text{litres mole}^{-1} \text{ sec}^{-1}$$

$$k_3 = 7.5 \times 10^{10} \exp(-11100/RT) \quad \text{litres mole}^{-1} \text{ sec}^{-1}$$

In older work by a similar method Schott and Kinsey (1958) found a slightly higher value for k_2 of $30 \times 10^{10} \exp(-17500/RT)$. They also studied the induction period for ignition, and Schott (1960) showed that the OH concentration greatly exceeded the equilibrium value and that a pseudo-equilibrium between [OH], [O] and [H] was quickly established by fast bimolecular exchange reactions.

Structure of Diffusion Flames. The oxy-hydrogen flat diffusion flame shows strong OH and O_2 absorption, and Wolfhard and Parker (1949, 1952) have used these absorptions, together with temperature measurements by the spectrum-line reversal method, to build up a detailed picture of concentrations and temperature through the reaction zone. OH and O_2 absorptions were compared quantitatively with those in premixed oxy-hydrogen flames of known composition and temperature in a region well above the reaction zone where equilibrium conditions could safely be assumed. The results are consistent with the maintenance of chemical equilibrium and thermal excitation throughout the reaction zone of the diffusion flame, and it seems that both O_2 and H_2 are raised to quite a high temperature (2000° C) before they reach the main reaction zone.

Excitation of Metal Spectra in H_2 /O_2 /N_2 Flames

When metal salts are sprayed into hydrogen flames the resonance lines of the metal atoms usually appear in emission, and in some cases band systems of oxides, hydrides or hydroxides are also observed. Sugden and colleagues at Cambridge between 1955 and 1960 published a long and valuable series of papers on measurements of the strengths of these metal lines and bands in hydrogen-oxygen-nitrogen flames, making quantitative measurements in flames of various compositions and temperatures and also studying the effect of height in the flame.

All flames were studied on what is essentially a modified Meker burner (Figure V.1) in which only the central region of the flame was coloured with added metal, this central region being protected from surrounding air and from wall-cooling by a ring of flame of similar gas composition but without added metal. An outer surrounding collar of flame also helped to improve stability. The metal grid of the usual Meker burner was replaced by a matrix of hypodermic needles.

Flames of general composition $aH_2 + O_2 + bN_2$ were studied and these were, in the papers of Sugden and colleagues, denoted by a capital letter representing the value of a (for a = 2½, 3, 3½, 4 and 4½ the letters F, K, P, U and Z were used) and by a subscript representing the value of b. The concentrations of added metal were obtained from the rate of consumption of liquid spray of known

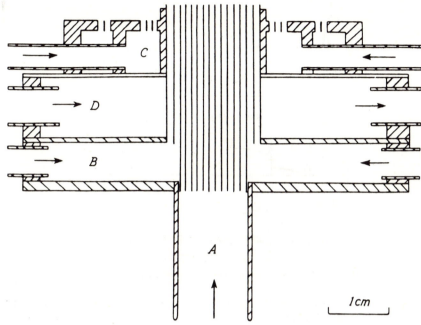

Fig. V.1. Section of burner used by Sugden and colleagues. A similar $H_2/O_2/N_2$ gas mixture was introduced at A, B and C, but only the flow A contained added metal salts. D was water cooling.

concentration and from measured gas flow rates. Temperatures of the flames were measured by the sodium-line reversal method.

For added sodium a relatively simple variation of excitation with composition and height is observed (Figure V.2). Padley and Sugden (1958) found that for hot flames (e.g. compositions P_3, K_3 and F_3) the excitation was thus dominated by thermal processes; the intensity was thus constant through the flame, apart from a narrow region very close to the base where combustion is incomplete and the full temperature is not reached. For the cooler flames (U_8 P_8) and for very rich mixtures (Z_5) the thermal emission from the main body of the flame is quite weak but at the base there is relatively strong emission due to the chemiluminescent reactions

$$Na + H + H = Na^* + H_2$$

and

$$Na + H + OH = Na^* + H_2O$$

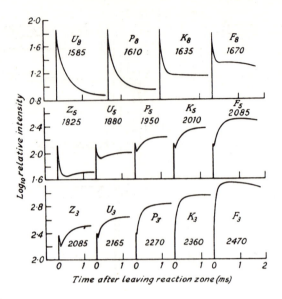

Fig. V.2. Relative strength of sodium D-line emission as a function of height (expressed as time above flame base in m sec) for flames of various composition, denoted by letter and subscript (see text). The numbers are measured reversal temperatures (deg K) in the equilibrium region.

This chemiluminescence is associated with an abnormally high population of free atoms and OH radicals in and close to the reaction zone; the high emission intensity is, of course, accompanied by an abnormally high reversal temperature in this part of the flame. For flames at intermediate temperature (U_5 P_5, Z_3) the sodium emission is high at the base of the reaction zone, falls to a minimum, and then rises again as the thermal emission increases towards equilibrium. The chemiluminescent excitation should be proportional to the square of the free atom concentration, whereas the departure from thermal equilibrium due to uncombined atoms will be proportional to the first power of the atom concentration.

For several other metals a similar pattern of excitation through the flame occurs. When the resonance line lies at long wavelength (red or near infra-red) the thermal emission is strong and tends to dominate, while for lines at short wavelength the thermal emission is relatively weak (as required by the Planck law) and the chemiluminescence shows up more strongly. Figure V.3 shows a comparison of excitation of lines of several metals in the same flame, from Padley and Sugden (1959*b*).

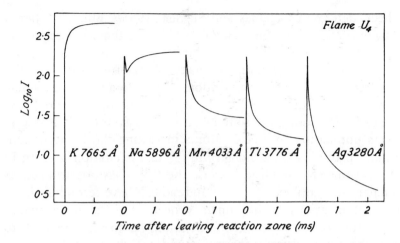

Fig. V.3. Variation of line intensity with height in flame (expressed as time in m sec) for various metals in the same flame.

From studies of the effect of mixture composition and determination of the concentrations [H], [OH] and [O] by methods to be discussed later, the chemiluminescent processes have been established. For Na the recombinations H + H and H + OH both contribute. For lead the process

$$Pb + H + OH = Pb^* + H_2O$$

is dominant, but for thallium

$$Tl + H + H = Tl^* + H_2O$$

is much more important than the H + OH recombination (Padley and Sugden 1959*b*).

For some metals the formation of molecules causes further complications, but may also give valuable information. With lithium, a stable hydroxide exists and a large fraction of the added metal is removed in this form, so that the free Li atom concentration is reduced. From studies in the equilibrium region the dissociation energy of LiOH and the equilibrium constant for the reaction

$$H + LiOH \underset{k_-}{\overset{k_+}{\rightleftharpoons}} Li + H_2O$$

in the form

$$K \equiv k_+/k_- = [LiOH] [H] / [Li] [H_2O]$$

have been determined (James and Sugden, 1955b, Bulewicz, James and Sugden, 1956). In some $H_2/O_2/N_2$ flames there is a region of strong Li emission just *above* the reaction zone, where the sodium emission is near its minimum intensity. This effect can be quite striking (see also Charton and Gaydon, 1958). In a flame containing both sodium and lithium, it may cause the flame to have a red base and yellow top — quite a simple lecture demonstration. It is due to an excess population of free H atoms increasing the Li atom concentration in this region although the thermal excitation of both Li and Na is somewhat reduced because the flame has not attained its full temperature until excess atoms and radicals have recombined.

These observations on lithium may be used to obtain a quantitative measure of the H-atom concentration. To do this one needs to measure the temperature in the region under study, either by Na-line reversal of from Na emission compared with that in the equilibrium zone (sodium does not have a hydroxide of significant stability to interfere), to know the equilibrium constant K at this temperature, and to calculate the Li concentration in the zone and in the equilibrium region, it being assumed that all added lithium not present as Li is present as LiOH. This is now a well-accepted method of measuring [H] in flames. Halstead and Jenkins (1969, 1970) have used the method, but with measurements of Li line absorption instead of emission, to determine the rate coefficients for the recombination reactions

$$H + H + M = H_2 + M \quad \text{and} \quad H + OH + M = H_2O + M$$

for various species of third body M.

Other methods of measuring the atomic hydrogen concentration are from measurement of CuH band intensity (Bulewicz and Sugden, 1956) and from study of the equilibrium Na + HCl = NaCl + H.

Concentrations of OH can be measured directly by absorption spectroscopy or from study of the intensity of the continuous spectrum emitted in the association

$$Na + OH = NaOH + h\nu$$

or the strength of CuOH band emission in the green (James and Sugden, 1958). In some cases atomic oxygen may be estimated from the yellow-green afterglow by titration with NO (see page 107, and James & Sugden, 1955b). These observations all tend to show a very high concentration of free radicals, compared with the equilibrium values, for the reaction zones of the lower temperature flames, and a rather slow recombination of this excess concentration. The results are also supported by mass-spectrometric studies (e.g. Dixon-Lewis and Williams, 1963). Even in and immediately above the reaction zone, where the atomic and OH populations are high, rapid bimolecular reactions

$$OH + H_2 \rightleftharpoons H_2O + H \text{ and } O + H_2 \rightleftharpoons OH + H$$

maintain a pseudo equilibrium between the free radicals. The removal of the excess population of these, however, requires slower termolecular reactions, mainly

$$H + H + M = H_2 + M \quad \text{and} \quad OH + H + M = H_2O + M.$$

Spectroscopic measurements of line and band intensities in these $H_2/O_2/N_2$ flames have been used to determine the dissociation energies of a number of metallic compounds, including the mono-hydroxides of Na, K, Li, Rb, Ga, Tl and Cs (Kelly and Padley, 1971), MnO and MnOH (Padley and Sugden, 1959*a*) SnO and SnOH (Bulewicz and Padley, 1971*a*) and CrO, CrO_2 and $HCrO_3$ (Bulewicz and Padley, 1971*b*).

Candoluminescence

The hydrogen flame, and to a less extent certain other flames containing hydrogen such as manufactured gas, possess the curious property of inducing luminescence in certain solids placed in contact with the flame. This phenomenon is usually referred to as cando-luminescence. It was first reported in 1842 by W. H. Balmain, and later investigated by J. Donan and by E. L. Nichols and colleagues at Cornell. E. C. W. Smith (1940*a*) carried out experiments to check Nichol's work and published an interesting review.

At one time it was thought that the radiation from certain substances heated to incandescence in a flame exceeded that from a black body at the surface temperature, but Smith showed that the apparent effect was really due to abnormal heating of the surface to a higher temperature than that which would be taken up by a black body in the flame. An extreme example is that of the Welsbach gas mantle; this consists mainly of thoria containing 1 per cent of ceria. This mixture has a high emissivity in the visible, especially in the blue, but low emissivity in the red. Thus the mantle may tend to become hotter than a black body because of the lower radiation losses resulting from the low infra-red emission, but the high blue emissivity enables it to give a good white light (Minchin, Densham and Wright, 1940). However, more recent work by Kondratenko and Sokolov (1966) shows that the main effect is due to heating of the surface by recombination of hydrogen atoms on it.

The candoluminescence exhibited by certain solids at below red heat is, however, undoubtedly a true luminescence. Smith (1940a) did not observe the effect with any pure substance, and showed that an activator was necessary. The oxides of Ca, Ba, Sr or Be are good parent substances, with various activators giving characteristic colours, bismuth (violet), manganese (golden yellow), lead (blue), antimony (bright blue) and praesodymium (pink). The spectra are characteristic of the activator rather than of the parent substance and consist of regions of continuum; praesodymium gives some diffuse lines. Boron nitride also shows strong candoluminescence.

To account for candoluminescence three possible explanations have been considered; (i) fluorescence in ultra-violet light from the flame; (ii) catalytic effects such as alternate oxidation and reduction at the surface; (iii) surface release of excess energy carried by active molecules or free atoms. Experimental work, summarized by Arthur and Townend (1954), shows that recombination of free hydrogen atoms is the true explanation. Similar luminescence can be obtained by placing the active substances in a stream of free hydrogen atoms from a low-pressure discharge.

Candoluminescence is most frequently observed with diffusion flames. Arthur and Townend have shown good photographs and demonstrated the effect by placing a porous block impregnated with the substance over the top of a burner. The gas is lit as it issues from the top of the block and a bright luminous ring is seen round the edge of the flame. Smith also showed that the effect occurs particularly strongly with hydrogen flames at reduced pressure, the luminescence spreading some mm from the flame. It seems, under these circumstances, to be associated with diffusion of the free atoms from the reaction zone of the flame into the cooler surrounding gas where their concentration must exceed the equilibrium value. Candoluminescence has only been observed from flames containing hydrogen and it is quenched by various additives. Arthur and Townend found that monatomic gases do not quench, diatomic gases quench slightly and polyatomic molecules, especially large organic ones, quench strongly.

Applications to Combustion Mechanism

The assignment of the main emission bands in hydrogen flames to the free radical OH by Watson led to a major advance in interpreting the reaction processes. Bonhoeffer and Haber (1928) included OH in their chain-reaction scheme, and this has been followed in practically all later variations.

Studies of the kinetics of initiation of explosions in closed vessels (e.g. Lewis and von Elbe, 1961; Minkoff and Tipper, 1962) show that surface initiation (probably H_2 + surface → H + H + surface, requiring 433 kJ/mole or 103 kcal/mole) and chain termination by formation of the relatively unreactive radical HO_2 (by H + O_2 + M = HO_2 + M) are important. H, OH and O are involved in the chain propagation and branching, and hydrogen peroxide also participates. However there is little evidence from optical spectroscopy for the formation of appreciable concentrations of HO_2 or H_2O_2 and even mass spectrometric detection is rather difficult and sometimes inconclusive because of the existence of other isotopes such as $O^{16}O^{17}$

In the gas phase, we have seen from shock tube studies (page 113) that the important reaction is

$$H_2 + O_2 = 2OH$$

The energy requirement for this is only 78 kJ/mole or 18 kcal, but the activation energy is actually 48 kcal. This reaction is likely to be important in detonations which are propagated by the accompanying shock wave and for which diffusion is unimportant because of the high speed of propagation.

In premixed flames (see Gaydon and Wolfhard, 1970) the chemical reactions are initiated by diffusion of H and OH from the reaction zone into the unburnt mixture, rather than by creation of new centres, as in spontaneous thermal ignition or detonation. The main chain propagation step is the slightly exothermic

$$OH + H_2 = H + H_2O + 61 \text{ kJ/mole (15 kcal)}$$

The slightly endothermic branching reactions

$$H + O_2 = OH + O - 69 \text{ kJ/mole (16 kcal)}$$

and

$$O + H_2 = OH + H - 8 \text{ kJ/mole (2 kcal)}$$

lead to a rapid build up to high concentrations of H, O and OH. A pseudo-equilibrium between these is rapidly established by bi-molecular exchange reactions. These radicals, especially the H atoms, diffuse from the reaction zone, both forwards into the unburnt mixture, thereby propagating the flame, and back into the burnt gases. Recombination of the excess radicals requires three-body collisions, mainly

$$H + H + M = H_2 + M + 435 \text{ kJ/mole (104 kcal)}$$

and

$$H + OH + M = H_2O + M + 497 \text{ kJ/mole (119 kcal)}$$

These are very exothermic and the flame does not attain its full adiabatic temperature until these reactions have established equilibrium. Under some conditions (especially at low pressure) the inverse predissociation

$$O + H \rightarrow OH(^4\Sigma^-) \rightarrow OH(^2\Pi) + h\nu$$

may contribute to the recombination processes.

A very detailed analysis of the structure of oxygen-hydrogen-nitrogen flames has been made by Dixon-Lewis (Dixon-Lewis and Williams, 1963; Dixon-Lewis, 1967, 1968); species concentrations have been measured mainly by mass spectrometry, but OH absorption and Na chemiluminescence have also been used, and the temperature profile has been measured with fine thermocouples. Calculations relate reaction rates (see page 113), transport properties (diffusion and thermal conductivity) with the burning velocity. The high concentrations of active species in the reaction zone have been measured and the frequently-made assumption of a pseudo-equilibrium between OH and H has been confirmed. This is the most complete analysis of flame structure and processes yet made.

The analysis by Dixon-Lewis also includes many reactions of HO_2 and H_2O_2, but as already stated these are not detected spectro-scopically. Hydrogen peroxide can be frozen out in considerable quantities from low-pressure explosions in chilled vessels (Egerton and Minkoff, 1947); some of this is due to recombination of OH radicals on the walls, but some may be formed in the gas phase. HO_2 might be formed either by $H + O_2 + M = HO_2 + M$ or possibly as a collision complex of appreciable life by direct association $H + O_2 = HO_2$ (Rosen, 1933; Rice, 1933); the further exothermic reaction $HO_2 + H_2 = H_2O + OH$ would continue the chain; here HO_2 may be considered as a 'sticky collision' which increases the chance of the three-body collision

$$H + O_2 + H_2 = H_2O + OH + 418 \text{ kJ/mole (100 kcal)}$$

Ozone is another conceivable intermediary, but studies of the absorption spectrum show that its concentration is never measureable.

In some flames and explosions the release of energy may not be complete, and in the past this latent energy was attributed to metastable molecules (David, Brown and El Din, 1932). O_2 does possess some metastable electronic states of low energy, but there is no evidence that H_2O does. These effects are usually studied in highly-dilute or near-limit flames and are again probably due to persistence of free H atoms (David and Mann, 1947) but selective diffusion may also alter the mixture strength and complicate interpretation of results (see Gaydon and Wolfhard, 1970).

The Hydrogen/Nitrous Oxide Flame

The visible and ultra-violet emission spectra of flames of hydrogen with nitrous oxide were first described by Dixon and Higgins (1928) and Fowler and Badami (1931), and the author (Gaydon, 1942b) extended observations into the near infra-red. The appearance and spectrum vary rather widely according to the type of flame.

With pre-mixed gases the flame has a non-luminous outer cone, which shows only the OH bands, and also a definite inner cone with a brightly luminous yellow cap; in this it differs from the hydrogen/air or oxygen flame. The spectrum of the yellow cap shows the γ bands of NO in the ultra-violet (see page 362), the NH band at 3360 Å (page 361) and again OH, while in the visible region there is a complex structure usually referred to as the ammonia α-band, now known to be due to the N_2 radical.

This α-band also occurs in the flame of ammonia and has been obtained particularly well in absorption following the flash photo-lysis of ammonia or hydrazine; Herzberg and Ramsay (1952) established the identity of the photolysis absorption and flame emission, while Dressler and Ramsay (1959) have succeeded in making a full rotational and vibrational analysis of this difficult spectrum, confirming the assignment to NH_2 and also studying the isotope ND_2. The excited electronic state is linear, HNH, and of type $^2 \Pi_u$, while the ground state is bent, with an angle of 103°, and is of type $^2 B_1$. The strongest bands, in the photolysis absorption, fall into a single progression $(0, v_2', 0) \leftarrow (0, 0, 0)$ with v_2' (the quantum number for the bending vibration) taking values from 3 to 18. In this progression the structure of the bands alternates, those bands with odd v_2' showing rotational structure due to vibronic sub-bands of types Σ, Δ aand Γ, while those bands with even v_2' show Π and Φ vibronic sub-bands. In emission the line structure is very compli-cated, but under small dispersion there are strong features around 5713, 6042, 6302, 6330, and 6652 Å.

In H_2/N_2O flames the OH is much stronger in the inner than in the outer cone, even for hot flames. Observations on low-pressure flames, at around 10 torr, by Gaydon and Wolfhard (1949b) show that the OH has a rather high rotational temperature, but the vibrational intensity distribution is quite different from that of hydrogen/oxygen flames, the (1,0) band being strong but the (2,1) quite weak. The NH band, too, is emitted from the base of the flame, while the NH_2 is strongest rather higher in the reaction zone (Gaydon and Wolfhard, 1949a).

In premixed flames with excess N_2O the yellow cap is less noticeable and the whole flame shows a greyish-green colour, the spectrum of which is continuous and may be attributed to reaction of atomic oxygen with nitric oxide.

The emission of NH and NH_2 bands from premixed flames of H_2 with N_2O is in marked contrast with the absence of these bands in $H_2/O_2/N_2$ mixtures. It indicates that the N_2O does not all decompose to N_2 and oxygen, but that nitrogen containing molecules actually react with the hydrogen. The difference in spectra of H_2/N_2O from $H_2/O_2/N_2$ flames is clear proof that the radiation is not just thermal emission from equilibrium products. The variation in intensity of OH and NH with mixture strength and with flame temperature (controlled by adding N_2 as diluent) has been studied by Ausloos and Van Tiggelen (1953a) who concluded that at least part of the OH radiation is not thermal in origin and that NH participates in the chain reactions but not in the chain branching steps.

Simple diffusion flames of H_2 burning in N_2O or of N_2O burning in O_2 do not show the characteristic banded spectrum of the premixed flame. Apart from OH they only show a continuous spectrum in the visible and near ultra-violet. This continuum is presumably due to the reaction between O and NO. It seems that N_2O is decomposed to nitrogen, oxygen and NO in the preheating zone, so that N_2O molecules do not reach the H_2 to give the NH or NH_2 emission.

Flames with NO and NO_2

Flames of H_2 with NO, although hotter even than those of H_2/O_2, are very difficult to light because of the high ignition temperature and large quenching distance, but can be maintained on large burners (see Gaydon and Wolfhard, 1970). Their spectra are quite different from those of premixed H_2/N_2O flames; the main emission is of the OH bands, with some O_2 Schumann—Runge and perhaps weak NO bands (Wolfhard and Parker, 1955); in preliminary work weak NH emission was observed, but this appears to be spurious, due to N_2O contamination. It seems that the NO decomposes thermally to $N_2 + O_2$ and the H_2 then burns with the O_2. NO is, however, very stable and is only decomposed at very high temperature, 3000 K and above. Studies of the effect of mixture strength on burning velocity, quenching distance and minimum ignition energy indicate that diffusion of hydrogen atoms is not important in the propagation mechanism which is controlled mainly by thermal decomposition of NO.

Hydrogen burns fairly readily with NO_2 but the flame has a much lower temperature than would be expected for complete combustion; for the stoichiometric flame the measured temperature is about 1820 K compared with an adiabatic value of 2930 K. The flame has a single yellowish reaction zone whose spectrum is mainly continuous (NO + O) and the OH radiation is very weak, only about 1/100 of the intensity of OH in H_2/O_2. Strong NO absorption bands are found in the gases above the reaction zone. These studies of the spectrum by Wolfhard and Parker (1953, 1955) lead to the conclusion that the NO_2 decomposes to NO and oxygen, the oxygen burns rapidly with the hydrogen, but the NO is not decomposed because the temperature attained in the first-stage flame, 1820 K or less, is insufficient to decompose the NO which then behaves as an inert diluent.

Chapter VI

The Carbon Monoxide Flame

The Spectrum

The flame of carbon monoxide burning with air or oxygen is bright blue, and the spectrum consists of a strong continuum with a weak system of numerous narrow bands superposed on it. The emission is strongest in the blue to near ultra-violet, 4500–3500 Å, but extends from the near infra-red to well down in the ultra-violet below 2400 Å. The banded part of the spectrum is clearest in the visible and becomes increasingly masked by the continuum at shorter wavelengths.

In early studies by Weston (1925), Kondratiev (1930) and Gaydon (1940) the bands were provisionally assigned to CO_2, and more recently Dixon (1963) has made a convincing analysis of the band structure and shown that it is indeed due to this molecule. In hot flames the continuum is entirely dominant, but the band system is relatively more prominent in flames at lower temperatures or at reduced pressure. A small dispersion spectrogram is reproduced in Plate 1e; this was taken of a low-pressure flame and a plate of high contrast was used to emphasise the band structure. Large-dispersion photographs of the band, by Dixon, Plate 7, were taken in an afterglow source.

For CO flames there does not seem to be any marked difference between the radiation from diffusion and from premixed flames. Premixed flames do show an inner cone, from which the radiation is more intense than from the region above, and the inner and outer cones can be separated in a Smithells' separator, this revealing that the interconal gases are practically non-luminous, while the blue inner and outer cones have similar spectra. For CO flames the reaction zone is much thicker than for the inner cone of a Bunsen flame. In diffusion flames there is no marked change in the spectrum with position in the flame, but bands of the Schumann–Runge system of O_2 also occur on the oxygen side.

While the continuum, the CO-flame bands of CO_2 and sometimes the O_2 bands are the only features of pure dry CO flames, the OH band at 3064 Å is normally a very strong and persistent feature, and it is rare to see a spectrum of a CO flame without this band. In the photographic infra-red there is some extension of the continuum and the CO_2 bands, but these become weaker towards longer wavelengths; but again with undried gases the vibration-rotation bands of H_2O are a surprisingly strong feature of the spectrum (Gaydon, 1942b).

Emission bands of O_2 from CO flames were first reported by Hornbeck (1948), Herman, Hopfield, Hornbeck and Silverman (1949) and Hornbeck and Hopfield (1949). They studied the very hot closed-vessel explosions of CO/O_2 mixtures in a spherical bomb. These spectra showed the Schumann–Runge bands, and, more remarkable, the forbidden "atmospheric" bands of O_2 in the extreme red. The Schumann–Runge bands (see page 104) also occur in steady flames of CO with O_2; at first there was some confusion between these bands and the true CO-flame bands but Wolfhard and Gaydon (1949) showed that the systems were separate, the O_2 bands occurring in very hot flames with excess O_2, the emission being presumably thermal, while the CO-flame bands are seen best in cooler flames. The atmospheric bands have a fairly open rotational structure and are degraded to longer wavelengths. They are due to the transition $b^1 \Sigma_g^+ - x^3 \Sigma_g^-$, which is forbidden by practically every selection rule. They are normally observed only in absorption by long paths of oxygen (seen typically in the solar spectrum through several miles of air) and their occurrence in emission from these explosion flames is quite unexpected. The strongest head of this atmospheric system is at 7593.7 Å (0,0); there are other fairly prominent heads 7683.8 (1,1), 7779 (2,2), 7879 (3,3) and 8597.8 (0,1). Griffing and Laidler (1949) suggested that the oxygen molecules were excited to the $b^1 \Sigma^+$ state by collisions with excited CO_2 molecules in a triplet state, thus conserving spin. However, the atmospheric bands have also been observed quite strongly in emission from an afterglow of pure oxygen (Broida and Gaydon, 1954) and can probably be explained as due to reactions of atomic oxygen or metastable oxygen molecules in a $^1 \Delta$ state which may also be present in the explosion flames.

The spectrum of the carbon monoxide flame with nitrous oxide is similar to that with oxygen, but the colour is a less clear blue and the continuum has a slightly different distribution of intensity. The difference is probably due to the superposition of some of the yellow-green NO + O continuum (page 106) due to the formation of a little nitric oxide from the N_2O.

The spectrum of burning solid carbon is similar to that of the CO flame, apart from the addition of the thermal continuum from the incandescent solid (Whittingham, 1950). The carbon-monoxide flame spectrum also appears when CO_2 is introduced into an oxy-hydrogen flame.

Carbon monoxide which has been stored in steel cylinders is often contaminated with iron carbonyl, and the spectrum of the flame then shows strong bands of FeO, mostly in the yellow and orange, and such flames usually have a yellow tip. The iron carbonyl can be removed by passing the gas through a heated tube packed with crushed porcelain, or through a bed of activated charcoal. Another persistent impurity in the spectra of CO flames is CuCl. These characteristic bands (see page 355) are strongest in the blue and blue-green; they were particularly troublesome during an investigation of the cool-flame spectrum of CO (Gaydon, 1943), and have also been observed in gun flashes, and frequently occur in the blue flames of coal fires, especially if salt is thrown on to a fire. The main bands of FeO and CuCl are listed in the appendix.

The Carbon Monoxide Flame Bands

As obtained in ordinary flames, the numerous narrow bands of comparable intensity, which constitute this system, are rather masked by the continuous spectrum and show little regularity. The first attempt at analysis was made by A. Fowler and is included in the first paper by Weston (1925); many of the stronger bands were arranged into arrays using intervals of around 570 and 370 cm^{-1}. Kondratiev (1930) arranged some of the bands in long series with intervals increasing from around 550 cm^{-1} to over 600 cm^{-1}, this interval being provisionally identified with the 667 cm^{-1} fundamental frequency of CO_2, known from the infra-red spectrum. These long series were not, however, completely confirmed by the author (Gaydon, 1940) who found that there were some disturbing irregularities in the size of the intervals and in the intensity distribution. The author studied the bands at the long-wave end of the spectrum in the flame at reduced pressure and found that many bands occurred in pairs, with a separation of around 60 cm^{-1} and were degraded to the violet; frequency intervals of around 345, 565 and 2065 cm^{-1} were noted and some vibrational arrays were proposed.

The system is clearly too complex for a diatomic molecule, and since the bands occur in the flames of pure dry CO with O_2 the emitter must be a polyatomic molecule containing only C and O. All the experimental evidence points to CO_2. Various other possible molecules, such as CO_3, C_3O_2 and O_3 have been considered and can be eliminated. The difficulty is the lack of agreement, either in wavelength of the bands or in the frequency intervals, with the known CO_2 spectrum. In a discharge tube at low pressure CO_2 shows only strong band systems of CO and some bands of the ionized molecule CO_2^+. In absorption CO_2 is transparent all through the visible and quartz ultra-violet region and the first known absorption bands lie below 1700 Å, with really strong bands lying at still shorter wavelengths, below 1200 Å. Herzberg and Herzberg (1953) studied CO_2 absorption in the photographic infra-red using path lengths of up to 5000 m but observed only bands of the vibration-rotation spectrum. Even liquid carbon dioxide is transparent to at least 2150 Å (Eiseman and Harris, 1932) and since the usual selection rules break down at least partially in the liquid state, this rules out the possibility of there being any low-lying electronic levels of CO_2. From infra-red and Raman data, the three fundamental vibrational frequencies for CO_2, reduced to infinitesimal amplitude, are the symmetrical stretching frequency ω_1 at 1388.2 cm^{-1}, the bending frequency $\omega_2 = 667.4$ and the asymmetrical stretching frequency $\omega_3 = 2349.2$ cm^{-1}. The connection between these fundamentals and the frequency intervals found for the carbon monoxide flame bands is far from obvious.

The author (Gaydon, 1940) suggested that the flame bands were indeed due to CO_2 and were part of the absorption system below 1700 Å. The big difference in wavelength of the emission and absorption bands was explained in terms of the Franck–Condon principle (see page 79). The ground state of CO_2 is known to be linear O—C—O and it was assumed that the excited state was bent; following the electronic transition the molecule would find itself displaced from its new equilibrium configuration, and would vibrate strongly about this configuration. Thus emission would correspond to transitions from low vibrational levels of the excited electronic state to levels of the ground state in which the bending vibration was strongly excited, i.e. to vibrational levels with large values of v_2''. Similarly in absorption transitions from the lowest vibrational levels of the ground state would go to high levels of v_2'.

The carbon monoxide flame bands have also been obtained in the spectrum of the afterglow of CO_2 in a discharge tube (Fowler and Gaydon, 1933) although not in the discharge itself. Other sources in which they have been observed are the cool flame, at around $600°C$, just below the ignition temperature (Gaydon, 1943), and the "atomic flame" of CO reacting with atomic oxygen from a discharge tube (Broida and Gaydon, 1953a). In all these three sources the emission is very weak compared with that of a normal flame, but the background continuum is weak or absent and the rotational temperature is lower and hence the bands are narrower and clearer. Dixon (1963) used the afterglow of CO_2 for his studies of the band structure and despite the weakness of this faint blue glow succeeded in getting some good plates in the second order of a fairly large grating spectrograph with an exposure time of five days (see Plate 7).

Dixon's analysis of the spectrum starts from the author's assumption that it is due to a transition from a bent excited electronic state of CO_2, which is responsible for the absorption bands below $1700 Å$, to the ground state. Vibrational levels of the ground state are known to fairly high energy from the vibration-rotation spectrum (e.g. in the atmosphere of the planet Venus). Although there are three fundamental vibrational frequencies, the bending vibration (v_2) can occur in two planes at right angles and is thus doubly degenerate, so that a second-order effect causes a splitting of levels and it is necessary to use a fourth quantum number l to specify the energy levels*. A further major complication is that ω_1 is very nearly equal to $2\omega_2$ so that there is strong Fermi resonance between these two vibrational modes. Thus for the higher vibrational levels it is not easy to assign separately the values of the quantum numbers v_1'' and v_2''; vibrational energy levels occur in groups ("Polyads") for which only the sum $(2v_1'' + v_2'')$ can be specified. Dixon has extrapolated from the known vibrational levels

---

*For even values of v_2, l takes the values 0, 2, 4, 6 etc up to v_2; for odd v_2, $l = 1, 3, 5$ etc to v_2.

of the ground state to much higher levels and has shown that the flame bands in the region 3100 to 3800 Å can be explained as transitions from two vibrational levels (the lowest vibrational level, and one with $v_2' = 1$) of an excited $^1B_2$ electronic state to vibrational levels of the ground $^1\Sigma^+$ state having $v_3'' = 0$ and values of $(2v_1'' + v_2'')$ between 22 and 28.

In the bent upper state the rotational energy levels require assignment of two rotational quantum numbers J and K, while for the linear ground state the rotational quantum number is J, but in addition the quantum number l (see above) effectively replaces K. The bands are attributed to a parallel-type transition, for which the selection rules are $\Delta K = 0$, $\Delta J = 0$, ± 1, but for $K = 0$ $\Delta J = \pm 1$ only. Bands are of two types, according to whether $(2v_1'' + v_2'')$ is even or odd. For even values, l (which is always equivalent to K) is also even and each band consists of a number of sub-bands; the first head is designated as of type Σ ($K = l = 0$) and should show only single P and R branches; the second head, which lies about 30 to 40 cm^{-1} to the short-wave side of the first, is designated Δ and has $K = l = 2$ and it is stronger than the Σ band; a third sub-band, designated Γ, with $K = l = 4$, is usually observed still further to shorter wavelengths by about 100 cm^{-1}. For odd values of $(2v_1'' + v_2'')$, l is also odd and these bands show a wider doublet separation of around 60 to 80 cm^{-1} between heads of similar strength denoted Π (with $K = l = 1$) and Φ (with $K = l = 3$); these wider doublets may be identified with Gaydon's violet-degraded pairs which were observed at longer wave-lengths; these bands with odd l may show a third sub-band, denoted H (with $K = l = 5$) around 130 cm^{-1} from the Φ sub-band. Thus in each group of bands, corresponding to transitions to successive levels in a polyad, the appearance alternates between those with the narrower (30–40 cm^{-1}) doublet separation and those with the wider separation (60–80 cm^{-1}); in the region studied by Dixon, 3100–3800 Å, the separation between bands of similar type is around 300 cm^{-1}. This structure may be seen in Plate 7.

No complete rotational analysis of a band has been made, but Dixon has compared the structure of some Σ-type bands and of one Δ-type with theory. The agreement is not entirely convincing if the diagrams are examined critically, but agreement is sufficient for Dixon to deduce values of the moments of inertia. In the excited $^1B_2$ state the molecule is found to be bent at an angle of $122 \pm 2°$, and has a C—O bond length of 1.246 Å. The vibrational analysis gives the system origin at 46700 ± 20 cm^{-1}. The upper electronic state is believed to be the same as that involved in the absorption bands which at room temperature lie below 1700 Å; this absorption has maximum intensity around 1475 Å but later work (Generalov and Losev, 1966) has shown that in gases shock-heated to high temperature there is absorption both at 2380 and 3000 Å. Walsh (1953) and Mulliken (1958) assigned this system to a transition from the ground state to a $^1B_2$ excited state. Dixon's analysis shows that an energy of approximately 22000 cm^{-1} (2.7 e.V) is required to straighten a molecule in the bent $^1B_2$ state.

The Excitation of the Flame Bands

The production of the flame bands can fairly certainly be attributed to reaction between atomic oxygen and carbon monoxide. However, as pointed out long ago by Herzberg (1932), the ground electronic state of atomic oxygen is ^3P and that of CO is $^1\Sigma$, so that spin conservation rules forbid direct correlation of ground state CO_2 ($^1\Sigma$) with $CO(^1\Sigma) + O(^3P)$. The ground state of CO_2 should correlate with $CO(^1\Sigma) + O(^1D)$. Gaydon (previous edition) suggested that the CO + O reaction led to formation of CO_2 in a bent excited triplet state, transitions from which led to the flame-band emission. This suggestion was supported by an analogy with the banded SO_2 afterglow spectrum (Gaydon, 1934) which does involve a triplet-singlet transition. However, from the work of Dixon there can be little doubt that the upper state of the flame bands is the bent $^1B_2$ state. Dixon's analysis puts the system origin at 5.79 e.V. (46700 cm^{-1}). The energy released by the association CO + $O(^3P)$ = CO_2 would be 5.45 e.V at 0 K or 5.52 e.V at 300 K, indicating that an activation energy of 0.27 ± 0.1 e.V is required. Although the flame bands are emitted in the reaction between unheated CO and atomic oxygen from a discharge tube (Broida and Gaydon, 1953a), Clyne and Thrush (1962) later studied the effect of temperature on the reaction and found a small temperature dependence, indicating an activation energy of 3.7 ± 0.5 kcal (0.16 ± 0.02 e.V), which may be considered reasonable agreement.

Prior excitation of atomic oxygen to the ^1D state need not be considered. The activation energy for that (1.16 e.V) is high, and Clyne and Thrush report that the CO flame bands are also produced when the atomic oxygen comes from the reaction $N + NO = N_2 + O(^3P)$.

Measurements by Clyne and Thrush showed that the light intensity from the flame bands, in the region 3200–3900 Å, was accurately proportional to the concentrations of CO and of atomic oxygen, and it was also independent of total pressure in the range 0.86 to 2.69 torr. However, although independent of pressure the light emission does depend on the nature of any diluent gas. At 293 K they found the following values for the constant I_c for the intensity $I = I_c [O] [CO]$

Diluent	$I_c \times 10^{-3} \, cm^2 \, mole^{-1} \, s^{-1}$
O_2	12.0 ± 0.8
N_2	11.2 ± 0.8
Ar	8.0 ± 0.8
Ne	5.6 ± 1.0
He	5.6 ± 0.8

This is best explained by assuming that both formation and quenching of excited CO_2 molecules depend on the presence of a third body and that relative cross sections for the forward rate and for quenching depend on the nature of this third body. As already noted, the overall reaction

$$CO(X^1\Sigma) + O(^3P) + M = CO_2(X^1\Sigma) + M + h\nu$$

involves a change in spin; the spin reversal could conceivably occur (i) during formation of the excited CO_2 by the third body (ii) by radiationless transition from a triplet to a singlet state after the three-body collision, or (iii) during the subsequent radiative transition to the ground state. However, although this third alternative was previously favoured by the author, Dixon's analysis seems to exclude it. Clyne and Thrush note that for case (i) oxygen, which has a triplet ground state, should be particularly efficient at stabilizing the CO_2 into a singlet state; it is not obvious that this is so, and they favour the alternative (ii), i.e. formation of CO_2 in a triplet state, probably $^3B_2$, followed by a radiationless transition to the $^1B_2$ and then the emission of the banded radiation.

The possible role of oxygen in assisting spin reversal is, however, interesting. The surprising occurrence of the strongly forbidden $b^1\Sigma_g^+ - X^3\Sigma_g^-$ "Atmospheric" bands of O_2 in emission from CO/O_2

explosion flames has been noted; they have also been reported from steady CO/O_2 flames and less clearly from the outer cones of hydrocarbon/oxygen flames (Hornbeck and Herman, 1951b). It is now known (e.g. Derwent and Thrush, 1971) that $^1\Delta$ oxygen molecules have a very long radiative lifetime and that one of their decay processes is

$$O_2(^1\Delta) + O_2(^1\Delta) = O_2(b^1\Sigma_g^+) + O_2(X^3\Sigma_g^-)$$

leading to "Atmospheric" band emission from the oxygen afterglow, the strength being proportional to the square of the $O_2(^1\Delta)$ concentration. If O_2 molecules participate either in the formation of excited $CO_2(^1B_2)$ or in quenching of the triplet $CO_2(^3B_2)$ then they might well end in the long-lived $^1\Delta$ state, and when the reaction rate is high, as in flames with O_2, but not in slower flames with air, the $O_2(^1\Delta) + O_2(^1\Delta)$ process could lead to emission of the "Atmospheric" bands.

The Continuous Spectrum: Pressure and Temperature Effects

The continuous spectrum is believed to have a different origin from the banded emission, although Kondratiev (1930) did suggest that it might be due to blending of many lines of the band system, the complexity of which increased at high temperature, and Clyne and Thrush (1962) also took this view. However, the author's observations show that as the temperature is raised the continuum becomes markedly stronger, but there is comparatively little change in the structure of the bands which remain at about the same intensity. Generally, low pressure or low temperature favour the banded emission, while in hot flames the continuum is dominant. Knipe and Gordon (1955) estimated that for explosion flames, initially at 100 to 200 torr, the continuum accounted for 90% of the light emission, while Kaskan (1959a) estimated that it was 95% in CO/air flames. In the cool flame of CO at around 600°C (Gaydon, 1943) the continuum is relatively weak, and in the CO/atomic oxygen glow (Broida and Gaydon, 1953a) it appears to be absent.

To separate the effects of pressure and temperature, Gaydon and Guedeney (1955) preheated the CO and air or O_2 in flat diffusion flames. The absolute strength of the bands was unaffected by temperature, showing incidentally that this emission was not thermal in origin, but the continuum increased in intensity with the preheating and also its distribution of intensity with wavelength showed a marked alteration. For CO/air, preheated to 750°C so that the adiabatic flame temperature at the stoichiometric point rose from 2025 to 2325 K, the short-wave end of the spectrum was strengthened most; I_{pr} and I_{un} are the relative intensities with and without preheat

$\lambda(\text{Å})$	5000	4500	4000	3000	2800
I_{pr}/I_{un}	1.5	1.7	1.9	2.2	2.5

Gaydon and Guedeney concluded that the continuum was due to a reaction between CO and O, but required an activation energy, probably because the collision took place on a repulsive potential energy surface. At higher temperature the colliding CO and O would approach closer, leading to a higher probability of emission and also to shorter wavelength radiation because transitions would then take place to nearer the potential-energy minimum of the ground-state CO_2 surface.

Kaskan (1959a) has studied mixed carbon monoxide plus hydrogen flames on a cooled porous-plate burner and produced strong quantitative evidence that under these conditions the continuum is indeed also due to a CO + O reaction. It could be either

$$CO + O = CO_2 + h\nu$$

or the set of reactions

$$CO + O + M = CO_2^* + M$$
$$CO_2^* + M = CO_2 + M$$
$$CO_2^* = CO_2 + h\nu$$

There is no conclusive evidence, but Kaskan, in agreement with the author, favours the simpler two-body process. In this mixed flame, Kaskan used absorption spectrometry to estimate the concentration of OH; as with hydrogen flames this is two to three orders higher than the equilibrium value in and just above the reaction zone. He assumes that the quasi-equilibrium

$$OH + OH \rightleftharpoons H_2O + O$$

is maintained by rapid two-body collision processes, and that [O] can be deduced from the equilibrium constant, denoted K_3, the square of [OH] and the water content; [CO] is calculated from the known stoichiometry and temperature. A logarithmic plot of light intensity per unit path length divided by [CO] against $[OH]/\{K_3/[H_2O]\}^{\frac{1}{2}}$ is indeed a good straight line of slope 2.

For these mixed flames Kaskan again found that the short-wave part of the continuum, below 4200 Å, was strengthened at high temperature, but at longer wavelengths he did not find an activation energy necessary, although remarking that one of two or three kcal/mole could have gone unnoticed. The rate constant (at around 1600 K) for $CO + O = CO_2 + h\nu$ is given as 2.7×10^{-19} cm^3/molecule sec, giving about 5×10^{-10} for the probability of radiation during a $CO + O$ collision; if the duration of a collision is assumed to be between 10^{-13} and 10^{-14} s, the transition probability for the radiative transition would be of the order* 10^4 sec^{-1}, indicating a slightly forbidden transition, possibly due to the change of multiplicity.

Combustion Processes for Carbon Monoxide

The combustion of carbon monoxide is very dependent on the presence of traces of moisture or hydrogen; this has been known since the work of H. B. Dixon in 1880. As the gases are progressively dried the flame speed decreases and the percentage of combustion taking place falls. For very dry mixtures ignition can only be produced using very much more energetic ignition sources than those which are sufficient for moist gases. Ubbelohde (1933) found that when CO was burnt in air or O_2 on an externally heated silica jet and the gases were progressively dried, then the flame enlarged and finally went out and could not be relit even with the jet heated to 1400°C. Obviously the stability of a flame on a jet will depend on the quenching distance and burner size, but pure dry CO is undoubtedly very difficult to burn. Similar effects were observed with traces of hydrogen instead of water, there being a critical range between a molar concentration of H_2 of 10^{-3} and 10^{-5} within which the flame length varied markedly with hydrogen concentration. Friedman and Cyphers (1956) measured burning velocities for moist $CO/O_2/N_2$ mixtures at 100 torr pressure using a flat-flame burner. With 0.0023 mole fraction of water vapour the burning velocity, S_u, was about 18 cm/sec, rising to 49 cm/sec for

*Kaskan states 10^3, but there appears to be a slip.

0.127 mole of water; a plot of S_u against $[H_2O]^{1/4}$ was quite linear and extrapolated back towards zero S_u for dry gas.

The spectrum of ordinary CO/air flames always shows strong OH emission, and the most careful drying is necessary to reduce this. Weston (1925) found that at high pressure (>5 atm.) OH emission was less pronounced, and it is known that the effect of drying on the combustion is also then less. Kondratieva and Kondratiev (1938) found that moisture also greatly reduced the strength of the blue CO-flame emission. In later experiments on diffusion flames Gaydon and Guedeney (1955) found that hydrogen also caused a marked reduction in the blue radiation, both the CO-flame bands and the continuum being equally affected; addition of a little H_2 would not greatly affect the flame temperature, and the reduced emission, by as much as a factor of 4, must indicate that the emission is of chemiluminescent origin and is altered by a change in reaction processes. The decrease in light yield in Kondratiev's experiments could not be accounted for by assuming quenching of electronically excited CO_2.

In the presence of moisture the flame propagation undoubtedly involves the chain reactions

$$OH + CO = CO_2 + H + 102 \text{ kJ/mole} \qquad (1)$$
$$H + O_2 = OH + O - 69 \text{ kJ/mole} \qquad (2)$$
and
$$O + H_2O = OH + OH - 68 \text{ kJ/mole} \qquad (3).$$

(1) is exothermic and chain propagating, while the two chain-branching steps (2) and (3) are slightly endothermic. Addition of nitric oxide to a carbon monoxide flame produces strong yellow-green continuous emission due to the NO + O reaction, showing the presence of atomic oxygen. Kaskan's measurements of OH absorption show considerable excess OH concentrations above the equilibrium value. Quantitative measurements (Kaskan, 1959b) of OH concentration and light emission at various heights in a flame show that reaction (1) is well equilibrated above 1500 K and that the fall in [CO] with height is linked with removal of excess radicals by three-body recombination processes rather than by combustion of CO. In other words, reactions (1), (2) and (3) are all bimolecular and fairly fast, leading to a build-up of an excess population of free radicals, especially oxygen atoms, which are only removed by relatively slow three-body processes or the spin-forbidden radiative recombination $CO + O = CO_2 + h\nu$.

Pure dry CO burns with great reluctance, and the combustion when it occurs is seldom complete. It is usually accepted that a dry reaction does exist, although the author is not quite convinced. Among discussions of the kinetics of the dry reaction are those by Hoare and Walsh (1954), Lewis and von Elbe (1961) and Minkoff and Tipper (1962). No simple bimolecular chain reaction can be written down. The direct reaction

$$CO + O_2 = CO_2 + O$$

is slightly exothermic (34 kJ/mole) but has, of course, a high energy barrier. Shock-tube studies (Sulzmann, Myers and Bartle, 1965) of the gas-phase reaction, using pure gases with not more than one part per million of moisture, show an induction period from which an apparent activation energy, varying with temperature, has been deduced. At high temperature a rate equation (with R in calories) is given as

$$k = 3.5 \pm 1.6 \times 10^{12} \exp\left(-51\,000 \pm 7000/RT\right) \text{cm}^3 \text{mole}^{-1}\text{s}^{-1}.$$

Possible chain propagation and branching steps which have been considered by various authors are:

the material chain
$$O + O_2 + M = O_3 + M \tag{4}$$
$$O_3 + CO = CO_2 + O + O \tag{5}$$
the direct energy chain
$$CO + O\,(+M) = CO_2^*(+M) \tag{6}$$
$$CO_2^* + O_2 = CO_2 + O + O \tag{7}$$
and the indirect energy chains
$$CO_2^* + O_2 = CO_2 + O_2^*$$
$$O_2^* + CO = CO_2 + O$$
and
$$CO_2^* + CO = CO_2 + CO^*$$
$$CO^* + O_2 = CO_2 + CO$$

The material chain has little to recommend it. Burgoyne and Hirsch (1954) when studying the inhibition of methane combustion did comment that prior to the ignition of pure CO/O_2 mixture there was a smell of ozone and that it could be detected in the products by reaction with KI. However, there is little evidence, e.g. spectroscopic, for detection of O_3 under flame conditions. Reaction (4) is slow because it is three-body and the $O_3 + CO$ reaction is also known to be very slow (Garvin, 1954).

The direct energy chain most probably involves electronically excited CO_2. For $^1B_2$, Dixon's analysis gives an excitation energy of 46700 cm^{-1} so that reaction (6) will be slightly endothermic (see page 133). Alternatively the $^3B_2$ state might be involved; if we accept Clyne and Thrush's view that $^1B_2$ is populated by radiationless transitions from $^3B_2$, then the latter will lie at slightly lower energy, so that (6) might be nearer thermoneutral; (7) is likely to be slightly exothermic in either case. There seems little real advantage in the indirect energy chains, although they could play a part.

Comparison of observed and calculated maximum temperatures in both explosion and stationary flames often reveals a significant discrepancy. Many observations of this temperature difference were made by the late Prof. W. T. David and colleagues at Leeds and attributed to a "latent energy" in the gaseous combustion products. Measurements of flame temperature by the sodium-line reversal method often agreed fairly well with calculated values, whereas measurements by platinum-wire resistance thermometry gave low temperatures. Such effects were reported for many flames but were particularly strong for carbon monoxide flames, often indicating around 20 per cent of latent energy; addition of moisture or hydrogen reduced this. Evidence for this type of disequilibrium is shown (Figure VI.1) by comparison of temperatures in CO/air explosion flames using bare platinum-rhodium resistance wires and similar wires coated with a thin protective layer of silica (David, Leah and Pugh (1941). For weak mixtures the temperature difference reached $300°$. Somewhat similar results have been obtained for stationary flames, the amount of latent energy varying with flame height, being greatest, of course, just above the reaction zone.

It was at first thought that this latent energy might be due to long-lived electronically excited molecules of CO_2; the author also considered the possibility of slow vibrational relaxation of ground state CO_2 molecules formed with excess vibrational energy in the transverse mode. Leah, Rounthwaite and Bradley (1950) realized that part of the effect was due to persistence of free oxygen atoms, which can be detected by the NO + O reaction, but still maintained that there was a residual discrepancy. However, more recent knowledge of the lifetime of electronically excited states of CO_2 and of vibrational relaxation times (from shock-tube work) rule out these possibilities. O_2 molecules in the metastable $^1\Delta$ state are long lived, but could not carry much energy. It seems that for pure CO flames all reactions are relatively slow and there is a build up of an excess population of free

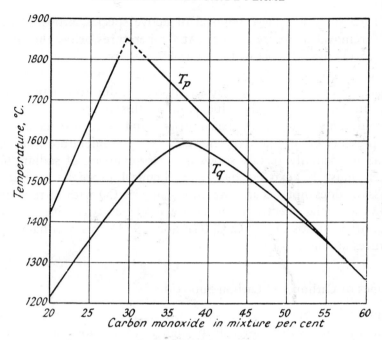

Fig. VI.1. Temperatures of CO/air explosion flames using bare platinum wires
(T_p) and quartz-coated wires (T_q)

atoms; this is even more marked than for hydrogen flames. These
excess free oxygen atoms are particularly persistent for fuel-weak
mixtures. With thermometry using bare wires there is some catalytic
recombination on the surface which reduces locally the discrepancy
between observed and calculated temperature. With silica-coated
wires this is not so and the translational temperature of the gas
molecules is measured. With sodium-line reversal measurements there
may be some chemiluminescent excitation of Na by recombining
oxygen atoms.

The slow combustion and spontaneous thermal ignition of carbon
monoxide depend on gas purity, vessel dimensions and surface
conditions. The ignition temperature is relatively very high (around
$800°C$ for the dry gas) at atmospheric pressure, but falls with
pressure. A graph of pressure versus ignition temperature shows a
marked "explosion peninsula"; results by Hoare and Walsh (1954),
for $2 CO + O_2$ under their experimental conditions, indicate a
minimum ignition temperature of around $530°C$ at around 20 torr.
The lower limit will be due to heat loss to the surface and the upper

limit to some gas-phase removal of active species, probably three-body removal of oxygen atoms. At temperatures below the ignition limit, and especially below the lower limit, there is a blue glow whose spectrum (Gaydon, 1943) shows relatively strong CO_2 bands and weak continuum, both, as already discussed, due to reaction of CO and atomic oxygen. The blue glow at this temperature must be chemiluminescent not thermal in origin, and its occurrence indicates the presence of atomic oxygen even at these temperatures. This cool flame incidentally readily shows impurity emission of sodium lines, which must be excited by some chemical process too, but OH emission does not occur. The cool flame of CO with nitrous oxide shows a similar spectrum and also a tendency to excite spectra of impurities, in this case of CuCl (Gaydon, 1943); this cool flame again involves atomic oxygen.

Flames of Carbon and Carbon Suboxide

Solid carbon gives a blue glow when heated in air or oxygen. Whittingham (1950) passed air through a carbon tube 1.5 cm internal diameter which was heated electrically to $850°C$ and observed a blue glow which filled the tube apart from a thin dark region close to the walls. The spectrum was similar to that of the cool flame of CO, showing both bands of CO_2 and some continuum; boron oxide bands occurred as impurity.

At much higher temperatures carbon begins to sublime and mass-spectrometric observations show that the vapour, in addition to atomic carbon, contains C_2, C_3 and C_5 molecules. Savadatti and Broida (1966) made spectroscopic studies of the flame or glow obtained when carbon vapour reacted with various species. They passed helium through a carbon tube heated electrically to up to $2800°C$ and led the gas at reduced pressure (10 torr) into an unheated reaction vessel where other species were introduced. With molecular oxygen a green glow showed Swan bands of C_2 and two systems of CO, the Triplet system $(d^3\Delta - a^3\Pi)$ and the Third Positive system $(b^3\Sigma - a^3\Pi)$. The C_2 seemed to show an enhancement of bands with $v' = 2$. Reaction with atomic oxygen, from a microwave discharge, showed the same bands of C_2 and CO, but more strongly, and also the 4050 Å "Comet-head" band of C_3. This C_3 is unlikely to be formed by the reaction of carbon vapour with atomic oxygen; Savadatti and Broida suggest that C_3 which is already present in the carbon vapour is excited by an energy transfer reaction

$$C_3 + M^* = C_3^* + M$$

They do not specify the nature of M, but it seems likely that it is CO, either electronically excited or carrying excess vibrational energy. The Triplet and Third Positive systems of CO do not usually occur in flame spectra, and Savadatti and Broida suggest that they are excited by

$$C + O + M = CO^* + M$$

which is one of the few reactions giving the necessary 10 e.V. It is interesting that there is no mention of the Fourth Positive system of CO, below 2400 Å which does occur in some flames; their observations are stated to extend to 2200 Å, but the published spectra do not go so far. Carbon vapour does not glow with molecular hydrogen, atomic hydrogen or molecular nitrogen, but atomic nitrogen gives strong CN emission.

Carbon suboxide, C_3O_2, is a stable gas although it polymerizes very readily. The author and G. Pannetier made some preliminary studies of its flame spectrum. Premixed with air it gives a Bunsen-like flame with a green inner cone which shows strong C_2 Swan emission while the outer cone shows the CO-flame spectrum. In the presence of moisture, bands of CH and OH appear. Rich mixtures give soot formation and Clusius and Huber (1949) have described effects of drying on the soot formation and combustion limits.

The spectrum of the flame of carbon suboxide burning with atomic oxygen has been reported by Becker and Bayes (1968) in a discussion of reactions proposed by Kydd and Foss to account for CO Fourth Positive emission in hydrocarbon flames. Becket and Bayes say that the Fourth Positive bands ($A\,^1\Pi - x\,^1\Sigma$) occur strongly in this flame and also the Triplet system ($d^3\Delta - a^3\Pi$) and Herman system $e^3\Sigma - a^3\Pi$. They suggests that C_2O is formed, followed by

$$O(^3P) + C_2O(^3\Sigma) = CO(A\,^1\Pi, d\,^3\Delta \text{ or } e^3\Sigma) + CO(x\,^1\Sigma).$$

These reactions may be important in organic flames, considered in the following chapter.

Chapter VII

The Spectra of Organic Flames

A preliminary description of the principal features of the spectra of Bunsen-type flames has been given in Chapter I. Here the various band systems encountered in flames will be described in detail and the evidence from which the emitting species are determined will be discussed; a full list of wavelengths of band heads and other data are given in the Appendix. These descriptions of the systems will be followed by sections in which the effects of fuel type, mixture strength and other flame conditions on the spectra are dealt with. A discussion of the processes of formation and excitation of the various radicals and of their role in the combustion reactions will be left to Chapters VIII and X.

The CH Bands

Three systems of CH are observed in flames, all having a common $^2\Pi$ final level. The systems all show the open rotational structure characteristic of a diatomic hydride, and the branches are double, indicating a molecule with an odd number of electrons. They are also obtained readily from discharges through pure hydrocarbon vapours, and the emitter can only be CH.

The 4315 Å system is the strongest (see Plate 3*b*) and is due to a $^2\Delta - {}^2\Pi$ transition. It is degraded, rather weakly, to shorter wavelengths so that the strong heads are those of the Q branch, Q_1 at 4314.2 and Q_2 at 4312.7 Å. Neither the P nor R branches form true heads, but the lines of the P branch of the (0,0) band close up together and under small dispersion appear to form a diffuse head near 4390 Å. The weaker (1,1) band is superposed on the (0,0) band and does not add any distinctive feature to the appearance of the band under small dispersion. A curious strong line-like feature at 4323 Å in front of the main Q heads is the piled up Q branches of the (2,2) band; the strength of this is rather variable, and probably changes with effective vibrational temperature, being most prominent in hot flames, such as that of oxy-acetylene. There is a very weak (0,1) band with Q head at 4941.2 Å.

The 3900 Å system (plate 3c) is also quite strong. This is due to a $^2\Sigma-^2\Pi$ transition and is fairly strongly degraded to the red so that its rotational structure appears more open than that of the 4315 system. The strong (0,0) band has a good R head at 3871.4 Å and a Q head of comparable strength at 3889 Å. Weak (1,0) and (1,1) bands with R heads at 3627.2 and 4025.3 Å also occur. Predissociation in the higher rotational levels of these bands usually causes the branches to break off abruptly beyond $N' = 7$; under some conditions (Durie, 1952a) this predissociation is suppressed and these bands are then relatively stronger.

The system at 3143 Å is the weakest of the three and is usually partly obscured by the strong rotational structure of the 3064 Å band of OH. It appears best in hot flames like oxy-acetylene, and is seldom obvious in flames with air. It is seen most clearly with rich mixtures, as masking by OH is then less serious; Plate 3d, shows the structure clearly as obtained in a rich low-pressure oxy-acetylene flame. It is due to a $^2\Sigma-^2\Pi$ transition and is hardly degraded at all either way so that the Q branches form a strong intensity maximum, like a slightly wide atomic line, at 3144 Å (under very large dispersion there are violet-degraded Q_1 and Q_2 heads at 3144.9 and 3144.2 and a red-degraded head at 3143.4). The P and R branches form nearly equally spaced line series on either side of the Q maximum, without forming a head.

The $^2\Pi$ state which is the lower state for all three systems is the ground state of CH, $x^2\Pi$. This is shown by the observation (McKellar, 1940) of CH absorption lines in interstellar space where the temperature is so low and collisions so infrequent that absorption from excited states could not possibly occur. CH is, however, very difficult to observe in absorption in flames, but since the previous edition of this book, it has been obtained using techniques of high sensitivity (see page 26). A very low-lying $^4\Sigma^-$ state of CH has been predicted theoretically but no band systems involving it have been reported.

Bands of C_2

The well-known Swan bands of C_2 form a compact system in the green region of the spectrum. The bands are degraded to the violet and show well-marked sequences, the (1,0), (0,0) and (0,1) sequences having outstanding heads at 4737.1, 5165.2 and 5635.5 Å. See Plate 6 b. Under large dispersion the bands show P and R branches, the branch lines all being triple. The transition is $^3\Pi-^3\Pi$. The emitter of the Swan bands was for long in doubt, the system being variously attributed to carbon or to a compound of carbon with oxygen or hydrogen. Detailed examination of the rotational structure shows an apparent staggering of the branch triplets; in a $^3\Pi-^3\Pi$ transition the lines should be double due to Λ doubling, but in the Swan bands alternate components of the Λ doubling are missing, producing the apparent staggering. This missing of alternate lines is characteristic of a homonuclear molecule, and this observation combined with experimental work by Pretty (1928) settled the emitter as C_2 beyond doubt.

Besides the strong Swan bands, $A\,^3\Pi-X\,^3\Pi$, several other systems due to C_2 are well known in the spectra of discharge tubes, the carbon arc or the King furnace, and these all occur weakly in flames. They are only visible using fairly large dispersion and with hot flames maintained by oxygen. Low-pressure flames, in which the continuous background is relatively weak, are especially suitable.

The Fox—Herzberg system was reported by Hornbeck and Herman (1950). This is also a triplet system involving the same lower state as the Swan bands and is due to the transition $B\,^3\Pi-X\,^3\Pi$. The bands are degraded to the red and the (0,3) and (0,4) heads at 2855 and 2987 Å are probably the strongest. The system extends from 2378 to 3283 Å in discharge tubes; in flames the OH bands tend to mask this system.

It is now known that the $X\,^3\Pi$ lower state of these systems is not really the ground state of C_2, although it is still denoted X. There are several singlet transitions and Ballik and Ramsay (1959) have shown that the lowest singlet level, now denoted $x^1\Sigma$ (formerly $a^1\Sigma$) lies just below $X\,^3\Pi$, by some 610 cm^{-1}. The most readily observed of the single systems in flames is Mulliken's band, $d^1\Sigma-x^1\Sigma$, with origin at 2312.7 Å. This is a headless band with simple rotational structure of P and R branches, and appears fairly open because, like all C_2 systems, it has alternate missing lines. It was first observed by the author in low-pressure C_2H_2/O_2 flames. Another weak singlet system, the Deslandres–d'Azambuja system, $c^1\Pi - b^1\Pi$ has also been recorded by Hornbeck and Herman (1949); the system is degraded to the red and has its (0,0) band at 3852.2 Å, and can sometimes be seen just in front of the CH 3900 band, but it is very weak indeed.

Phillips infra-red system, $b^1\Pi - x^1\Sigma$, is probably quite a strong system in the inner cones of all organic flames because it requires the lowest excitation energy of the singlet systems, but being fairly well down in the photographic infra-red it is seldom observed in practice. Hornbeck and Herman (1949) reported it in flames. The bands are degraded to longer wavelengths and the main (3,0), (2,0), (1,0) and (0,0) sequence heads are at 7714, 8751, 10135 and 12070 Å.

Vaidya's Hydrocarbon Flame Bands (HCO)

These bands were first studied in detail by the late Dr Vaidya (1934) using an ethylene/air flame, but they can be observed in flames of all hydrocarbons, including methane, acetylene and benzene. The system lies in the region 2500–4000 Å, with the strongest bands lying between the OH at 3064 and the CH at 3900 Å. They occur best in flames of very weak mixtures and in low-temperature flames such as the author's chilled flame (Gaydon, 1942a) and that on Hornbeck and Herman's low-temperature burner. In hot flames the bands are relatively weak and masked by strong OH and CH. In the acetylene/atomic oxygen glow the heads show up fairly clearly because of the lower rotational temperature, and this source is useful for studying their structure.

The bands are degraded to the red. Under small dispersion the heads appear fairly sharp, but with larger dispersion the complex rotational structure becomes visible and the exact positions of the heads are difficult to locate; thus small-dispersion measurements of heads, e.g. by Vaidya, are systematically displaced by about 10 cm^{-1} towards the red as compared with large-dispersion measurements. Several attempts have been made to analyse the vibrational structure (see later) and a number of fragmentary arrays have been established. Since the bands cannot be represented by a single array characteristic of a diatomic molecule, we may assume that we are dealing with a polyatomic emitter.

A number of possible emitters have been considered. Vaidya was led, on the suggestion of the late Prof. A. Fowler, to attribute the bands to HCO. Vaidya supported this hypothesis by comparing the vibrational frequencies obtained from his partial analysis with those of the ground and excited $B\,^2\Pi$ states of the iso-electronic molecule NO. The real reason for suggesting HCO as the emitter was that this seemed a likely fragment in the break-down of a peroxide formed by direct introduction of oxygen into an acetylene molecule; it has indeed been shown (Gaydon, 1942a) that peroxidic substances are present in the chilled flames which give the bands so well.

Other emitters which have been considered are formaldehyde, an isomeric formaldehyde HC·OH, CH_2 and CHO_2. A number of experiments, under controlled conditions or in other sources than flames, are of relevance in determining the emitter. The bands do not occur under normal conditions in a discharge tube although both CH and OH occur readily; this makes emission from a molecule formed by association of CH and OH unlikely. The author has also failed to observe the bands in either a discharge through glyoxal CHO·CHO or in a flame of burning glyoxal. Their production seems to involve some combustion process such as the breakdown of a peroxide.

The only other known source of the hydrocarbon flame bands is the fluorescence spectrum of formaldehyde excited by very short ultra-violet radiation (Dyne and Style, 1947). The pressure dependence of this fluorescence shows that it is not due to secondary effects and that the emitter must be either formaldehyde itself or some decomposition product. The only possible decomposition products are HCO and CH_2; we shall see later that CH_2 can be excluded, and the actual fluorescence process appears to be

$$HCHO + h\nu_1 = H + HCO^*$$

$$HCO^* \rightarrow HCO + h\nu_2$$

The hydrocarbon flame bands can be obtained from flames in which the hydrogen of the hydrocarbon is replaced by deuterium. Flames of deutero-acetylene have been studied by Vaidya (1951) and others, and the bands are found to show a large isotopic displacement, thus proving that the emitter contains hydrogen. Spokes and Gaydon (1959) compared the flame spectra, using a low-temperature dilute flame (page 49), of normal acetylene, C_2H_2, deuteroacetylene, C_2D_2 and partly deuterated acetylene C_2HD; the partly deuterated acetylene showed a spectrum which consisted of a superposition of the bands from C_2H_2 and from C_2D_2 but no additional bands of the type which would have been expected from a molecule containing one deuterium and one hydrogen atom. This experiment appears to rule out CH_2 (which should show bands of CH_2, CD_2 and CHD) and formaldehyde CH_2O as emitter, and taken with Dyne and Style's fluorescence observations leaves only HCO as a possible emitter of the hydrocarbon flame bands.

Another band system, certainly due to HCO, has been observed at the red end of the visible region (strongest bands 6138 and 5624 Å) in absorption when glyoxal, aldehydes and formates are flash photolysed (Herzberg and Ramsay, 1955). Detection of absorption by hydrocarbon flame bands, which would probably lie at shorter wavelengths than the emission bands, in the same source has not been possible because the parent compound (glyoxal etc.) usually absorbs strongly at these wavelengths. Analysis of these flash-photolysis bands shows that HCO is bent in the ground state, with an angle of around 120°, and that the upper state of this transition has a linear configuration. Values of the rotational constants have been obtained and the ground state is found to be of type $^2A'$, and is denoted $\tilde{X}^2A'$. Ground-state vibrational frequencies are not easy to assign from these absorption experiments, but Johns, Priddle and Ramsay (1963) find a bending frequency ω_2 of 1083.0 for HCO and 847.4 for DCO; a value of $\omega_3 = 1820$ for HCO is doubtful as the band may be due to the isotope $H^{13}CO$.

HCO can also be studied in a solid matrix at low temperature (14 K) by association of H atoms (from photolysis of HBr etc.) with CO. The vibrational frequencies of the radical in the matrix have been observed by infra-red absorption (Ewing, Thompson and Pimentel, 1960; Milligan and Jacox, 1964). For HCO we have $\omega_1 = 2488$, ω_2 1090 and $\omega_3 = 1861$; for DCO $\omega_1 = 1937$, $\omega_2 = 852$ and $\omega_3 = 1800$. These frequencies will be slightly different from those in the gas phase but are very valuable in later discussion of the vibrational analysis.

The rotational structure of the hydrocarbon flame bands is complex but fairly open, as would be expected for a molecule containing hydrogen. Hornbeck and Herman (1954) studied the bands on a grating instrument with resolving power 150,000; they comment that the gross structure shows heads which might be quadratic K structure. Dixon (1969) studied some bands on an Ebert grating spectrograph with a resolving power of 100,000; although both he and Hornbeck and Herman express their measurements to

0.01 cm^{-1} there is, for the 3377 Å band, practically no agreement! However, Dixon's published photograph of this band (see Plate 8) confirms the K-type structure, and partial analysis of three bands has been made, this being consistent with that of a type a (parallel) band. Values of the rotational constants B and A for the lower states of these bands are close to those obtained by Herzberg and Ramsay for their absorption bands from the ground state; the hydrocarbon flame bands involve higher vibrational levels of the ground state, so exact agreement is not to be expected, but this comparison confirms that the ground electronic state of HCO is the final state of the flame bands.

Vibrational Analysis. A number of attempts at a vibrational analysis have been made, but there is still some doubt about it. A non-linear HCO molecule will have three fundamental vibrations, which, following Herzberg (1966), we shall denote as ν_1 (of frequency ω_1 cm^{-1}) for the C–H stretching mode, ν_2 (ω_2) for HCO bending, and ν_3 (ω_3) for C–O stretching. Vaidya called the C–O stretching ν_1 and Dixon called it ν_2; in Table VII.1 Herzberg's notation has been used in making the comparison between various authors. For the isotopic molecule DCO we shall expect a big change in ω_1, a moderate change in ω_2 and only a small change for the C–O stretching ω_3.

The main regularity is the occurrence of fairly long progressions of bands separated by intervals, in the progression, varying from ~ 1870 to 1720 cm^{-1}. Vaidya assigned this interval, which he took as 1883 cm^{-1}, to the ground state C–O stretching, here denoted ω_3''. This assignment appears safe and has been generally accepted. It agrees with the solid matrix value of 1861, although not with the

TABLE VII.1

Proposed vibrational assignments for the hydrocarbon flame bands of HCO and DCO

Authors	HCO					DCO			
	ω_1''	ω_2''	ω_3''	ω_2'	ω_3'	ω_1''	ω_2''	ω_3''	ω_3'
Vaidya		760	1883	465	1072			1859	970
H & H	2851		~1782						
M & S		1532	1880	1070	1219				
Dixon		1078	1976						
J, P & R		1083	1820?				847		
M & J	2488	1090	1861			1937	852	1800	

doubtful flash-photolysis value of 1820 cm^{-1}. It is the relative arrangement of these progressions into arrays and assignment of other frequencies which has led to difficulty.

Proposed vibrational frequencies in the ground state ('') and excited state (') for HCO and in some cases for DCO are summarized in Table VII.1. Data are from Vaidya (1934, 1964), Hornbeck and Herman (1954), Murphy and Schoen (1951) and Dixon (1969); ground-state data from flash photolysis (Johns, Priddle and Ramsay, 1963) and infra-red matrix absorption (Milligan and Jacox, 1964) are included.

Vaidya called his $0,0,0 \rightarrow 0,0,v_3$ progression A_0, his $0,0,1 \rightarrow 0,0,v_3$ progression A_1 etc, and other progressions were labelled A', A'', B_0 and B_1. The B progressions involve bands at rather shorter wavelengths and these bands are only observed in relatively hot flames and it seems most likely that these B bands involve a higher electronic state than that responsible for the A progressions.

Dixon noted that whereas bands of the A_1 progression had sharp lines, bands of the A_0 progression were slightly diffuse, having line widths of 1 to 5 cm^{-1}. He then says that "such a diffuseness in a low-pressure emission spectrum could hardly arise from pre-dissociation of the upper levels, since the rate of radiative decay would not be high enough to compete with the predissociation. The diffuseness must therefore be associated with predissociation of the lower levels. Hence the sharp A_1 bands and the slightly diffuse A_0 bands cannot have common lower levels, although they may have common upper levels." He then goes on to assign the frequency interval of around 1078 to ω_2'' instead of to ω_3'. This argument seems quite unacceptable. On Dixon's assignment the A_0 bands involve transitions $0,0,0 \rightarrow 0,1,v_3''$; the dissociation energy of ground-state HCO is, as Dixon says, around 1.2 e.V and for most of the A_0 bands the final vibrational levels lie well below this energy and could not possibly be predissociated; it would also be surprising if the extent of the predissociation were unaffected by the value of v_3''. Vaidya's A_2 progression involves rather weak bands, but appears to be genuine and if so its separation from the A_0 and A_1 progressions leads to a reasonable value of $x\omega$ on Vaidya's assignment, but not if the 1078 interval were assigned to the ground electronic state. For the isotopic molecule DCO a similar array to that involving the A_0, A_1 and A_2 progressions of HCO can be found (Spokes and Gaydon, 1959; Vaidya, 1964) and the frequency interval of around 970 cm^{-1}

TABLE VII.2

The array of A bands of HCO. Wavenumbers of the first heads of the bands are shown in ordinary print. The intervals between bands are printed in italics. Measurements from Dixon (1969) except where indicated.

$v_3'' \backslash v_3'$	0		1		2		3		4		5		6		7		8
0	39486*	*1876*	37610*	*1856*	35754	*1836*	33918	*1823*	32095+	*1784*	30311	*1760*	28551	*1744*	26807	*1724*	25081* A_0
			1077		*1071*		*1064*		*1077*		*1067*		*1059*		*1060*		*1067*
1			38687	*1862*	36825	*1843*	34982+	*1810*	33172	*1794*	31378+	*1768*	29610	*1743*	27867	*1719*	26148 A_1
			1033		*1045*		*1059*								*1050*		*1058*
2			39720*	*1850*	37870*	*1829*	36041*								28917*	*1711*	27206* A_2
																	1045
3																	28251* A_3

+Measurement from Hornbeck and Herman. *Measurement from Vaidya $+10 \ cm^{-1}$; Vaidya's measurements are systematically about $10 \ cm^{-1}$ lower than Dixon's, probably because he measured the stronger second or Q_{R_1} heads, while Dixon measured the first weaker Q_{R_0} heads.

can then only be assigned to the upper electronic state, not to the ground state as the matrix absorption shows that ω_2'' for DCO is about 852 cm^{-1}.

The diffuseness of bands of the A_0 progression does present a problem. With thermal excitation under equilibrium conditions, predissociation would not weaken the emission, as it does in non-equilibrium sources like discharge tubes, because loss by pre-dissociation will be compensated for by the inverse process of association. In ordinary chemiluminescence any predissociation in the excited state is likely to reduce the intensity of the affected bands. If excitation involves an association, such perhaps as $CH + O = CHO^*$ then the predissociated levels could even be strengthened, as in OH.

Table VII.2 shows the bands of the A array in the form given by Vaidya, but with improved measurements for most of the heads. The author still considers this the more likely analysis. More numerical data are given in the Appendix. Obviously there is room for further work on this system. The late Dr Vaidya started work using the isotope ^{18}O, but it was not completed; this would perhaps be the best approach, and might also enable a more definite assignment of the system origin ν_{000} as well as the vibrational frequencies. Shifts with DCO are so large that it is difficult to relate corresponding bands of the two isotopes. With ^{18}O or ^{13}C this should be easier.

The 4050 Å 'Comet Head' Group (C_3)

A group of apparently diffuse bands around 4050 Å, that is between the main CH bands at 4315 and 3900 Å, has long been known in the spectra of the heads of comets, and has also been obtained in discharge tubes containing flowing hydrocarbon vapours. First observations in flames were by Durie (1952b) who obtained them in diffusion flames of hydrocarbons burning with fluorine, and by Gaydon and Wolfhard (1952) who found them in the spectra of hydrocarbons burning with moist atomic hydrogen. Kiess and Bass (1954) and Kiess and Broida (1956) found them in the mantle or feather of rich oxy-acetylene flames, a source in which they are strong although often partly obscured by continuum. They also occur strongly in absorption after flash photolysis, in carbon furnaces and in absorption by flames (Jessen and Gaydon, 1969).

For many years it was thought that the bands were due to CH_2

but Monfils and Rosen (1949) found that there was no isotope shift on replacing hydrogen by deuterium. Douglas (1951) confirmed this and by using carbon enriched with ^{13}C found that the main head was split into six components and he showed that the emitter was a linear triatomic carbon molecule, C_3. The full analysis of the bands by Gausset et al (1965) shows that they are due to a $^1\Pi - ^1\Sigma$ transition in a linear molecule; the bands are of perpendicular type with R and Q heads and the vibronic structure has been fully investigated.

The strongest feature is always the red-degraded R head of the 0,0,0—0,0,0 band at 4049.8 Å and a second head at 4072.4 Å due to the 0,0,0—0,2,0 transition is usually fairly clear (see Plate 5e). In flames there is usually a lot of associated continuum, although Kiess and Broida's (1956) large dispersion spectrograms suggest that this may be due to rotational fine structure which is unresolved with smaller dispersion. The C_3 bands often appear to be associated with the formation of incipient solid carbon particles and usually require the presence of hydrogen. The difficult problem of the formation of these C_3 molecules, both in comets and stars and also in flames, has been discussed by Marr (1958) who has considered reactions such as $C_2 + CH = C_3 + H$ and $C_3H + CH = C_3 + CH_2$ but has not reached any definite conclusion.

Emeléus's Cool Flame Bands (CH_2O)

The spectrum of the cool flame of ether is quite different from the spectrum of ordinary hot flames and consists of a number of complex bands in the blue and near ultra-violet. The strongest bands cover the region 3700—4800 Å and are fairly evenly spaced. They are usually obtained only with small dispersion, because of the very weak emission from such cool flames, and then they appear diffuse and slightly degraded to the red. The spectrum was first reported by Emeléus (1926) who used very long exposures (250 hours) and a very wide slit, with the result that his values for the wavelengths were of low accuracy, making identification difficult. However Kondratiev suggested that the bands might be due to formaldehyde, and following the study of the resonance fluorescence spectrum of this molecule by Herzberg and Franz (1931) it was shown by Pearse (see Ubbelohde, 1935a), by direct comparison of spectrograms that the cool flame spectrum was indeed identical with the fluorescence spectrum of formaldehyde; this was checked by Kondratiev (1936) who remeasured the centres instead of the edges of Eméléus's bands. Bands of this type, with complex structure and no clear heads, are best identified by comparison of photographs rather than from measurements. Band centres are listed in the Appendix and a photograph is shown in Plate 2e.

These bands have since been obtained in a brighter source, a Tesla-coil discharge through flowing formaldehyde vapour, and this has enabled more precise measurements to be made and the rotational and vibrational structure of the bands has been studied in detail. Most of the bands fall into three short progressions and have complex rotational structure, being perpendicular type bands of HCHO. The analysis by Brand (1956) of the cool-flame bands and of the absorption of formaldehyde between 3550 and 2600 Å shows that they are all part of a $^1A''-^1A_1$ transition, but that in the $^1A''$ upper electronic state the molecule is non-planar. This complicates the vibrational selection rules so that the transition between the lowest vibrational level of the ground state and the lowest vibrational level of the excited state is forbidden and the (0,0,0–0,0,0) band is not usually observed. In absorption the bands are mostly due to transitions from the lowest vibrational level of the ground state to excited vibrational levels of the upper electronic state, while in emission transitions from the lowest vibrational level of $^1A''$ only take place to higher vibrational levels of the ground $^1A_1$ state. Thus there is practically no correspondence between the individual emission and absorption bands of formaldehyde.

The blue cool flames of higher paraffins, ethers and aldehydes all show this spectrum of formaldehyde, and it has been obtained during auto-ignition processes in engines. It is not obtained from normal hot flames, with one exception, the flame of methyl alcohol; Guénault (1934–37) reported the cool flame bands, together with other usual features such as CH and OH, in the methanol/air flame, and later Vaidya obtained the bands from the flame of methanol burning with atomic oxygen, while Gaydon and Wolfhard found them in the low-pressure flame of methanol with oxygen.

Band Systems of CO

Carbon monoxide is, of course, a major constituent of flame gases, especially of the interconal region in premixed flames. In low-pressure discharges the Angström bands in the visible region and the Third Positive bands of CO in the near ultra-violet are the most readily excited and are very strong. However, in flames there is insufficient energy for thermal excitation of these systems and, rather surprisingly to anyone familiar with discharge-tube spectra, they do not appear to have been recorded at all. Indeed there is very little emission from CO in flames, but the Fourth Positive System and three other systems have been reported and the Fourth Positive emission is of special interest.

The Fourth Positive bands lie in the far ultra-violet and consist of simple red-degraded bands with fairly close but regular rotational fine structure. The (0,0) band lies right down at 1544 Å, beyond the range of quartz instruments or the transmission of air, and all the strong bands lie below 2000 Å. The system is due to the transition $A\,^1\Pi - X\,^1\Sigma$, i.e. to the ground state of CO, and requires an energy of over 8 e.V for its excitation. Bands at the long-wave end of the system, between 2500 and 2000 Å were first noted by Gaydon and Wolfhard in low-pressure oxy-acetylene flames, using a quartz spectrograph. Hornbeck and Herman (1951b) also recorded bands of this system down to 1920 Å in the inner cone of an acetylene flame at atmospheric pressure. Hand (1962) extended observations to 1810 Å by shielding an oxy-acetylene flame with a flow of helium and using a vacuum grating spectrograph, while Kydd and Foss (1967) used a similar technique, with a lithium fluoride window, on flames at reduced pressure and measured strong Fourth Positive CO emission down to 1410 Å from flames of several hydrocarbons including methane. The strongest heads are at 1597.4 (0,1), 1653.0 (0,2)*, 1712.2 (0,3), 1729.5 (1,4) and 1792.4 (1,5). The vibrational intensity distribution seems fairly normal, the strongest bands being those with low v' but bands to $v' = 4$ (which require an excitation energy of 8.7 e.V) are fairly strong and in the low-pressure flames Gaydon and Wolfhard found very weak bands in the quartz ultra-violet up to $v' = 11$. In view of the very high excitation energy required, it seems that excitation must be chemiluminescent, and possible mechanisms have been discussed by Kydd and Foss, who attempted to correlate the strength of the CO Fourth Positive emission with that of other species such as CH and C_2; many suggested reactions were eliminated and no definite correlations found, although reactions

*Kydd and Foss incorrectly assigned this as the (3,4) band.

CH* + O = CO* + H and CH* + OH = CO* + H$_2$ were considered possible. Bayes (1967) and Becker and Bayes (1968) studied CO emission from carbon suboxide flames and suggested the reaction C$_2$O + O = CO* + CO to account for Fourth Positive emission in both the C$_3$O$_2$ and in hydrocarbon flames.

This CO Fourth Positive emission has an interesting application, because the bands lie at short wavelengths where normal thermal emission is low and beyond the solar spectrum or the transmission of ordinary glass. Thus the emission from burning hydrocarbons, even of flames with air, can be detected by "solar blind" detectors sensitive only to far ultra-violet light even in sunlight or in artificial light, and can serve for fire detection.

Three other systems of CO have been reported in flames of acetylene with atomic oxygen and also in carbon suboxide flames by Becker and Bayes (1968) and Oldman, Norris and Broida (1970). They are the Triplet system $d\,^3\Delta - a\,^3\Pi$ and Herman system $e\,^3\Sigma^- - a\,^3\Pi$ in the visible and the weak Cameron inter-multiplicity system $a\,^3\Pi - x\,^1\Sigma$ between 2600 and 2200 Å. They may occur weakly in ordinary hot flames but are difficult to observe because of other masking spectra.

Other Band Systems in Flames

In flames with air the nitrogen does not usually contribute appreciably to the emission, but in some of the hotter flames, such as stoichiometric acetylene/air, the CN Violet band system is weakly present. This system is degraded to the violet and has very well marked sequences. The rotational structure is close but fairly simple, showing only double P and R branches for each band, the transition being $^2\Sigma - ^2\Sigma$. The (0,0) sequence at 3883 Å is the strongest; this comes between the strong Q and weaker R heads of the 3900 CH band; in this sequence of CN the (0,0),(1,1) and (2,2) heads are close together and this little group of heads was referred to in some of the old literature on hydrocarbon flame spectra as "the three". The (0,1) sequence of CN is usually masked by CH unless it is very strong, but the (1,0) sequence at 3590 Å is clear from major overlapping and often shows in the spectra of hot flames containing nitrogen.

The CN Red system is not usually observed in the visible region in flames with air, although the (0,0) and (1,0) sequences in the infra-red do occur. In flames containing combined nitrogen, that is flames of nitrogenous fuels or flames supported by oxides of nitrogen, these CN bands, including those in the visible region, are very strong. In the infra-red there are obvious sequences, but in the visible region there are long progressions of equally spaced bands degraded to the red. The bands are due to a $^2\Pi - {}^2\Sigma$ transition and show double P, Q and R branches, but under small dispersion appear to have close triple heads.

Another band system involving nitrogen which appears weakly in hot flames with air is that of NH. The (0,0) band at 3370 Å has a strong piled up Q branch of unresolved lines, not clearly degraded either way, with open triplet P and R branches spreading out on either side of the Q maximum (see Plate 3e); however in flames with air it is always weak and largely masked by the much stronger OH band. In flames with oxides of nitrogen or of fuels containing combined nitrogen, the NH band is much stronger and the ammonia α-band due to NH_2, strongest in the yellow and green, and bands of NO in the ultra-violet, are also observed. These are discussed in a later chapter when dealing with flames containing combined nitrogen.

An unknown band structure in the red region of the spectrum was obtained by Gaydon and Wolfhard (1949a) in low-pressure flames. The structure was observed best in flames of ether or acetaldehyde burning with excess oxygen, but probably occurs also with hydrocarbons and other organic compounds, although overlapping by other band systems and the CO-flame continuum makes observation difficult. The structure is relatively open, suggesting a hydride, and extends from 5200 Å to the long-wave limit of plate sensitivity around 6500 Å. The complexity of the structure suggests a polyatomic emitter, and possibilities are CH_2, CH_3 or HCO; it occurs under similar conditions to the known hydrocarbon flame bands and could be due to emission bands of the same system of HCO as Herzberg and Ramsay obtained in flash photolysis. Another possibility is that it is an extension into the visible of the infra-red vibration-rotation spectrum of a stable emitter such as methane. Agnew and Agnew (1956) also report a similar banded structure in such flames.

The Carbon Line, λ2478

This line at 2478.57 Å was first observed in low-pressure flames, but also occurs at atmospheric pressure in hot flames of carbon-rich fuels. Slightly rich oxy-acetylene flames show it best and it is prominent in the tracings of spectra by Hornbeck and Herman (1951b).

It is due to the transition $3s\,^1P_0 \rightarrow 2p^2\,^1S$ and requires an excitation energy of 61982 cm^{-1} or 7.7 e.V. This is slightly less than required to excite the Fourth Positive System of CO which occurs under similar conditions. At 3000 K only about one atom in 10^{13} would, in equilibrium, be excited to this high energy. It is unlikely that excitation is by direct chemiluminescence, but it is probably excited by collision with another energy-rich molecule like CO formed by chemiluminescence in the flame gases. As this line does not involve the ^3P ground state of atomic carbon it cannot easily be used to deduce the carbon atom concentration.

Premixed Flames

Mixture Strength. The relative and absolute intensity of the various band systems varies considerably with the mixture strength. In general, the C_2 bands are strongest with very rich mixtures, the CH bands at slightly less rich mixtures, while OH has a flat maximum only slightly on the rich side of the stoichiometric point. The hydrocarbon flame bands appear best with weak mixtures; their absolute maximum intensity may be near the stoichiometric point but they maintain their strength at weak mixtures better than the other systems and so are relatively more prominent. In normal rich flames the HCO bands appear to be absent, but at the extreme limit, using an Egerton–Powling burner, Spokes and the author (Spokes, 1959) found that flames of ether and butane showed HCO emission in a fore-zone below the main reaction zone.

It is difficult to express results quantitatively because of changes in flame shape and structure, as discussed in Chapter II. However some measurements have been made of light emission by the various species as a function of the mixture strength λ. λ is defined as the ratio of the actual oxygen concentration in the mixture to the oxygen concentration at the stoichiometric point, for complete combustion to CO_2 and H_2O; thus $\lambda = 0.5$ is rich with only half the required amount of oxygen, $\lambda = 1.0$ is stoichiometric and $\lambda = 2.0$ is fuel-weak with twice the required amount of oxygen; λ is the reciprocal of the equivalence ratio, ϕ, used by some authors. Ausloos and Van Tiggelen (1953b) have studied acetylene/air flames and shown curves relating intensity to the percentage of acetylene in the mixture for the major emitters; these curves indicate that C_2 has maximum intensity at $\lambda = 0.65$, CH at $\lambda = 0.68$, OH at $\lambda = 0.85$ and HCO at $\lambda = 0.91$.

For flames with oxygen the range of mixture strengths is widened, especially on the rich side, but because of the big range in flame conditions measurements at 1 atm are difficult. In low-pressure flames Gaydon and Wolfhard (1950) were able to make some measurements of total C_2, CH and OH light emission from the reaction zone (i.e. the inner cone). For oxy-acetylene the C_2 had its maximum at $\lambda = 0.42$, CH was strongest at $\lambda = 0.59$ and OH near $\lambda = 0.7$. For other hydrocarbons the intensity maxima occur at rather less rich mixtures than for acetylene. For low-pressure ethylene/oxygen the C_2 has its maximum at about $\lambda = 0.57$, CH at 0.75 and OH shows a flat maximum around 0.9. The curves for C_2 and CH are shown in Fig. VII.1; these measurements were made with a flame on a 28 mm burner with, for rich mixtures ($\lambda < 1.0$) a constant oxygen flow and varying fuel supply, and for weak mixtures ($\lambda > 1.0$) a constant fuel supply and varying oxygen flow; the pressure was about 8 torr at the stoichiometric point, but rather higher for rich or weak mixtures, rising to 55 torr for the richest mixture. See also Bulewicz (1967).

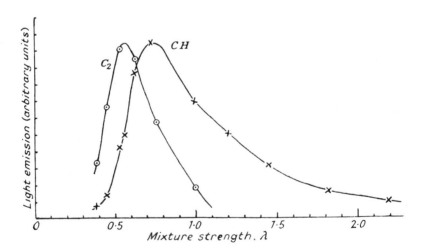

Fig. VII.1. Variation of C_2 and CH emission (arbitrary units) in a low-pressure ethylene/oxygen flame on a 28 mm diameter burner.

Fuel types and Light Yield Measurements. Fuels with a high carbon content, such as acetylene and benzene, give the most luminous inner cones and show the strongest C_2 bands. Saturated hydrocarbons show the C_2 rather less strongly than the unsaturated ones, and in methane the C_2 bands are much weaker and hardly visible with lean mixtures. Fuels containing oxygen, such as alcohols, ethers and aldehydes still show strong OH emission, but the C_2 and CH are less strong than with hydrocarbons. Methyl alcohol shows some CH emission, besides the strong OH, and is alone among normal flames in giving Emeléus's cool flame bands of formaldehyde; C_2 is barely detectable in methyl alcohol flames. Flames of formaldehyde and of formic acid show strong OH emission and some CO-flame spectrum (both the bands and continuum), but not C_2 ; CH is present very weakly in the formaldehyde flame. In all these organic flames the OH emission from the reaction zone is much stronger than from the burnt (i.e. interconal) gases above.

Quantitative comparison of the spectra of various fuels is difficult, especially at atmospheric pressure. For flat low-pressure flames there is less trouble from emission by the interconal gases and the outer cone and there is less self-absorption of radiation and Gaydon and Wolfhard (1950) made measurements of the C_2 emission for several flames. The total emission from the flame, at known distance, was compared with that from a calibrated strip lamp, the emission being integrated over the region of C_2 emission, a deduction being made for any background continuum. Results were expressed as the number of light quanta emitted for each O_2 molecule reacting, it

TABLE VII.3

C_2 light yield, in number of photons emitted per O_2 molecule consumed, at the stoichiometric point (st.) and for maximum emission.

Fuel	Light yield at st. $(\times 10^{-5})$	Pressure for st. mixture (torr)	Light yield at max. $(\times 10^{-5})$	λ for max. yield	Pressure at max. (torr)
CH_4	0.049	13	1.05	0.59	30
C_2H_2	11.3	5.5	112	0.42	14
C_2H_4	2.9	8.5	16	0.57	18
C_2H_6	1.4	15	4.6	0.64	26
C_3H_8	0.78	12	6.1	0.55	24
C_6H_6	14.3	7.6	33.5	0.66	16

being assumed that for weak mixtures the excess oxygen above the stoichiometric amount acts as a diluent and does not react. Results for the several hydrocarbons studied are summarized in Table VII.3.

For CH similar, but relative instead of absolute, measurements of light yield were also made, and are listed in Table VII.4. The light yield is on an arbitrary scale of 100 for the stoichiometric oxy-acetylene flame.

No data for OH are available because superposition of radiation from the burnt gases, variation in effective rotational temperature from fuel to fuel and self absorption make measurements particularly difficult.

Effects of pressure, temperature and dilution. The general character of the spectrum and the variation of the strength of the various band systems with mixture strength is similar at low and at atmospheric pressure, but the total light yield rises as the pressure is reduced. For oxy-acetylene the C_2 light yield rose from about 3.6×10^{-5} photons emitted per O_2 molecule reacting at 760 torr to 13.5×10^{-5} at 3.5 torr (Gaydon and Wolfhard, 1950). For prop-ane/oxygen the rise in C_2 with decreasing pressure was rather greater; for the stoichiometric mixture the yield rose from 0.04×10^{-5} at 760 torr to 0.78×10^{-5} at 12 torr; the position of maximum C_2 emission remained at $\lambda = 0.55$ and the yield rose from 0.4×10^{-5} at 760 torr to 6.1×10^{-5} at 24 torr. Kydd and Foss (1967) have also made some measurements on oxy-acetylene flames at reduced pressure, in the range 40 to 95 torr; they used a modified Meker burner and the flame was shielded by a flow of helium. They speak of the CH and CO Fourth Positive emission intensity varying as $1/(\text{pressure})^2$. This is surprising and suggests a major discrepancy

TABLE VII.4

CH light yield (arbitrary units) at the stoichiometric point and for maximum emission for some low-pressure flames with O_2.

Fuel	Light yield at stoich.	Light yield at max.	λ at max.
CH_4	1.2	2.7	0.67
C_2H_2	100	170	0.59
C_2H_4	20	48	0.75
C_2H_6	13	16	0.84
C_6H_6	61	78	0.85

from the results of Gaydon and Wolfhard, because if CH varied so strongly and C_2 so much less strongly with pressure, the low-pressure flames should be much bluer than those at higher pressure, which is not the case. In the author's work the burner diameter was increased at low pressure to reduce wall quenching effects, while Kydd and Foss apparently made all measurements on the same burner; however, the direction of the discrepancy seems the wrong way for it to be due to increased quenching at low pressure. Further careful quantitative measurements are obviously needed. Measurements by Bulewicz (1967) are especially difficult to interpret as she used a focused image of the flame on the spectrograph slit and so really only measured the brightness of the reaction zone, not the integrated intensity which is the product of brightness and size. Her apparent result that the concentration of excited species varied between the 2.0 and 2.5 positive power of the pressure is not easily related to the emission per reacting fuel molecule.

Hsieh and Townend (1939) noted that flames of ether/air and other fuels became very green at reduced pressure, apparently due to strong C_2 emission. All enclosed flames of rich mixtures are green, although often the effect is partly masked in unenclosed flames by the blue emission from the outer cone. The author has not noted any relative strengthening of C_2 at low pressure, although flames with air do tend to be greener than flames with oxygen because addition of nitrogen reduces CH emission more than C_2 emission.

Some interesting measurements of the effect of dilution and also of replacement of part of the hydrocarbon by hydrogen or carbon monoxide have been made in the low-pressure flames (Gaydon and Wolfhard, 1950). Dilution with nitrogen caused a relatively minor reduction in C_2 light yield, but excess O_2 had a strong quenching action. Results for flames of the type

$$C_2 H_2 + 2\tfrac{1}{2}O_2 + nX$$

where X is the diluent (N_2, etc.) are shown in Fig. VII.2.

When part of the hydrocarbon supply is replaced with hydrogen or carbon monoxide in the correct amount to keep the mixture stoichiometric there is a surprisingly rapid fall in the C_2 and CH light emission. This is not so marked with the unsaturated fuels like acetylene, but is strongest with saturated hydrocarbons. Thus in mixed ethane/hydrogen flames the C_2 intensity varies roughly as the

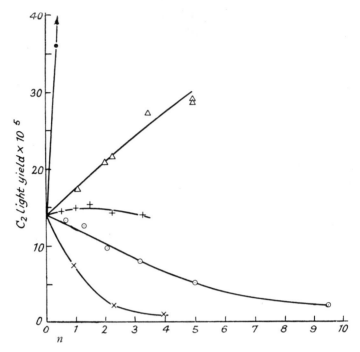

Fig. VII.2. Effect of dilution of low-pressure stoichiometric oxy-acetylene flames with various gases. ● C_2H_2, Δ H_2, + CO, ⊙ N_2, x O_2.

square of the ethane concentration (see Figure VII.3) while the CH varies about as the cube of the ethane proportion. For ethane/carbon monoxide mixtures the C_2 emission and CH emission are both roughly proportional to the fourth power of the ethane concentration. It should be noted that in these flames with ethane there is very little change in the theoretical flame temperature on replacing ethane by H_2 or CO so that the effects on C_2 and CH emission are not thermal but must be due to the chemical processes producing these species. The OH emission is not, however, so sensitive to this partial replacement of the hydrocarbon by another fuel; for ethane/hydrogen it appears to be proportional to the ethane concentration; it is about thirty times stronger in pure ethane than in pure hydrogen flames. In methane flames, the reduction in C_2 and CH by partial replacement of CH_4 with H_2 or CO is particularly marked; in CH_4/H_2 mixtures even the OH emission varies as the square of the CH_4 concentration.

Kydd and Foss (1967) have made similar measurements of relative

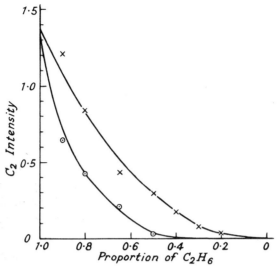

Fig. VII.3. Effect on C_2 emission of replacement of ethane by H_2 and CO in stoichiometric low-pressure flames. x H_2, ⊚⊚⊚CO.

intensities for mixed hydrocarbon/hydrogen/oxygen flames at reduced pressure. For stoichiometric mixtures they have plotted log (intensity) against log (total carbon) for bands of C_2, CH, OH, HCO (hydrocarbon flame bands) and CO (Fourth Positive system). The curves approach straight lines and the slope gives the mean power-law dependence of emission on carbon concentration. Their results are summarized* in Table VII.5.

TABLE VII.5

Slope of log(I) v log [C] curves for various emission bands in mixed hydrocarbon/hydrogen/oxygen flames.

Fuel	Emitter C_2	CH	OH	HCO	CO
C_2H_2	2.1	1.6	0.4	1.0	1.4
C_2H_4	2.9	2.8	0.5	1.9	2.2
C_3H_8	2.6	2.8	0.5	1.7	2.2
CH_4	3.4	3.1	0.7	1.5	1.8

*Kydd and Foss express their results to three significant figures, but their standard deviations and graphs show that this is not justified. The power-law only holds approximately.

Effects of dilution with nitrogen have also been studied by Nenquin, Thomas and Van Tiggelen (1956), who have also measured the flame temperatures. They assume that the emission is chemiluminescence and depends on the rate of reaction and thus on the activation energy and temperature; they have thus derived some values for the activation energies required to excite various band systems, especially in acetylene/oxygen/nitrogen and in acetylene/nitrous oxide flames. The interpretation of the results is difficult because the strength of emission depends on species concentrations, activation energies and also on quenching processes. Thomas (1951) attempted to study the effect of preheating on flame spectra and did observe some initial increase in C_2 and CH emission for small preheating, but changes in flame shape and stability prevented quantitative evaluation of the observations.

Structure of the Reaction Zone. The low-pressure flames are useful for studying the detailed spectroscopic structure of the reaction zone and Gaydon and Wolfhard (1949a) have described a number of flames. Usually the C_2 radiation is strongest near the base of the reaction zone, while CH comes higher, so that the base appears green and the top more violet (see for example the colour photograph in *Flames*, Gaydon and Wolfhard, 1970). Photographs of the spectrum, using a focused image of the flame on the spectrograph slit show the effect clearly (see Plate 6d). For methane there is no obvious resolution into green and violet zones, but in mixed methane/hydrogen flames the separation is again very clear. In some flames of very rich mixtures there is a faint bluish zone at the base of the main reaction zone, this showing CH bands. The OH radiation extends right across the reaction zone. In weak mixtures the HCO bands occur and these are fairly strong at the base, definitely below the CH region. In the formaldehyde flame the main features are OH and the CO-flame bands and these occur at the base of the reaction zone, with weak CH radiation coming much higher. In the methyl alcohol flame the cool flame bands of formaldehyde are at the base, with OH and CH higher in the flame.

Broida and Heath (1957) have made a more precise study of a stoichiometric oxy-acetylene flame at 2.2 torr, using photoelectric recording and large dispersion so that rotational temperature measurements could also be made. At this low pressure the reaction zone is around 3 cm thick and their logarithmic plots show (Fig. VII.4) that the C_2 has its maximum intensity lowest, the OH just a little

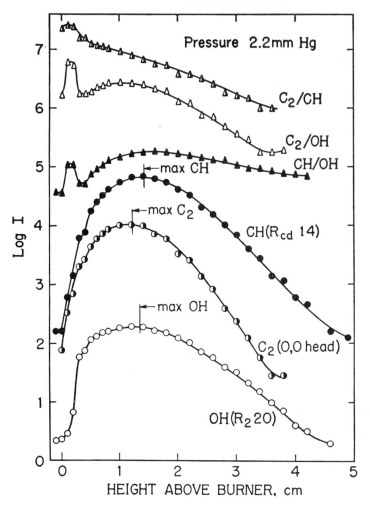

Fig. VII.4. Logarithmic plots of intensities of band systems, and their ratios, in low-pressure oxy-acetylene. From Broida and Heath.

higher and the CH about 2 or 3 mm above that of C_2. The curves for CH and C_2 show slight kinks near the base of the flame; these are more apparent in plots of the ratios CH/OH and C_2/OH They appear to be real anomalies as rotational temperatures also show additional peaks in this region. No reason has been assigned, but it is likely that two processes are involved in CH excitation, one accounting for the main CH emission and occurring most strongly in the upper part of the flame, and the other being more important in very rich mixtures and also causing the kink in the plot of intensity against height for CH in the flame of the stoichiometric mixture studied by Broida and Heath.

At atmospheric pressure the flame front is usually much too thin for detailed study, but in some slow-burning flames like butane/air, or flames near the limiting mixture strength, some details are visible. The flat flames on an Egerton-Powling burner (see page 42) are fairly thick but usually show little structure; however, rich flames of ether show a well defined cool-flame layer giving formaldehyde emission bands, below the main flame front (Thabet, 1951), and the spectra of first and second-stage ignition have been studied in this way by Agnew and Agnew (1956). In rich flames the yellow luminosity, usually attributed to carbon, occurs well above the reaction zone where C_2 and CH are emitted; Agnew and Agnew comment that this region of yellow luminosity does not show continuous emission in the infra-red as would be expected for carbon particles. Ordinary conical flames of acetylene/oxygen and ethylene/oxygen which are rich but just free from definite carbon formation show a whitish mantle* above the inner cone, this sometimes extending for a few cm. This mantle emits C_2, CH and often CN, and also strong C_3 at 4050 Å, on an apparently continuous background; Echigo, Nishiwaki and Hirata (1967) have, however, indicated that the emission, especially in the infra-red, may be banded not continuous.

Effects of Turbulence in the Gas Flow.   Flame propagation depends on conductive heat transfer and diffusion of active species, and under laminar flow conditions the burning velocity is a well defined quantity. Under turbulent flow conditions there is a considerable increase in effective burning velocity (see Gaydon and Wolfhard, 1970). Turbulence may have two main effects; it may break up the flame front (wrinkled flame concept) so that its area is increased and the volume rate of consumption of gas mixture shows a corresponding increase; or it may, by stirring within the reaction zone, cause increased heat transfer and radical diffusion and so alter the structure of the flame front itself. The first effect certainly occurs, but there has been some uncertainty about the importance of the second effect, although Fox and Weinberg (1962) have shown that the increase in flame area due to wrinkling is insufficient to account for the change in effective burning velocity; in their

---

*This is sometimes known as the acetylene "feather".

experiments with propane/air the propagation rate of the flame front itself was increased by a factor of 2½.

Gaydon and Wolfhard (1954) compared the spectrum of an ethylene/air flame under laminar flow conditions with that of a similar flame made slightly turbulent by the introduction of a grid; there was no change in the relative intensity of the main emission bands of CH, OH, C_2 and HCO, but the intensity of all systems fell by up to 20 per cent. John and Summerfield (1957) made similar studies of open and enclosed propane/air flames using higher levels of turbulence and found a decrease of up to 35 per cent in radiation under turbulent conditions. In Gaydon and Wolfhard's work the grid probably caused fairly large-scale turbulence and the small decrease in radiation was thought to be due to breaking up of the flame front into fragments, some of these fragments being quenched out and so reducing the radiation. However, the rather larger effect observed by John and Summerfield was attributed to modified transport processes within the flame front. In a study of the noise from turbulent flames Hurle, Price, Sugden and Thomas (1968) used the light intensity as a measure of chemical reaction rate; in their conclusion they say that emission intensities of C_2 and CH are directly proportional to volume flow rate for both laminar and turbulent flames, this supporting the wrinkled flame concept; however, in the text they note that above Reynolds numbers of 13,000 the C_2 and CH emission intensities were not linear with flow rate, and that above a Reynolds number of 17,000 the intensities actually fell with increasing flow. The experimental results in these three investigations all seem to agree, but not the conclusions drawn from them.

Diffusion Flames

The simple flame of a jet of fuel gas issuing from a round tube and burning in air is usually yellow and highly luminous. The spectrum of this is continuous and due to thermal emission of hot carbon particles. Owing to the carbon particles being smaller than the wavelength of light, their emissivity increases to shorter wavelengths, so that the intensity distribution in the continuum differs somewhat from that of a black body at the flame temperature; the colour temperature of such a flame tends to be rather higher than the true gas temperature. This type of radiation, and also scattering effect of carbon particles, are more fully discussed in *Flames*. Nearly all organic fuels give luminous flames of this type; the exceptions are carbon monoxide, methyl alcohol, formaldehyde and formic acid which give clear blue or bluish flames.

At the base of luminous diffusion flames of this type there is a bluish green region which shows C_2 and CH emission, and there may be a blue-violet outer region which shows the spectrum of burning carbon monoxide. In the candle flame these regions are fairly clearly defined, and there is also a dark inner region close to the wick from which no light is emitted. The detailed structure of diffusion flames is, however, best studied with the flat-flame burner developed by Wolfhard and Parker or the flat counter-flow diffusion flame (see Chapter III).

Most work has been done on flames with oxygen, as these have thicker reaction zones than diffusion flames with air (although in premixed flames it is the other way round). Away from the burner, the flame may be divided into three well-defined zones, the yellow luminous zone on the fuel side, where the carbon particles are present, a dark central zone whose spectrum, however, shows OH bands in the ultra-violet, and a blue zone on the oxygen side which emits the CO-flame spectrum and sometimes also the Schumann—Runge bands of O_2. The blue-green region, showing C_2 and CH bands, only occurs at the base, near the burner rim, and adjoins the yellow carbon zone, lying between this carbon zone and the dark central zone. There is some indication that the CH bands spread slightly further towards the oxygen side than the C_2 bands. The structure of this type of flame has been discussed fully in *Flames*; the general picture is of a wedge of combustion products separating the oxygen from the fuel, with the fuel breaking down thermally to carbon and hydrogen, the hydrogen diffusing rapidly across to the oxygen and the carbon being attacked by water vapour to give carbon monoxide and more hydrogen which again diffuse across. In the main part of the flame there is a close approach to both thermal and chemical equilibrium, but with a small region at the base in which there is some interdiffusion of fuel and oxygen and where there may be departures from both chemical and thermal equilibrium; it is this region which shows the C_2 and CH bands.

In flat diffusion flames between hydrocarbons and air the reaction zone is too thin for convenient study. The yellow carbon zone is less bright than for flames with oxygen, but the strength of the blue-green zone which shows C_2 and CH is well maintained and this zone extends rather higher. Further dilution with nitrogen gradually extinguishes the carbon luminosity but leaves the blue-green region surprisingly bright. Vaidya's hydrocarbon flame bands of HCO are not usually seen in diffusion flames, but the author has observed them in ethylene/air flames diluted with nitrogen; at the limit, just before the nitrogen extinguishes the flame, they are one of the strongest features of the spectrum. It has also been noted that in these diffusion flames the CH bands show the usual breaking off in rotational structure due to predissociation, indicating non-thermal excitation.

The effect of pressure on diffusion flames is rather similar to that of dilution with nitrogen. As the pressure is lowered the carbon zone weakens and contracts, while the blue-green zone expands and becomes much brighter, until finally a bright green flame is left, very similar in appearance to that of a premixed flame of a rich mixture at low pressure.

The true diffusion flame is only obtained with slow flows under so-called laminar flow conditions. With faster flows there is a change to turbulent conditions. First, vortices at the edge of the flame cause partial mixing of fuel and air so that we change towards a premixed type of flame. Thus as the flow rate is increased there is a brightening of the C_2 and CH emission. With sufficiently fast flow, the flame lifts some distance above the burner and at this stage there is a sudden drop in carbon formation and the base of the flame appears as a bright blue-green ring showing strong C_2, CH and OH emission (Gaydon and Wolfhard, 1954).

Flames at liquid surfaces, including flames on wicks and round large droplets, are basically of the diffusion type and show similar spectra.

Cool Flames and Preignition Glows

For most of the hydrocarbons, ethers and aldehydes the graph of ignition temperature against pressure takes a rather curious form whose general chracter is illustrated in Figure VII.5. At high pressure the ignition temperature is quite low and does not vary much with pressure. At low pressure the ignition temperature is much higher and tends to rise as the pressure is reduced. In the intermediate region the curve may take an irregular form, and just to the low-pressure side of the ignition point pale blue cool flames may be

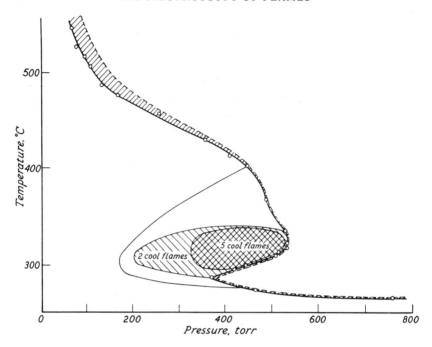

Fig. VII.5. Variation of ignition temperature with pressure for a typical higher paraffin (propane/oxygen).

observed to pass through the mixture. The experimental observations may be carried out in either a static or a flow system. In the static system the prepared mixture is admitted quickly to the heated reaction vessel until the required pressure is reached; in the cool-flame region one or more (often up to five) cool flames may be seen to traverse the mixture successively; there is an induction period, often of the order of a minute, sometimes a very long one of perhaps half an hour, before either the passage of cool flames or ignition. In the flow system the mixture is pumped steadily through a heated reaction tube; in the cool flame regime a glow may be stabilized in this tube. The ignition point depends not only on pressure but also on mixture strength, vessel dimensions, surface conditions and the induction period (or, for flow systems, on the flow rate). Thus curves of the type shown in Figure VII.5 depend very much on the actual experimental conditions and have little absolute significance.

Usually cool flames occur in the temperature range 200° to 450°C and for the higher hydrocarbons, ethers and aldehydes at rather below atmospheric pressure. For the lighter hydrocarbons the optimum pressures are higher; with propane/air rather above one atmosphere and with ethane and ethylene at several atmospheres. Methane does not usually give a cool flame, but has a high ignition temperature; however, Vanpee (1956) has found that a rich methane/oxygen mixture ($2\,CH_4 + O_2$) will give cool flames at around 500°C and 700 torr.

The spectra of all the cool flames studied are found to show Eméléus's formaldehyde bands. Using Topps and Townend's conical tube (see page 49) it has been possible to stabilize first and second stage cool flames. Usually both first and second stage flames show the same spectrum of formaldehyde. Agnew and Agnew (1956) studied first and second stage flames of ether/air for very rich mixtures on an Egerton-Powling burner; for the richest mixtures both first-stage and second-stage flames showed formaldehyde only, but as the mixture became less rich there was a smooth change in the spectrum of the second-stage flame from the cool-flame type to the full-ignition type for a rich mixture in which CH and HCO were emitted.

In addition to the cool flame regime, a glow is frequently observed just prior to ignition in the high-temperature regime. This "preignition glow" is indicated in Figure VII.5. Spectra of this preignition glow for propane/air differ from that of the cool flame (Gaydon and Moore, 1955). For weak propane/air the spectrum of the glow is very similar to that of the hot flame, showing strong OH, HCO and some CH, with a background probably due to the CO-flame spectrum. With rich mixtures, the spectrum of the preignition glow is intermediate between that of the hot flame and the cool flame, showing strong hydrocarbon flame bands of HCO, some CH and some formaldehyde emission, but OH is weak and C_2 absent. The normal hot flame through these mixtures shows strong C_2, CH and OH and only weak HCO. Plate 2, strips e, f and g illustrate this. The results for propane are probably typical of those for higher hydrocarbons. Mixtures of methane and carbon monoxide with oxygen give a preignition glow, the spectrum showing formaldehyde bands and CO-flame spectrum (Gaydon and Moore, 1955).

Absorption Spectra

Premixed Flames. The advantages and limitations of studying absorption spectra have been discussed in Chapter II, page 24. The

174

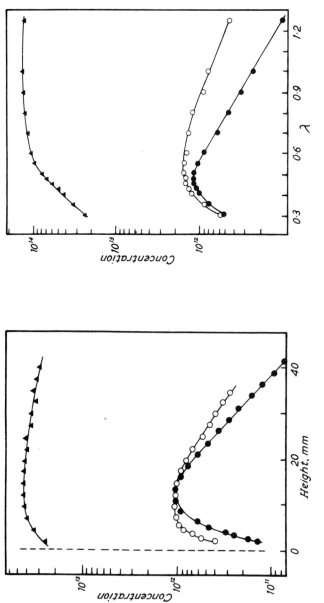

Fig. VII.6. Concentrations of radicals in low-pressure oxy-acetylene flames (*a*) with height above burner top, (*b*) with mixture strength λ. Results are for the $v = 0$ vibrational levels. For the flame used for (*a*) the pressure was 5 torr and λ = 0.5.

▲ $OH(x^2\Pi)$ ○ $CH(x^2\Pi)$ ● $C_2(x^3\Pi)$.

Redrawn from Bulewicz, Padley and Smith.

interconal gases of Bunsen-type flames at atmospheric pressure show fairly strong OH absorption, and the heads of the main bands at 3064 and 2811 Å are easily visible with small-dispersion instruments, despite the basic difficulty of observing bands with resolved fine structure under these conditions. The Schumann—Runge bands of O_2 at short wavelengths are also obtained fairly readily, especially in hot flames with oxygen when the absorption extends to longer wavelengths, perhaps to 2600 Å (Plate 4c). These bands show open rotational structure without prominent heads and are best detected with a spectrograph of fairly high dispersion. The NO γ bands at 2269 (0,0) and 2155 Å (1,0) can also be observed in absorption in these hot flames.

Repeated early attempts to observe absorption by the radicals CH, C_2 and HCO in the reaction zones of ordinary flames were unsuccessful. However, with the more refined techniques it has now been possible to detect, and in some cases to measure the concentrations of C_2 and CH. First observations in absorption were made on low-pressure flames (Gaydon, Spokes and van Suchtelen, 1960; Bleekrode and Nieuwpoort, 1965; Bulewicz, Padley and Smith, 1970). For CH the band at 3143 Å, although relatively weak in emission, is the one most easily detected in absorption because of its strong Q branch of superposed lines, but all three systems have been found. For C_2 the emission and absorption spectra are quantitatively different because the rotational temperature is abnormally high in emission but not in absorption; this is discussed later (page 198). The HCO bands are not observed in absorption; the strongest absorption bands would be expected to lie below 2000 Å and so not be readily accessible with quartz instruments. Herzberg and Ramsay's bands of HCO in the visible should occur if the concentration of HCO is high enough, but they do not seem to have been reported.

In low-pressure oxy-acetylene flames Bulewicz, Padley and Smith were able to measure the concentrations of C_2, CH and OH from the strength of the absorption bands. Results, as a function of position in the reaction zone and of mixture strength λ are shown in Figure VII.6. All these concentrations, particularly those for C_2 and CH, were orders of magnitude above the equilibrium values for these low-pressure flames. For the flame in Figure VII.6a the total number of flame gas molecules was 2.4×10^{16} ml^{-1}.

Bulewicz et al also measured the strength of emission of various bands and found strong confirmation of the reaction

$$C_2\,(x\,^3\Pi) + OH(x\,^2\Pi) \rightarrow CH(A\,^2\Delta \ or\ B\,^2\Sigma) + CO$$

which had been previously proposed by the author. This was shown by plots of CH emission intensity (for the 4315 Å band) against the product of the C_2 and OH concentrations. These plots are shown in Figure VII.7.

In the low-pressure flames of methane, dimethyl ether and acetaldehyde, Gaydon, Spokes and van Suchtelen found a diffuse band at 2157 Å which appears to be due to CH_3, having been previously found in flash photolysis by Herzberg and Shoosmith (1956); weaker bands, also possibly due to CH_3, were found at 2188 and 2080 Å. In flames of ethane and of diethyl ether a rather similar band with maxima at 2228 and 2242 Å has been provisionally attributed to the C_2H_5 radical.

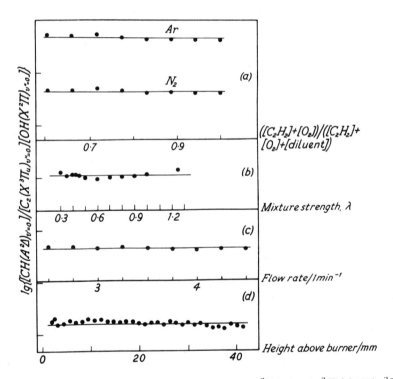

Fig. VII.7. Variation of the function $Y = [CH(A\,^2\Delta]/[C_2(x\,^3\Pi)]\,[OH(x\,^2\Pi)]$ with (a) Ar and N_2 dilution, (b) mixture strength, (c) flow rate and (d) height above burner. For details see Bulewicz, Padley and Smith (1970).

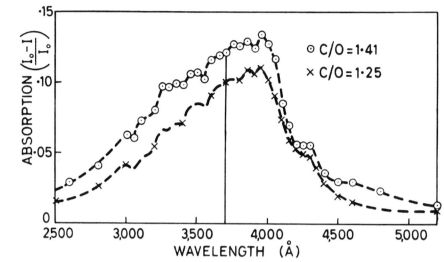

Fig. VII.8. Tracings of absorption in the tips of two rich oxy-acetylene flames under conditions of incipient soot formation. A multiple-reflection system was used to give eleven traversals through the flame at a height of about 2 mm above the tip of the inner cone.

In rich oxy-acetylene flames at atmospheric pressure the C_2 bands and all three systems of CH have also been observed in the reaction zone (Jessen and Gaydon, 1967), and in the mantle or "feather" just above the reaction zone the C_3 band at 4050 Å and some accompanying structure of rather uncertain origin between 3000 and 4000 Å has been obtained (Jessen and Gaydon, 1969); a tracing of this for two flames of slightly different composition is shown in Figure VII.8.

In these rich oxy-acetylene flames in the mantle above the reaction zone the concentration of C_2 is about 5×10^{14} molecules per millilitre, which is within a factor of 2 equal to the calculated equilibrium concentration; C_3 at about 10^{15} and CH at around 10^{14} are also near equilibrium. In the reaction zone itself, however, the concentrations of C_2 and CH are much higher than the equilibrium values; quantitative measurements in the thin reaction zone are difficult, but the concentrations may be around 50 times that for equilibrium. C_3 has not been detected in the reaction zone; it is probable that its concentration does not exceed the equilibrium value.

Diffusion Flames. In the flat diffusion flames of hydrocarbons burning with oxygen, as studied on the special burner by Wolfhard and Parker (see page 42), the blue region on the oxygen side shows O_2 Schumann–Runge absorption, the main dark central zone shows OH bands strongly, and the luminous carbon zone shows a continuum due to absorption and scattering by carbon particles; this absorption by carbon is almost complete. Just to the fuel side of the carbon zone, Wolfhard and Parker (1950) found apparently continuous absorption which increased in strength towards shorter wavelengths. This part of the flame has a very steep temperature gradient and schlieren effects are troublesome. Garton and Broida (1953) failed to find this continuous absorption in their studies of a similar flame further in the ultra-violet. Wolfhard and Parker found a similar continuous absorption, increasing in strength towards shorter wavelengths when hydrocarbon fuels were pyrolysed in a heated tube.

More recent experiments (Laud and Gaydon, 1971) using a flat counter-flow diffusion flame have shown that this absorption on the fuel side of the carbon zone is not strictly continuous, but shows quite a lot of diffuse structure. A tracing of the spectrum is shown in Figure VII.9. A re-examination of the absorption on a Wolfhard–Parker burner showed that this continuum also possessed similar

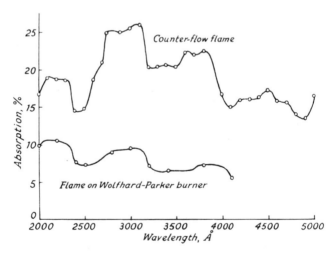

Fig. VII.9. Absorption, at the position of maximum strength, for ethylene flat diffusion flames on counterflow and Wolfhard–Parker burners. For the counter-flow flame the fuel flow was 10 ml/s C_2H_4 + 35 ml/s N_2, and the oxidant flow was 20 ml/s O_2 + 25 ml/s N_2; the effective path length was 55 mm. For the Wolfhard–Parker flame, pure ethylene and oxygen were used with a path length of 25 mm.

structure, which the original qualitative observations by photographic methods had failed to reveal. The origin of this diffuse structure is unassigned, but it must be due to some molecules formed in the pyrolysis or oxidation and which probably precede soot formation; it cannot be due to incipient particles as these should give a more uniform continuum. There is little evidence from the quantitative measurements that this absorption strengthens to shorter wavelengths; indeed it is stronger in the near ultra-violet than in the far ultra-violet.

When the fuel gas itself has discrete absorption bands it is possible to follow the decomposition of the fuel in the preheating zone of diffusion flames, by noting the disappearance of the absorption bands. Quantitative results are difficult to obtain because of the reduction in gas density due to thermal expansion and because the higher temperature produces changes in the structure and strength of the absorption spectrum. However, Wolfhard and Parker were able to show that when a little benzene was introduced into the ethylene side of an ethylene/oxygen flame, the benzene decomposed long before it reached the carbon zone. In these diffusion flames a wedge of combustion products separates undecomposed fuel from oxygen.

The Slow Combustion of Hydrocarbons. The absorption spectra of various hydrocarbons and related compounds undergoing slow combustion below the normal ignition temperature were first studied by Egerton and Pidgeon (1933). The hydrocarbons themselves are transparent throughout the quartz ultraviolet, but strong absorption was observed during their slow combustion in air. With the lower hydrocarbons, such as propane, there was an induction period of several minutes, followed by the development of continuous absorption (referred to as "extinction") spreading from the short-wave end of the spectrum. This was followed by the appearance of a banded spectrum in the region 3200 Å to 2800 Å which was identified with the known absorption of formaldehyde; bands are listed in the Appendix and shown in Plate 4*d*. The initial extinction at the ultraviolet end, preceding the formaldehyde, might be due to either acids or peroxidic substances, which are all known to show continuous absorption in the far ultra-violet. Estimates of the intensity of absorption compared with that produced by known quantities of acids found to be present by titration, indicated that the main part of this extinction was due to acids; with butane, however, the extinction was rather stronger than could be accounted for by the known amount of acid present, suggesting that during the early stages another substance, such as a peroxide, might be present.

In addition to the extinction in the ultra-violet and the formaldehyde absorption, it was found by Egerton and Pidgeon, and confirmed by Ubbelohde (1935b) that during the combustion of butane and higher paraffins there was a strong region of continuous absorption around 2600 Å, referred to as the X band; this appeared at the end of the induction period and died out later as the formaldehyde absorption developed. This X band was also observed in the slow combustion of butyl and higher alcohols and with isopentane, but not with propane, iso-butane, butyl or other aldehydes, butylene, amylenes or diethyl ether. Under conditions when a cool flame occurred there was a sharp drop in this absorption at 2600 Å, as the cool flame passed but a rise in the continuous absorption further in the ultra-violet. In later work Burt, Skuse and Thomas (1965) found that normal hot ignition occurred the instant the strength of absorption at 2600 Å reached a certain critical value.

Quite a lot of work has been done to identify the origin of this absorption at 2600 Å. Egerton and Ubbelohde examined the spectra of many organic compounds and compared them with the X band. Barusch and colleagues (1951a, 1951b) and Thomas and Crandall (1951) attributed the absorption to β-dicarbonyls, i.e. to the conjugated groups

$$\underset{\text{—C—C—C—}}{\overset{\text{O   O}}{\underset{\|\quad\|}{}}} \quad \text{or} \quad \underset{\text{—C—C—CH}}{\overset{\text{O   O}}{\underset{\|\quad\|}{}}}$$

Further work on n-paraffins by Burt, Skuse and Thomas (1965) led them to doubt this conclusion and they favoured branching intermediates or peroxy radicals. However, Fish, Haskell and Read (1969) have since made more quantitative measurements and also studied the combustion products by gas chromatography and have shown that unsaturated ketones and β-dicarbonyls are responsible. For the slow combustion of methyl pentane they find that the absorption can be accounted for quantitatively by the combined presence of several unsaturated ketones, especially 2 methylpent-2-2en-4-one, with β-dicarbonyls, like 4-methylpentan-3-onal, making a minor contribution.

In addition to formaldehyde absorption, bands of propionaldehyde were observed by Newitt and Baxt (1939) during the early stages of slow oxidation of propane. In the slow combustion of heptane, Egerton and Pidgeon reported a condensation product forming on the windows and showing absorption bands at 2600, 2530, 2470, 2420 and 2350 Å; these are due to benzene. This work has not been repeated, but the indication that ring closure to form aromatics may occur in slow combustion is interesting. Spokes (1959) also reported benzene absorption in low-pressure flames. These bands of benzene are strong and easily detected, but care is necessary to avoid any contamination and further work is desirable.

Chapter VIII

Measurements of Effective Temperature and Studies with Special Sources

In the three preceding chapters, observations on the emission and absorption spectra of normal stationary flames have been described. In this chapter we shall be concerned with additional observations aimed at elucidating the detailed state of the gases in the reaction zone. These will include quantitative measurements on the relative intensities of lines and bands, or on the contours of individual lines, with the results expressed as effective temperatures. The various methods of measuring flame temperatures have been described fully in the recent edition of *Flames* (Gaydon and Wolfhard, 1970); here the principles of the methods will be referred to only briefly, but results will be given. Other observations in special sources such as flash photolysis, atomic flames and shock tubes, and also tracer observations using stable isotopes for the study of flame spectra, give information about basic chemical processes and excitation mechanisms and are relevant to the understanding of flame spectra. Work on some special flames, such as those supported by fluorine instead of by oxygen, is also valuable for a discussion of the reaction and excitation processes in normal flames and is therefore included here.

Electronic Excitation Temperatures

It has already been pointed out that the radiation from the primary reaction zone or inner cone is often much stronger than from burnt gases above. This is partly because some radicals like CH and C_2 are only present in the reaction zone, but the OH emission is also often much stronger from the reaction zone than from above. When metal salts are present in flames there is sometimes abnormally strong excitation of the atomic lines of the metal in the reaction zone. This emission of visible and ultra-violet radiation all involves excited electronic states of the atoms or molecules. Effective electronic excitation temperatures are usually measured by two methods, the spectrum-line reversal method and the line-ratio method.

In the spectrum-line reversal method the brightness of a background source giving a continuous spectrum is adjusted until the spectrum line in the flame is just at the reversal point, that is, it appears neither in absorption as a dark line against the bright background, nor in emission superposed on the continuum. It can be shown from Kirchhoff's law that at this point the effective excitation temperature of the spectrum line is the same as the brightness temperature of the background source, which can be measured with an optical pyrometer. In equilibrium the ratio of the population of atoms in the upper and lower energy states corresponding to the particular spectrum line will be $e^{-E/kT}$, where k is the Boltzmann constant, T the temperature and E the energy difference between the two states. Under non-equlibrium conditions the population in the upper state may be too high, giving an abnormally strong emission of the line and necessitating a higher temperature of the background source to reverse the line. The reversal temperature then gives an effective electronic excitation temperature for the particular electronic transition involved, and this temperature is really just a measure of the ratio of populations in the two electronic states.

With flames at atmospheric pressure the reaction zone is very thin and it is difficult to restrict the light beam to this zone and to prevent interference by the interconal gases or surrounding cooler layers of the flame. Thus work has mainly been done with low-pressure flames (Gaydon and Wolfhard, 1951a) using a special burner in which only the centre of the flame was coloured with added metal. Most complete measurements were made on acetylene/air flames, and results are summarized in Table VIII.1. This flame had an adiabatic temperature of 2340 K, but owing to heat losses to the burner the true temperature was probably rather lower; the reversal temperature in the interconal gases just above the reaction zone was about 2000 K but as there would be a little radiation depletion at this pressure the true temperature was probably about 2050 K (see *Flames*, page 252).

Generally, the reversal temperatures increase with increase of excitation energy of the line, and for lines in the ultra-violet the temperatures are far above the equilibrium flame temperature. The

TABLE VIII.1

Reversal temperatures for various metal lines in the reaction zone of a stoichiometric acetylene/air flame at a pressure of 24 torr

Line	Reversal temp, K	E/kT	Line	Reversal temp, K	E/kT
Na 5890—6	2010	12.1	Fe 2983.5	3050	15.7
Fe 3859.9	2565	14.4	Fe 2966.9	3068	15.7
Fe 3824.4	2580	14.45	Fe 2936.9	3100	15.7
Tl 3775.7	2350	16.2	Pb 2833.0	2970	17.0
Fe 3719.9	2650	14.5	Fe 2719.0	3310	15.9
Fe 3679.9	2630	14.8	Fe 2522.8	3365	16.8
Fe 3440.6	2880	14.4	Fe 2501.1	3410	16.75
Fe 3020.6	3058	15.5	Fe 2483.3	3450	16.7

The 3064 band of OH absorbed too weakly for accurate measurement, but the reversal point for low rotational levels was about the same as that for the Fe line 3020 (i.e. about 3058 K); higher rotational levels had a higher reversal temperature. The non-resonance line of Tl at 5350 reversed at 2750 K.

reality of the departure from equilibrium is shown by the observation that non-resonance lines of Pb show in emission even against the full carbon arc as background (~ 3800 K), although the resonance line at 2833 Å appears clearly in absorption on these same plates (Plate 6c). As a more quantitative statement of the anomaly, the population in the upper level of the Fe line at 2483 Å is 2700 times higher than that for equilibrium at the adiabatic flame temperature (2340 K) or 10^5 times that for the actual temperature (around 2050 K). The extent of the anomaly varies with the added metal; it is greater for iron than for thallium or lead. For Fe a plot of E/kT against E gives a fairly good straight line whose slope corresponds to the very high temperature of 8000 K; the line does not, of course, pass through the origin as it should for equilibrium.

Although the anomalous excitation is easiest to measure in low-pressure flames, it occurs also in flames up to at least atmospheric pressure, and probably does not change much with pressure, as shown by the results in Table VIII.2.

Reversal measurements on the reaction zones of low-pressure flames show that the anomalously high electronic excitation of added metals occurs strongly in premixed flames of hydrocarbons with air or oxygen, especially for acetylene. Methane shows the

TABLE VIII.2

Reversal temperature for the Fe line at 3020 Å in the reaction zone of stoichiometric acetylene/air flames at various pressures

Pressure, torr	Burner diameter mm.	Reversal temp., K
13.5	54	2900
35	27	3000
120	7.5	3010
200	3.6	3060
700	(0.6)	≥2990

effect, but less strongly than other hydrocarbons. Methyl alcohol also shows the anomaly quite definitely. Formaldehyde and carbon monoxide show little if any abnormal excitation, and low-pressure hydrogen flames seem free from it, the Fe lines and also the OH bands being uniform through the reaction zone and into the hot burnt gases. The effect is not, however, restricted to organic flames. Ammonia/oxygen and hydrogen/nitrous oxide flames show the anomaly quite strongly. Flames of cyanogen also appear to show the anomaly, but those of carbon disulphide do not show any obvious sign of it. In acetylene flames, rich mixtures tend to show the effect slightly more strongly than weak mixtures. In mixed flames such as methane/hydrogen/oxygen the extent of the anomaly appears to be proportional to the methane concentration; this contrasts with the light-yield studies which showed that the CH intensity varies as nearly the fourth power of the methane concentration.

In diffusion flames with oxygen at atmospheric pressure there was no obvious sign of abnormal excitation of added metals, although at low pressure the effect appeared strongly. In diffusion flames of hydrocarbons with air at 1 atm. there is probably a slight excess excitation near the base of the flame.

The line-ratio method involves the comparison of the intensity in emission of two or more lines of the same element, the excitation energies and transition probabilities for the lines being known. Self-absorption tends to weaken the emission and so strong resonance lines are unsuitable. The method is also experimentally rather more exacting; it is most suitable for high temperature sources (> 5000 K) where numerous lines are excited and reversal measurements are difficult because of the absence of a suitable background source. It is not generally to be recommended for measurement of flame temperatures. It does, however, have some advantages for the study of thin reaction zones, where the excitation temperature may be quite high and the thinness of the zone may restrict reversal measurements.

Iron is again the best metal for these line-ratio measurements, and Broida and Shuler (1957) studied several flames, introducing the iron as ferrocene $(Fe(C_5H_5)_2)$. For stoichiometric oxy-hydrogen at 1 atm. they found 2845 K above the reaction zone, that is a little below the adiabatic flame temperature of 3100 K, and slightly more than this, about 3325 K, in the reaction zone; possibly the hydrocarbon part of the ferrocene could have had some influence. For stoichiometric oxy-acetylene at 1 atm., the adiabatic temperature should be 3320 K, and they found 3200 K above the reaction zone and in the reaction zone excitation was non-thermal with line-ratios for lines of excitation energy below about 34000 cm^{-1} giving a temperature of 3600 K, but lines of higher energy being much more strongly excited. For low-pressure (2.1 torr) oxy-acetylene the anomaly in the reaction zone was even more marked, highly excited lines of Fe indicating temperatures around 10 000 K. These results thus tend to confirm those by the reversal method.

For the radicals C_2, CH and OH, which are present in the reaction zones, measurements are difficult, but there is plenty of evidence that the electronic excitation is also very high; this is one of the reasons why C_2 and CH are so difficult to observe in absorption.

For OH we have already seen that in C_2H_2/air the reversal temperature near the band head is about the same as that of Fe lines in the same spectral region, i.e. ~3058 K. Gaydon, Spokes and van Suchtelen (1960) found that in oxy-acetylene flames the low rotational lines of OH reversed at about 2990 K, and high rotational lines at up to 3440 K.

For C_2 in rich oxy-acetylene at low pressure Gaydon, Spokes and van Suchtelen compared the emission in the band head with that from a calibrated lamp, and used the values of the absorption to deduce an excitation temperature of 3200 K, compared with a flame temperature around 2500 K; it now seems likely that inadequate allowance was made for the limited resolving power of the spectrograph, in which case the excitation temperature could be higher. Bleekrode and Nieuwpoort (1965) measured the lower state

C_2 concentration* as $[C_2(x^3\Pi)] \sim 3 \times 10^{13}$ and at the flame temperature of 2500 K this should lead, in thermal excitation, to $[C_2(A^3\Pi)] \sim 6 \times 10^8$ whereas they calculate from Gaydon and Wolfhard's light-yield measurements on a similar flame that the actual $[C_2(A^3\Pi)]$ was between 10^{10} and 10^{11}; this would correspond to an excitation temperature around 4000 K. Similarly they found for CH that $[CH(x^2\Pi)] \sim 2 \times 10^{12}$ which would lead on thermal excitation to $[CH(A^2\Delta)] \sim 4 \times 10^6$, whereas they deduce between 10^9 and 10^{10} from the light yield, a value which would indicate a very high CH excitation temperature around 5100 K.

These very high excitation temperatures for the C_2 bands in the green and CH in the blue contrast with excitation temperatures of metal atoms which, for lines in the visible region, are not abnormally high. It seems likely that the strong excitation of C_2 and CH, and perhaps also OH, is due to these radicals being formed in the excited electronic state as a result of the chemical reactions, that is true chemiluminescence, whereas the excitation of metal atoms, which leads to abnormally strong emission of lines in the ultra-violet, involves some other process such as their acting as third body in some exothermic process or receiving excess vibrational energy from some newly formed species. These processes will be discussed again in Chapter X.

Translational Temperatures from Doppler Broadening

The factors which influence the breadth and shape of spectrum lines have been discussed at the end of Chapter IV. The Doppler effect and the random thermal motion of the molecules leads to a line breadth which depends on the square root of the absolute temperature. However, under ordinary flame conditions at atmospheric pressure, pressure broadening, which leads to wider wings to the lines, masks the Doppler broadening and prevents temperature measurements being usefully made from line contour. Self-absorption adds to the difficulty.

---

*These concentrations are higher than the more recent and probably better values by Bulewicz et al; this may be partly due to the higher pressure and partly due to the use of different oscillator strengths for C_2 and CH.

In low-pressure flames both pressure broadening and self absorption should be much smaller and Gaydon and Wolfhard (1949c)
explored the possibility of determining the translational temperature
from the line breadth for the reaction zone of oxy-acetylene flames.
A quartz Littrow-type spectrograph was crossed with a Fabry—Perot
interferometer to give the high dispersion necessary for the study of
line contour. Experimental details, methods of computation and
results are given in the original paper.

Most work was done with CH, especially the Q_2 (½) line of the
3900 Å system as this is one of the few lines free from overlapping.
For the brightest part of the reaction zone the line breadth
corresponded to a temperature of 4000 ± 400 K, compared with an
adiabatic flame temperature of 2700 K. A typical spectrum is shown
in Plate 5c. Measurements on oxy-acetylene flames between 1.9 and
14 torr gave about this value, and similar high temperatures were
indicated for the 4315 Å band. The translational temperature varied
slightly with mixture strength and also through the reaction zone
from 3000 K at the base to 4500 K at the top. Although the results
clearly indicated lack of translational equilibrium, the contours did
not depart measurably from the pure Doppler form. Recently G.
Adomeit (Combustion Institute European Symposium (Sheffield),
Ed. F. J. Weinberg, Academic Press, 1973, p. 47) has confirmed these
results.

In contrast to these results at low pressure, some measurements on
an oxy-acetylene flame at atmospheric pressure gave a lower
apparent temperature of only 2600 K; under these conditions a
closer approach to translational equilibrium must be established but
pressure-broadening effects are likely to modify the contour, and
later work by Harned and Ginsburg (1958) and Rank, Saksena and
Wiggins (1958) does indeed indicate fairly narrow lines but with a
non-Doppler contour due to pressure broadening at atmospheric
pressure.

From the high translational temperature at low pressure and the
more normal one at atmospheric pressure, it appears that the CH
radicals are formed in the excited electronic state and with high
translational energy. For this to lead to a broadening of the lines the
excited molecules must radiate before their excess kinetic energy has
been lost by collision. Assuming a normal gas-kinetic collision
cross-section this requires that the radiative lifetime of CH should be
quite short, around 10^{-7} sec.

For C_2 the rotational structure is too complex for studies with the resolving power available to Gaydon and Wolfhard. For OH at 3064 Å the reflectivity of the Fabry—Perot aluminized mirrors was too low to permit an accurate determination of contours. At low pressure the widths corresponded to a temperature of about 3500 K, again suggesting some lack of translational equilibrium, but possibly due to other broadening effects or lack of resolving power. At atmospheric pressure the rotational lines were definitely wider, corresponding to temperatures up to 4750 K, but this width is almost certainly due to self-absorption and collision broadening.

Measurement of Rotational Temperatures: OH

The interchange between rotational and translational energy of molecules is usually very rapid, ultra-sonic and shock-tube measurments of rotational relaxation times corresponding only to between 5 and 10 times the interval between collisions. The relative intensities of the lines in the rotational fine structure of a band can be used to derive the rotational energy of the emitting or absorbing molecules, and one might expect that this would give a good measure of the flame temperature. However, we shall see that for emission bands of molecules formed in the reaction zones this is often not so, presumably because the molecules are formed in the excited electronic state and with abnormal rotational energy and emit before the rotational energy can relax; in some cases collisions, instead of relaxing the rotational energy, will quench the electronic excitation and so prevent observation of emission from the rotationally relaxed molecules.

The intensity of an individual line of the rotational fine structure of a band depends on the transition probability P for that particular line, on the population of molecules in the initial state, and of course on instrumental conditions. We may write the intensity I of a particular line as

$$I = C \cdot P \cdot \nu^4 \, e^{-E_r/kT} \quad \text{for emission*, or as}$$

$$I = C \cdot P \cdot \nu \, e^{-E_r/kT} \quad \text{for absorption}$$

Here C is a constant which depends on the instrumental conditions; it usually varies only slowly with wavelength and is often

*This ν^4 factor is for the total intensity of the line; the central intensity varies as ν^3.

assumed to be constant over a single band. v is the wave number of the line, E_r is the rotational energy of the initial level of the line, k is the Boltzmann constant and T the temperature.

The transition probability P is the product of rotational, vibrational and electronic contributions. For a single band the electronic and vibrational contributions are usually assumed to be constant so that we need only consider the rotational part, although in some cases (e.g. OH, see page 100) there is in fact appreciable interaction between rotational and vibrational terms so that a correction is needed. The rotational intensity factor can be derived theoretically if the type of electronic transition is known; the formulae for a $^2\Sigma - ^2\Pi$ transition like that of OH have already been given (page 100) and full formulae are given by Kovacs (1969). The experimental methods and ways of computing the temperature are described fully in *Flames*.

Line intensities may be measured photo-electrically or photographically. The photo-electric method is usually simpler and more accurate but requires the flame to be kept at steady intensity throughout the time taken to scan the spectrum. For photographic work it is necessary to calibrate the plate, and this needs care as the variation of photographic density with intensity is often different for a continuous spectrum from that for narrow spectrum lines; flames always give some continuous background spectrum and this must be deducted from the intensity of the lines; if this lies below the threshold sensitivity of the plate it may cause serious error, and it may be necessary to take additional long exposures to bring the background above threshold so that it can be measured.

The transition probability P in any main branch of a band increases with rotational quantum number. The population term $e^{-E_r/kT}$ decreases rapidly with rotational quantum number because $E_r = B \cdot J(J + 1)$, or in case b coupling $E_r = B \cdot N(N + 1)$. Thus as we go, with increasing rotational quantum number, along a branch the line intensity increases, reaches a maximum and then falls. It is possible to calculate the temperature from the position of maximum intensity but the method is insensitive.

The equation for line intensity can, for emission, be rewritten in the form

$$\log_e I = \log_e C + \log_e(P \cdot v^4) - E_r/kT$$

so that if $\log_e I - \log_e(P \cdot v^4)$ is plotted against E_r we should obtain a straight line whose slope will give kT and hence T. This is the method which has been used most frequently. It has the advantage that any departure from a Maxwell–Boltzmann distribution of rotational energies shows up as deviations from the straightness of the line in the plot. A number of lines, not necessarily in the same branch, are used so that any errors due to blended or otherwise poor lines can be detected and eliminated. The method may be subject to errors due to self-absorption, and these are discussed later.

Another method, used by Dieke and Crosswhite (1962) and by Broida is known as the iso-intensity method. If in any branch the intensity is plotted against rotational quantum number we can select one weak line at the beginning of the branch and find another line towards the end of the branch which has the same intensity; it is possible to interpolate between lines so that the intensity coincidence may correspond to a hypothetical line of non-integral quantum number. If we denote the two lines by the subscripts a and b, then from the intensity formula, since the lines have equal intensity, we may, for emission, write

$$P_a \cdot v_a{}^4 e^{-E_{ra}/kT} = P_b \cdot v_b^4 e^{-E_{rb}/kT}$$

or

$$(E_{ra} - E_{rb})/kT \times \log e = \log (P_a v_a^4) - \log (P_b v_b^4)$$

from which T can be calculated. The method has two big advantages; it depends on intensity matching instead of measurement so that if the lines are not far separated in wavelength the results are independent of plate calibration and are unaffected by background continuum; to a first approximation it is unaffected by self-absorption because lines of equal intensity suffer equally from self-absorption. The method depends on estimates of intensity of only a few lines and may therefore be affected by errors due to blending in these lines. Departures from a Maxwell–Boltzmann distribution are less easily detected by this method than by the plotting method; when there is such a departure there are often two effective temperatures, one for rotational levels of low energy and one for high energy and then these temperatures can only be determined by the plotting method.

The most serious errors in measurement of rotational temperature are due to self-absorption and to non-uniform temperature across the flame.

In the plotting method the strongest lines in the centre of the graph are weakened most by self absorption and so fall below the correct straight line. Thus there is a dip in the centre of the curve and the slope of the first (low rotational energy) part of the line tends to be increased and to give too low a temperature, while the slope for higher rotational energy is reduced to give an apparent temperature above the true value. Results by the iso-intensity method are to a first approximation unaffected by self-absorption; however emission intensity is proportional to $P\nu^4$ and absorption to $P\nu$ so that if the lines of equal intensity have different ν the absorption error will not cancel. If the comparison is made between lines in the beginning and tail of a head-forming branch, like the R_2 branch of OH, the values of ν are nearly the same and this error is negligible. If the effective temperatures in emission and absorption are not the same then the two iso-intensity lines will be affected by different amounts.

When the temperature is not uniform through the flame, then if there are no complications from self-absorption, the rotational temperature for an emission band will be a mean value in which the hottest region of the flame will be weighted strongly because of its higher intensity in emission.

If we have both self-absorption and non-uniform temperature, it is very difficult to be sure of avoiding error. The hot central regions of the flame emit most strongly those lines of high rotational energy, while the cool outer layers absorb and weaken the lines of low rotational energy. Under these conditions both the plotting and iso-intensity methods fail unless self-absorption can actually be measured and corrected for.

The molecule which has been most studied for rotational temperature measurements is OH and the basic data are available in the paper by Dieke and Crosswhite (1962); this type of data has been used for most measurements in the past, but it is necessary for accurate work to correct the rotational transition probabilities for rotation-vibration interaction using data from Learner (1962) and Anketell and Learner (1967) as discussed on page 101. Most work has been done in emission and gives an effective rotational temperature for the $v = 0$ level of the $A^2\Sigma^+$ state. For steady flames there should be little difficulty in using the light-chopper technique to measure the band in absorption and obtain the temperature for the ground $X^2\Pi$ state.

To avoid difficulties due to self-absorption, which is strong in flames at atmospheric pressure, and to take advantage of the thicker reaction zone, measurements were made on various low-pressure flames by Gaydon and Wolfhard (1948, 1949a, 1949b) and by Broida, Shuler and colleagues.

The low-pressure hydrogen flame does not show a clearly developed reaction zone, but the weak OH emission was studied in both the flame front and above, in the "interconal" gases. The plotting method gave a good straight line, indicating a thermal energy distribution at a temperature just a little below the adiabatic value of 2520 K; there was no appreciable dip due to self-absorption.

For various organic fuels the OH emission is much stronger from the reaction zone than from the interconal gases. In all cases studied the interconal gases gave a straight-line plot corresponding to a temperature around or a little below the adiabatic flame temperature. For a few organic flames (formaldehyde, formic acid, methyl alcohol, moist carbon monoxide) the plot for OH emission in the reaction zone also was straight and indicated a reasonable temperature. Such a plot for formic acid is shown in Figure VIII.1. However, for most organic compounds the plot showed definite evidence of an abnormal intensity distribution. In some cases, such as acetylene/oxygen, (see Figure VIII.1) the plot was still straight but had a slope corresponding to an abnormally high temperature. In other cases, such as methane, the plot was curved, the part at low values of N', the rotational quantum number*, giving a moderate temperature, and the part at high N' giving again an abnormally high temperature. In some cases, such as the mixed acetylene/hydrogen flame shown in Figure VIII.1 the plot showed two straight sections, the first at low N' giving a temperature near the adiabatic value, and the second at high N' giving a very high value. Results for a number of flames are summarized in Table VIII.3. The striking difference in rotational structure of the OH band which occurs when a little acetylene is added to an oxy-hydrogen flame is shown in Plate 6, strips a and b.

---

*For doublet states in case b coupling, as for OH, the rotational energy is determined mainly by N rather than by J. This quantum number N was denoted K in the original papers and the previous edition of this book; N is now internationally accepted.

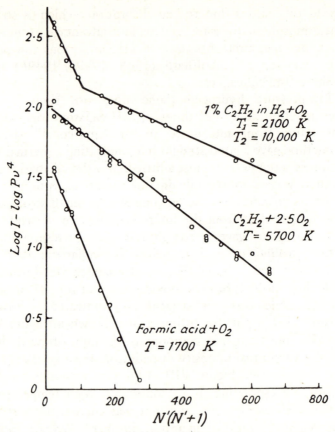

Fig. VIII.1. Plots of log I − log($P \cdot \nu^4$) against $N'(N' + 1)$ for OH emission in various low-pressure flames. $N'(N' + 1)$ is, with sufficient accuracy, proportional to the rotational energy E_r.

The variation of effective temperature with pressure is interesting and is summarised for stoichiometric oxy-acetylene in Table VIII.4.

In these low-pressure flames of acetylene the rotational temperature is higher at the base than at the top of the reaction zone; this was noted by Gaydon and Wolfhard, and by Broida and Heath (1957) who studied the temperature distribution through the reaction zone at 4.5 torr and found a narrow temperature maximum of 9000 K at the base, a flat region around 5500 K in the main part of the luminous zone and a fall to a much lower value at the top; Bulewicz, Padley and Smith (1970) included similar measurements.

It was suggested at one time (Penner, 1952) that these anomal-

TABLE VIII.3

Effective rotational temperatures for OH emission in the reaction zone of low-pressure flames (stoichiometric mixtures)

Fuel	Supporting gas	Pressure torr	Temp at low N' K	Temp at high N' K	Form of curve
H_2	O_2	7	2300	2350	Fairly straight
H_2	Air	39	2060	2060	Straight
C_2H_2	O_2	13	5700	5600	Straight
C_2H_2	O_2	1.5	(4600)	8750	Straight except at very low N'
C_2H_2	Air	27	2300	5700	Two straight parts
CH_4	O_2	8½	2800	7500	Curved smoothly
n-heptane	O_2	4	(2400)	5850	Two straight parts
CH_3OH	O_2	5	1900	2300	Fairly straight except v. high N'
CH_2O	O_2	6½	2140	2140	Straight
CHOOH	O_2	9	1720	1720	Straight
C_2H_5OH	O_2	4½	2760	6400	Two straight parts
Acetone	O_2	4½	(2200)	(4800)	Curved
Ethylene oxide	O_2	3	2840	7000	Two straight parts
CO (moist)	O_2	32	2440	2440	Straight
H_2	N_2O	9½	(2000)	3550	Curved
C_2H_2	N_2O	10	2840	3960	Curved
CH_4	N_2O	11	2760	2760	Fairly straight
CH_3OH	N_2O	10	2330	2330	Fairly straight
n-heptane	N_2O	14	2150	3300	Curved
Acetone	N_2O	14	(1870)	3100	Curved

ously high rotational temperatures for OH might be due to the self-absorption effect, but it is now quite clear that this is not so; direct measurement of absorption by a two-path method and against a background source giving fine OH lines have both conclusively proved (Broida and Kostkowski, 1955, 1956) that absorption is not the cause of the effect.

TABLE VIII.4

OH rotational temperatures in emission for the reaction zones of oxy-acetylene flames at various pressures

Pressure, torr	1.5	2.5	5.5	13	760
Temperature, K	8750	7000	6200	5700	5400

In flames at atmospheric pressure the danger of self-absorption is more real, but even so some striking results have been obtained by the iso-intensity method. In premixed flames with oxygen the observed rotational intensity anomalies are usually fairly small, but there is evidence of high temperatures in the inner cone, although these are often partially masked by thermal emission from the interconal gases. In oxy-acetylene the lower part of the inner cone is seen through a thin layer of interconal gas and a high temperature is recorded, but at the tip of the flame, which is seen through several mm of interconal gas, the temperature is nearer the adiabatic one. For flames with air the rotational temperature anomaly is easier to observe; for methane/air Broida found a temperature of 5200 K in the inner cone.

In low-pressure oxy-acetylene flames Gaydon and Wolfhard found that dilution with hydrogen increased the OH rotational temperature in emission. Some similar effects of dilution have been found by Kane and Broida (1953) at atmospheric pressure. Thus when oxy-acetylene is diluted with up to 90 per cent of argon and nitrogen, temperatures are around 5500 K. With an oxy-hydrogen flame containing 5 per cent acetylene and then diluted with 90 per cent of argon the temperature is 10 000 K. In a flame diluted with N_2 it has been shown that the radiation from the inner cone is 10^4 times that to be expected thermally. The effects appear to be partly due to the dilution suppressing the thermal radiation so leaving the chemiluminescence easier to observe, but special deactivation processes may be necessary to explain some of the results. Deactivation by normal $OH(x^2 \Pi)$ radicals would certainly be less in dilute flames. Dilution with oxygen does not have the same effect as dilution with N_2 or Ar.

There is obviously more than a single process responsible for the strong OH excitation. Thus formic acid and formaldehyde flames show strong OH but not high rotational temperature. Flames with N_2O show a different rotational intensity distribution from flames with O_2. The main chemiluminescent process, in organic flames with O_2, which gives the high rotational temperatures in $OH(A^2 \Sigma^+)$, appears to be associated with CH emission, which is always strong in these flames. The results can probably be explained (Gaydon and Wolfhard, 1948) by formation of electronically and rotationally excited OH by the reaction

$$CH + O_2 = CO + OH^*$$

followed by independent deactivation of the electronic and rotational excitations. The electronic deactivation occurs less frequently and then, of course, in a single step. The loss of rotational energy, however, may be only partial at each collision. At very low pressure, where there is no deactivation, the excited OH radicals radiate from the energy levels in which they were initially formed, displaying a very high effective temperature. It may appear a little curious that the plot gives such a good straight line, indicating a Maxwell—Boltzmann type of distribution of rotational energy, but it is likely that at high flame temperatures the reactions will lead to some distribution of energy in the products rather than to a selective formation in particular levels, and there seems no reason why this distribution should not be a probability distribution similar to the Maxwell—Boltzmann type. At high gas pressure most excited molecules will be quenched before they radiate, but for those that do radiate the average rotational energy will depend on how much of the initial rotation has been lost by collision before radiation, and this will depend on the relative efficiencies of electronic and rotational deactivation. At high pressure the rotational temperature should be independent of pressure. At pressures where the chance of electronic deactivation is comparable with the radiative lifetime, the effective temperature will decrease with increasing pressure. This pressure range seems to be around 1 to 10 torr.

Some fluorescence quenching studies on OH in low-pressure oxy-acetylene flame gases showed that the quenching efficiency was very high. First reports (Broida and Carrington, 1955, Carrington 1959a) appeared to indicate quenching at every collision, thus ruling out the above explanation of the pressure dependence. However later work (Carrington, 1959b) necessitated some reinterpretation of the earlier quenching experiments. When the flame gases are exposed to light from the Bi resonance line at 3067.7 Å the OH is excited to the $N = 11$ level of $A^2\Sigma^+$ but it was found that radiation from other rotational levels also occurred due to rotational energy changes during collisions; the rate at which molecules leave $N = 11$ by rotational transfer is found to be about twice the rate at which they are quenched electronically; this appears to be consistent with the above explanation of the pressure dependence of rotational temperature.

Values of rotational temperature of OH determined from the absorption spectrum give much lower values. Gaydon, Spokes and van Suchtelen (1960) made some preliminary measurements, by the iso-intensity method, for stoichiometric oxy-acetylene at 2 torr pressure and obtained a rotational temperature for OH $(x^2\Pi)$ of 2300 ± 300 K, which was a little below the adiabatic flame temperature and appeared to indicate that a true flame temperature was being measured. However, more careful measurements by Bulewicz, Padley and Smith (1970) using corrected rotational transition probabilities from Learner gave temperatures up to 3000 K for rich oxy-acetylene at low pressure. This throws some doubt on whether even the ground-state rotational temperature is strictly in equilibrium with translational energy; further work is desirable. OH is known to be an important chain carrier and it could just possibly be that it is formed with high rotational energy by an exothermic process and that its chemical removal in chain reactions is sufficiently rapid for its average rotational energy still to be slightly high.

Absorption measurements away from the reaction zone should, of course, give the true flame temperature. Measurements of this type have proved useful for detonations; thus Miyama (1962) has found a corrected OH absorption rotational temperature in good agreement with the calculated temperature of 3715 K behind detonations through an acetylene/hydrogen/oxygen mixture.

Rotational Temperatures for CH, C_2, NH etc.

The strong CH emission bands at 4315 and 3900 Å show open rotational structure, but because of the spin splitting and Λ-doubling it is necesary to use quite large dispersion for studies of rotational intensity distribution. The higher rotational levels of both bands are also affected by predissociation and are therefore unsuitable for measurements of temperature*. The 3900 Å band is affected most and many authors have restricted their measurements to the 4315 Å band. The results of early measurements (Gaydon and Wolfhard, 1949c) are summarized in Table VIII.5.

These values appear to be only slightly above the adiabatic flame temperatures, and, for low-pressure flames, values in general agree-

*Criticism by Dieke (1955) of results by Gaydon and Wolfhard is unjustified as only rotational levels below the predissociation limit were used; Dieke's empirically determined transition probablities for the predissociated levels seem inadmissible as the extent of the predissociation will depend on the degree of departure from equilibrium which varies with flame conditions.

TABLE VIII.5

Effective rotational temperatures for CH in emission

Flame	Pressure	T_{rot} 3900 Å K	T_{rot} 4315 Å K
Oxy-acetylene welding fl.	1 atm.	3200	3600
$C_2H_2 + 2\frac{1}{2}O_2$	1.8 torr	2460	2750
$C_2H_2 + 12$ air	12 torr		2420
Acetylene/N_2O	8 torr	2800	
Oxy-methane	18 torr	2670	
Oxy-methanol	8 tor		2400

ment with these have been confirmed (Broida and Heath, 1957; Bulewicz, Padley and Smith, 1970). Ndaalio and Deckers (1967) made a particular study of low-pressure methane flames and found that although the CH($A^2\Delta$) gave a statistical distribution of energy corresponding to a temperature of 2450 ± 150 K, this should not be regarded as a flame temperature because it was independent of pressure (2 to 35 torr), dilution (40 per cent N_2) or mixture strength; however, in ethylene flames there was some dependence on these parameters and they concluded that the excited CH was produced by a single process in the methane flame, but in the ethylene flame either two processes occurred, or the energy distribution in the initial reactants had some influence on the excited CH product.

In the mantle of rich oxy-acetylene flames Marr (1957) found a rotational temperature of 3350 K for CH 4315 Å, which is not far from the flame temperature.

Absorption by CH in low-pressure flames (Bulewicz, Padley and Smith, 1970) gave temperatures between 2000 and 2300 K, in good agreement with the adiabatic flame temperature and appreciably lower than the CH emission, and Bulewicz et al concluded that the excited CH was formed by C_2 + OH = CO + CH but the ground state CH($x^2\Pi$) was not formed in this way.

For the weak band of CH at 3143 Å the few measurements available, by Gaydon and Wolfhard, indicate a much higher rotational temperature in emission e.g. 5070 K in rich oxy-acetylene at 11 torr. This suggests a different excitation process from that for the other two bands.

For the Swan bands of C_2 the rotational structure is relatively close and the transition is between triplet states with a small spin splitting. However, the alternating missing lines due to the homo-nuclear character of the molecule simplify the spectrum a little by removing the Λ-doubling, so that with high dispersion the energy distribution among the higher rotational levels can be studied.

Measurements by Gaydon and Wolfhard (1950) indicated a very high temperature for C_2 in emission, this varying between 4950 K for oxy-acetylene at 1 atm. to rather lower values of 3800 K at low pressure. Later work has confirmed that the rotational temperature is high, but gives a different pressure dependence. Broida and Heath (1957) found temperatures between 3800 and 5000 K in low-pressure oxy-acetylene, the variations being with mixture strength and position in the reaction zone. Bleekrode (1966) found over 6000 K at very low pressure. Ndaalio and Deckers (1967) found 6500 K, independent of fuel and mixture strength, for various flames at below 3 torr, but found that the rotational temperature fell with increasing pressure. The latest work by Bulewicz, Padley and Smith (1970) shows that the plot of $(\log I - \log Pv^4)$ against $N'(N' + 1)$ for C_2 is not straight, but can be divided into two fairly straight regions, one below $N' = 50$ and the other above $N' = 50$ and that the slopes and temperatures for these parts of the line are different and have a different pressure dependence. Table VIII.6 summarizes their results.

This reveals a rather complicated picture. Most probably the situation is basically similar to that for OH. Direct chemi-luminescence results in formation of electronically excited C_2 with a very high but non-thermal distribution of rotational energy, which is revealed at low pressure. At higher pressure collision processes modify this rotational intensity distribution but, because the

TABLE VIII.6

Effective rotational temperatures for C_2 in emission
from oxy-acetylene flames

Pressure torr	Temperature for low rotational levels, $N' < 50$	Temperature for high rotational levels, $N' > 50$
3	7100 K	4800 K
7.5	6300	5100
13	6300	5200
20	6100	6000
760	4100	6300

collision processes for conversion of rotational and translational energy are more effective at low than at high energy, the distribution of rotational energy changes in this rather strange way.

Bulewicz et al were also able to measure the rotational temperature from the absorption bands in low-pressure rich oxy-acetylene flames. Temperatures appear to vary between 2900 K for very rich ($\lambda = 0.3$) and 2400 K for less rich flames ($\lambda = 0.7$); these

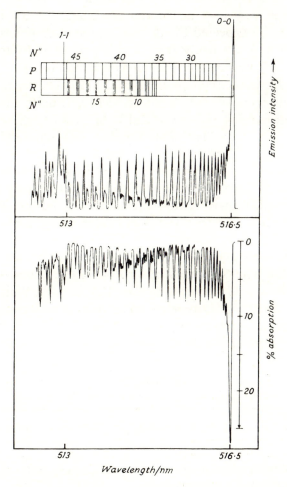

Fig. VIII.2. Traces of the emission and absorption spectrum of C_2 for a rich oxy-acetylene flame ($\lambda = 0.5$) at a pressure of 5 torr. Note that in emission lines P(40) to P(45) are much stronger than R(13) to R(17) but in absorption these lines are of comparable strength due to the lower effective temperature.

values are not far from the adiabatic flame temperature, but the variation with mixture strength is rather puzzling. The real difference between temperatures in emission and absorption is shown in their traces of the C_2 (0,0) band, reproduced in Figure VIII.2

In the mantle of rich oxy-acetylene flames at 1 atm. the C_2 emission temperature is not far from the adiabatic flame temperature of about 3300 K; Marr (1957) found 3000 K and Jessen and Gaydon (1969) found 3200 K. The latter also found that the C_2 concentration, although very high in the reaction zone, for the mantle was about that to be expected for thermal equilibrium with solid carbon particles, so it seems likely that in this part of the flame thermal excitation of C_2 has replaced the chemiluminescence.

In some flames containing nitrogen compounds, or supported by oxides of nitrogen, the NH band at 3360 Å is quite strong. The Q branch is unresolved, but the P and R branches form well resolved triplet branches (see Plate 3e). Gaydon and Wolfhard measured effective rotational temperatures in emission for several flames and some results are given in Table VIII.7. The values are all rather above the adiabatic flame temperatures, although not by very much; it is not clear whether there is true chemiluminescence or whether the high excitation is due to transfer of vibrational energy from other molecules.

Some of these flames also show the γ system of NO and the Violet system of CN. The NO bands lie far down in the ultra-violet where plate sensitivity and contrast is low; Gaydon and Wolfhard made some approximate measurements which appear to indicate again an anomalously high temperature, above 4000 K, in the reaction zones. Rotational temperatures for CN emission have frequently been used

TABLE VIII.7
Effective rotational temperatures for NH emission

Flame	Mixture strength	Pressure	Temp. K
C_2H_2/N_2O	Stoich.	1 atm.	3760 ± 300
C_2H_2/N_2O	Stoich.	11.5 torr	3460
C_2H_2/N_2O	Rich, $\lambda = 0.6$	11 torr	3760
C_2H_2/N_2O	Weak, $\lambda = 1.6$	18 torr	3100
H_2/N_2O	Stoich.	1 atm.	3500
NH_3/O_2	Stoich.	35 torr	3060
NH_3/N_2O	Stoich.	32 torr	2950

for arc and plasma sources and in astrophysics, but little has been done in flames. For the cyanogen/oxygen flame at 1 atm. Thomas, Gaydon and Brewer (1952) found that the CN rotational temperature in the reaction zone was only 3150 K, compared with C_2 emission at 4700 K and an adiabatic temperature for this exceptionally hot flame of 4850 K; this is a rare example of an effective rotational temperature in the reaction zone being below the adiabatic temperature. However, in $C_2 N_2 /O_2 /N_2$ flames at low pressure Bulewicz, Padley and Smith (1973) have found a rotational temperature for CN in emission of 5500 to 6500 K, about the same as the C_2 rotational temperature and the CN and C_2 vibrational temperatures; in absorption the rotational temperature for the CN ground state was close to the flame temperature.

The Schumann—Runge bands of O_2 have open rotational structure and should be suitable for temperature studies. Their emission is usually rather weak and comes from the hot product gases rather than the reaction zone, so measurements would probably yield a genuine flame temperature rather than information about non-equilibrium processes in the reaction zone.

Effective Vibrational Temperatures

Quantitative measurements of vibrational intensity distribution and effective vibrational temperature are more difficult than the corresponding rotational studies. The basic methods and limitations of determining vibrational transition probabilities have been discussed in Chapter IV (page 79). In nearly all actual band systems there is appreciable overlapping of bands so that it is not possible to pick out bands due to particular vibrational transitions and to measure their total intensities. In practice one may either measure the intensity of individual lines of the rotational fine structure using high dispersion, or, alternatively, the apparent intensity under small dispersion of some particular feature of each band, such as the band head or the maximum intensity in a close Q branch. In general the former method will be the more reliable, but difficulty may arise if the effective rotational temperatures in the vibrational levels being studied are not the same; thus Broida and Heath (1957) found that the rotational temperature in the CH (0,0) band was, for low-pressure oxy-acetylene, usually about 2700 K, but that of the (1,1) band only 2100 K; in this case the full rotational structure was resolved and they were able to measure the effective vibrational temperature in the upper electronic state and this usually came close to the rotational temperature of the (0,0) band. When comparison is made between band features, such as heads or Q maxima, theoretical intensity factors cannot easily be used; either experimentally determined factors must be employed or the theoretical factors must be corrected for the distribution of rotational structure in the feature, as this will vary from band to band and perhaps also with rotational temperature. Despite these major difficulties a number of values of effective vibrational temperature have been published since the previous edition of this book.

For OH we have seen (page 110) that the predissociation and its inverse preassociation produce population anomalies in the vibrational levels 2 and 3 of the $A^2\Sigma^+$ state. These anomalies are particularly important in hydrogen flames at low temperature. Shuler (1950) found a reasonable vibrational temperature, in agreement with the adiabatic flame temperature, for the outer cone of a methane flame at atmospheric pressure, but for the inner cone of an acetylene flame Shuler and Broida (1952) reported an excess population in $v' = 3$ and some underpopulation of $v' = 1$ compared with that to be expected at the flame temperature.

For C_2 and CH there seems to be a general tendency for the effective vibrational temperature to be very close to the rotational temperature for the same electronic state. This is not really surprising; in an exothermic reaction the excess energy tends to be fairly distributed between the various possible forms. Thus for CH in low-pressure oxy-acetylene Broida and Heath (1957) and Bulewicz, Padley and Smith (1970) found a vibrational temperature for $CH(A^2\Delta)$ of about 2800, in agreement with the rotational temperature, and the latter authors also found for the ground state $CH(x^2\Pi)$ a vibrational temperature of 2300 ± 100 K. They also found that for C_2 upper state vibrational temperatures tended to lie between the two values, at high and low N', for rotational temperature, and for the ground state vibrational temperatures were between 2800 and 30000 K; they attribute very high values over 10 000 K quoted by Bleekrode to use of an unsatisfactory method involving integrated band intensities. At atmospheric pressure Jessen and Gaydon (1969) found a high vibrational temperature (~ 7000 K) for upper state C_2 in the reaction zone, and slightly high (3900 K) in absorption for the ground state. In the mantle region the vibrational temperature for C_2 (A $^3\Pi$) was 300 to 600K above the adiabatic flame temperature, but the ground state C_2 (x $^3\Pi$) gave 3200 K in agreement with the adiabatic value.

Vibrational temperatures for CN emission have been measured in the very hot cyanogen/oxygen flame by Thomas, Gaydon and Brewer (1952). The full temperature, close to the adiabatic value, of 4800 K was only attained about 2 mm above the visible flame front. Close to the reaction zone the effective vibrational temperature was only 3150 K, the same as the CN rotational temperature in this region.

Predissociation

The use of predissociation as a test for departures from equilibrium has been noted in Chapter IV, page 84.

Observation of predissociation indicates a departure from local thermodynamic equilibrium. Suppression of the predissociation may indicate either that equilibrium is established and excitation is "thermal" or that some other process, such as collision deactivation is dominant in controlling the relative populations in the excited states so that the predissociation has a negligible effect in disturbing the population distribution.

Higher rotational levels of all the main CH bands show pre-dissociation. In the (0,0) band of the 3900 Å system it sets in at $N' = 15$, and in the spectra of discharge tubes and low-pressure flames all lines arising from $N' \geqslant 15$ are weak or absent. In the inner cones of organic flames at atmospheric pressure the phenomenon is shown less strongly and lines are usually observed out to $N' = 17$. Durie (1952a) found that in a rich ($\lambda = 0.4$) hydrogen/air flame to which a little ethylene or other organic fuel had been added, the predissociation was much less noticeable, lines being observed out to $N' = 20$. In the (1,0) and (1,1) bands of this system the pre-dissociation usually causes termination of the branches beyond $N' = 7$, but Durie found lines to $N' = 13$ in the hydro-gen/air + ethylene flame. These rich hydrogen/air flames are now known to contain an excess population of H atoms, and this presumably suppresses the predissociation in some way.

Clearly in organic flames CH excitation is chemiluminescent and the population in excited electronic states is much greater than for equilibrium; the life for predissociation is much shorter than the radiative lifetime so those levels affected by predissociation are weakened and radiate less. At atmospheric pressure collision processes are important, the collision lifetime being less than the radiative life but still longer than the time for predissociation. In the mantle of rich oxy-acetylene flames we have seen that there is a closer approach to local thermodynamic equilibrium and Marr (1957) notes that in this region the CH predissociation is partly but not completely suppressed. Even here CH is not in complete equilibrium.

The predissociation in OH has already been mentioned several times. In the reaction zone of organic flames Gaydon and Wolfhard (1951b) found that the highest rotational levels of $v' = 1$ and all levels of $v' = 2$ were weakened by predissociation. Again the effect is strongest in low-pressure flames, but still observable at atmospheric pressure. The radiative lifetime for OH is known to be about 10^{-6} sec and at low pressure the collision lifetime is also of this order, but at atmospheric pressure the collision lifetime becomes around 10^{-8} sec and is then obviously becoming comparable with the lifetime for this weak predissociation. However, it is still clear that even at atmospheric pressure the OH excitation in the reaction zone is non-thermal and must be due to chemiluminescence.

Flash Photolysis

The technique of flash photolysis has been described in Chapter III, page 54. Since the previous edition of this book numerous papers on flash photolysis have appeared and valuable new information about the spectra of simple radicals, about energy transfer processes and about flash-initiated combustion reactions have been obtained. Here some results on combustion of organic compounds will be discussed.

Explosions of acetylene and other hydrocarbons with oxygen, sensitized by NO_2 and initiated by flash photolysis have been studied by Norrish, Porter and Thrush (1953, 1955) and Norrish (1953). After an induction period of a few milliseconds the first absorption bands to appear were always those of OH. This indicated that OH radicals took part in the initial chain reactions in which the other radicals were not formed in significant quantities. This was followed by a period in which all the other radicals observed — C_2, CH, CN, NH, and C_3 — increased rapidly in concentration. The CN radical gave the strongest absorption bands and was followed in explosions

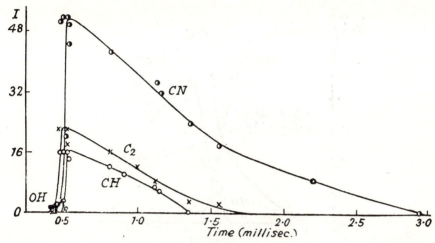

Fig. VIII.3. Time dependence of radical concentrations in the explosion of oxy-acetylene initiated by flash photolysis. Mixture: C_2H_2 13 torr, O_2 10 torr, NO_2 1.5 torr.

of acetylene, ethylene, ethane and methane. C_2 and C_3 were only detected with acetylene explosions, and CH only with acetylene and ethylene. Figure VIII.3 shows the time dependence of some of these radical concentrations in a rich acetylene/oxygen explosion (from Norrish, 1953). Generally it seems that the C_2, CH, CN and C_3 varied together, although in this curve there is some indication of C_2 preceding CH. We may recall that in low-pressure flames the C_2 *emission* definitely precedes CH.

In rich mixtures (richer than equimolecular for C_2H_2) the carbon radicals gave strong absorption while the OH was quite weak, but in weaker mixtures the carbon radicals were suppressed and the OH was strong and sometimes continued to increase in strength after the carbon radicals had passed their peak; this was presumably due to further reaction of oxygen with hydrogen which was formed in the break-up of the hydrocarbon. It may be noted that the OH absorption in acetylene explosions had about the same strength as in oxy-hydrogen explosions; this is further proof that the strong OH emission from acetylene flames, compared with that in hydrogen flames, is due to chemiluminescence and not to excess concentration of OH. Figure VIII.4 shows the variation of the maximum concentrations of the radicals in oxy-acetylene as a function of mixture strength.

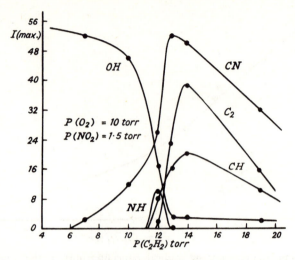

Fig. VIII.4. Effect of mixture strength on maximum radical concentrations in flash-initiated oxy-acetylene explosions.

It seems that the OH concentration could not rise high while carbon radicals were present, but that as soon as there was an excess of oxygen the carbon radicals were rapidly removed and the OH concentration rose. Addition of water or CO_2 produced similar effects to those of adding oxygen, one molecule of H_2O or CO_2 being equivalent to nearly $\frac{1}{2} O_2$.

Appreciable continuous absorption accompanied the radical absorption and its intensity distribution is shown in Figure VIII.5. Immediately after the flash there was weak continuous absorption, rather stronger around 4000 Å than at shorter wave-lengths; this always occurred and was probably due to residual NO_2. This was followed by an absorption with peak intensity around 3800 to 3900 Å, which reached its maximum strength at about the same time as the carbon radical absorptions and was probably related to C_3 and to the continuous emission (Marr and Nicholls, 1955) and absorption (Jessen and Gaydon, 1969) in the mantles of rich oxyacetylene flames. This was followed by a more uniform continuous absorption which increased in intensity towards shorter wavelengths; this was associated with carbon formation, but it is not known whether it is actually due to small carbon particles or whether it is related to the pyrolysis continuum observed by Wolfhard and Parker and by Laud and Gaydon (1971) in diffusion flames. These

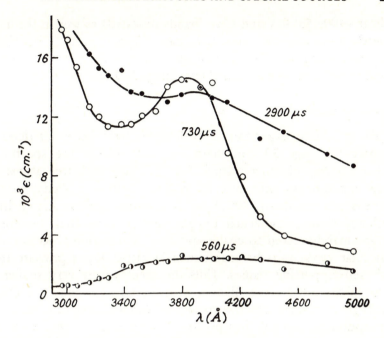

Fig. VIII.5. Continuous absorption against wavelength observed at three times during flash initiated oxy-acetylene explosions. Mixture: C_2H_2 19 torr, O_2 10 torr, NO_2 1.5 torr.

last two continua were strongest with acetylene explosions, but were also observed, with similar characteristics, in explosions of rich mixtures of other hydrocarbons.

The flash-initiated combustion of formaldehyde, formic acid and acetaldehyde with oxygen have been studied (McKellar and Norrish, 1960) without the need to use NO_2 as sensitizer. The absorption spectra show bands of formaldehyde, OH, the HCO radical (Herzberg and Ramsay's bands at 5624 and 6138 Å) and a continuum 2200–2450 Å. By controlling the initial temperature of the mixture and the mixture strength it was possible to study both slow combustion processes and explosive reaction. In the slow combustion the HCO bands were absent or very weak due to rapid removal by reaction of HCO with O_2. In explosive combustion of rich mixtures the HCO absorption was stronger but of short duration due to thermal decomposition of HCO to H + CO.

In a recent study (Petrella and Sellers, 1971) of the flash initiated ignition of styrene, $C_6H_5CHCH_2$, absorption bands of the C_6H_5

radical 4400—5300 Å and CH_2 (bands unstated) as well as the usual radicals C_2, C_3, OH and CH, have been reported, but little information is given. Flash-photolysis studies of the role of lead ethyl in reducing knock are discussed later (page 290).

Atomic Flames

Again the basic experimental techniques have been described in Chapter III, page 50, and here we shall discuss the spectra of hydrocarbons and other organic fuels reacting with free atoms. In early experiments with atomic oxygen (Geib and Vaidya, 1941; Gaydon and Wolfhard, 1952) the free atoms were generated by a discharge through molecular oxygen, and it seems that the atomic oxygen mainly served to enable ordinary combustion processes with molecular oxygen to be maintained at rather lower pressure than with self-supporting flames. Thus the spectra were very similar to those of the low-pressure flames, with the OH, CH and C_2 emission bands giving similar effective rotational temperatures to those in the corresponding normal low-pressure flames; and high electronic excitation of added metal atoms, such as Fe, was again obtained. The occurrence of these flames indicated that reaction of atomic oxygen with fuels, other than methane, could proceed with low activation energy. Chemical sampling (Ferguson and Broida, 1955) revealed mainly normal combustion products (CO, CO_2, H_2O, H_2) with traces (especially in rich mixtures) of intermediary products such as formic acid, formaldehyde and methane; there was a little waxy residue which was soluble in water and which released iodine from potassium iodide and might therefore be a peroxide. There was no sign of polymerization of hydrocarbons.

Using a titration method ($NO + N = N_2 + O$) to produce atomic oxygen which was practically free from molecular oxygen, Krishnamachari and Broida (1961) found that the flame with acetylene still gave strong C_2 and CH emission but that the OH emission was now very weak, and that on deliberate addition of a little molecular oxygen the OH emission was considerably strengthened (in their experiments by a factor of about 7), while the CH and C_2 were weakened by a factor of around ten, i.e. a relative stengthening of OH compared with CH of about 70 times. This is strong support for chemiluminescent excitation of OH by

$$CH + O_2 = CO + OH^*$$

In the flame with atomic oxygen which was free from O_2, the effective rotational temperatures of C_2 and OH were still very high. The vibrational temperature for C_2 emission was also very high (8500 K); the rotational temperature of CH (1900 K) and vibrational temperatures of CH (1500 K) and OH (1100 K) were also far above the gas temperature; the CN rotational temperature was only 400 K.

Observations on $O + C_2H_2$ flames have been extended into the infra-red by Clough, Schwartz and Thrush (1970) with the important result that highly vibrationally excited CO molecules have been found. They mainly studied the first overtone bands, $\Delta v = 2$, around 2.5 μm. With excess of atomic oxygen ($O/C_2H_2 = 20$) they found emission from all vibrational levels up to $v = 14$ of ground-state CO and attributed this to the reaction

$$O + C_2H_2 = CO + CH_2 + 200 \text{ kJ/mole.}$$

With a higher proportion of fuel ($O/C_2H_2 = 1.8$) they found excitation of still higher vibrational levels, up to $v = 33$, requiring an energy of 680 kJ/mole. This was attributed to the reaction

$$O + CH = CO + H + 736 \text{ kJ/mole.}$$

Their tracings of the infra-red spectrum are reproduced in Figure VIII.6. This direct observation of the formation of carbon monoxide

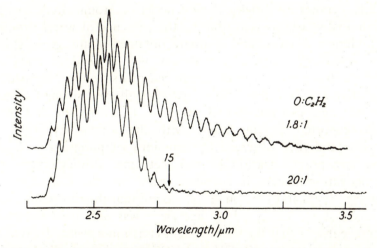

Fig. VIII.6. Infra-red chemiluminescence of CO in the atomic flame $O + C_2H_2$. The O/C_2H_2 ratios are indicated. The arrow on the lower trace indicates the origin of the $15 \to 13$ band.

molecules with very high vibrational energy probably provides an explanation of high metal-excitation temperatures. Since the reaction $O + CH = CHO^+ + e^-$ has been used to explain chemi-ionization, the link between the two phenomena is also indicated.

These flames of atomic oxygen and hydrocarbons (especially C_2H_2) also lead to emission of far ultra-violet Fourth Positive bands of CO. Marmo, Padur and Warneck (1967) used flames of acetylene with atomic oxygen produced both by discharge through $O_2 + Ar$ and from the $NO + N$ titration and found that the vacuum-ultra-violet spectrum showed bands of the $A^1\Pi - X^1\Sigma$ system with v' up to 7, although the strongest bands had low v'. There was also some ultra-violet emission down to 1260 Å requiring an energy of nearly 10 e.V. Observations on other flames of atomic oxygen with carbon suboxide (Becker and Bayes, 1968) and with C_2F_4 (Fontijn et al, 1970) also show this CO Fourth Positive emission, which has been attributed to the reaction

$$C_2O + O = CO(A^1\Pi) + CO(X^1\Sigma)$$

In acetylene flames the C_2O may be produced by

$$O + C_2H_2 = C_2O + H_2$$

and, for excitation to nearly 10 e.V., some vibrational energy may be retained by the C_2O until it reacts with O.

Early reports of bright flames between atomic hydrogen and various hydrocarbons were due to the presence of water vapour in the hydrogen. Gaydon and Wolfhard obtained a faint green glow at times when apparently dry hydrogen was allowed to react with acetylene, but were not quite convinced of its reality. The spectrum of the flame with slightly moist hydrogen showed strong C_2 bands, some CH and also the 4050 Å group of C_3. Ferguson and Broida (1955) reported a faint flame with dry atomic hydrogen and acetylene and they observed C_2 emission; sampling again failed to reveal any polymerization of the acetylene, although very rapid polymerization occurs if acetylene gets into the discharge tube. However, Bayes and Jansson (1964) found that the recombination of hydrogen atoms catalyzed by acetylene was definitely not chemi-luminescent and that the residual light emission usually observed was due to oxygen atoms. O, OH and HO_2 were individually generated and added to the $H + C_2H_2$ and it was found that only the oxygen atoms caused luminescence. With moist atomic hydrogen the

PLATES

DESCRIPTION OF PLATE 1

(a) Spectrum of flat diffusion flame of hydrogen burning with oxygen, on a medium quartz spectrograph. This is of the region rather to the hydrogen side of the stoichiometric point and shows only OH bands.

(b) As above, but towards the oxygen side. This shows weak Schumann-Runge bands of O_2 as well as OH.

(c) Spectrum of flame of O_2 burning in an atmosphere of H_2 at a quartz tube. Taken with a large-aperture glass prism spectrograph on an Ilford long-range spectrum plate. Comparison spectrum below of Ne lamp containing He.

(d) Premixed hydrogen/nitrous oxide flame at 10 torr pressure taken with small quartz spectrograph. This shows the " ammonia α-band " of NH_2 and the unusual vibrational intensity distribution of OH.

(e) Premixed carbon monoxide/air flame at reduced pressure. Medium quartz spectrograph and plate of high contrast. This shows the CO-flame bands (CO_2). Fe arc comparison below.

PLATE 1

DESCRIPTION OF PLATE 2

(*a*) Spectrum of small jet of H_2 burning in air; small quartz spectrograph. This shows the (1, 0), (2, 1) and (3, 2) bands of OH with nearly equal strength.

(*b*) Premixed H_2/O_2 flame at 7 torr pressure. This shows unusually strong OH (2, 1), especially at the top of the flame.

(*c*) Electric discharge from a transformer through the above flame. This shows a normal development of the OH bands, with the (1, 0) band relatively strong.

(*d*) Spectrum of blue glow of CO reacting with atomic oxygen (from a discharge tube). Quartz Raman spectrograph. This shows the CO-flame bands well developed and the continuum absent.

(*e*) Cool-flame type glow of rich propane/air in tube heated to about 560° C. Raman quartz spectrograph on Zenith plate. This shows Eméleus's cool-flame bands of formaldehyde. Comparison spectrum below of Ne lamp containing He.

(*f*) As above, in preignition glow region at about 610° C. This shows hydrocarbon flame bands (HCO) and a little CH and OH as well as formaldehyde, but no C_2.

(*g*) Normal propagating hot flame through same propane/air mixture as above two spectra. This shows strong C_2 Swan bands, CH and OH, with only weak HCO and no formaldehyde.

PLATE 2

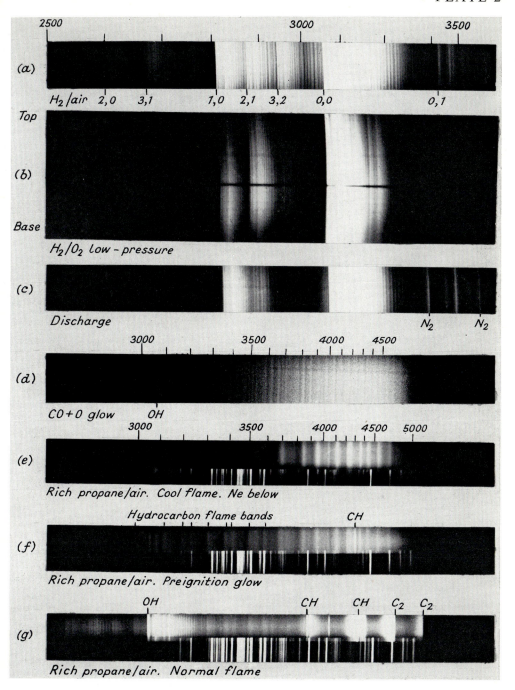

DESCRIPTION OF PLATE 3

This plate shows the rotational fine structure of the diatomic hydride bands found in flame spectra. All spectra are of the reaction zone of low-pressure flames and were taken with the large $1\frac{1}{2}$-prism Littrow quartz spectrograph.

(a) The (0, 0) band of **OH** at 3064 Å. This is due to a $^2\Sigma - {}^2\Pi$ transition. Some lines or heads of the six main branches are indicated. Since the $^2\Pi$ state is near Hund's case a coupling, the lines are indicated by their J values, rather than by N values, as in Dieke and Crosswhite's tables.

(b) The $^2\Delta - {}^2\Pi$ band of **CH** near 4315 Å. This is near Hund's case b coupling and the N values of a few lines are indicated. The P, Q, and R branches are each split into four components by the spin doubling and the Λ doubling. At low rotational quantum numbers, as in P4, the Λ doubling is negligible but the spin splitting is appreciable, while at higher N as in P12, the Λ doubling is considerable while the spin doubling is unresolved. Fe arc comparison above.

(c) The $^2\Sigma - {}^2\Pi$ band of **CH** near 3900 Å. The N values of a few lines are marked. The spin doubling is clearly resolved in lines of low N, such as Q5 and P4, but becomes small at higher N. Note the sudden termination of the P branch at P16 due to predissociation.

(d) The $^2\Sigma - {}^2\Pi$ band of **CH** at 3143 Å. This is hardly degraded at all, so that the Q branch appears as a single intensity maximum. N values of a few lines are indicated. There is some overlapping **OH**.

(e) The $^3\Pi - {}^3\Sigma$ band of **NH**. The spin splitting into triplets is clearly visible; it decreases with increasing N. The Q branch is unresolved. Fe arc comparison below.

PLATE 3

DESCRIPTION OF PLATE 4

(a) Inner cone of slightly rich premixed ethylene/air flame, photographed with a small quartz spectrograph.

(b) Very weak acetylene/oxygen flame diluted with CO_2. Medium quartz spectrograph. This shows Vaidya's hydrocarbon flame bands of HCO relatively strongly, with OH and CH weak and C_2 almost suppressed.

(c) Absorption spectrum, taken with high-pressure xenon lamp as background, of the interconal gases of a slightly weak ethylene/oxygen flame. Large $1\frac{1}{2}$ prism quartz Littrow spectrograph. This shows many lines of the complex Schumann-Runge system of O_2. Fe arc comparison below.

(d) Absorption spectrum, taken with hydrogen discharge tube as background, obtained during the slow combustion of propane. Medium quartz spectrograph. The absorption bands are due to formaldehyde, and there is some continuous absorption at shorter wavelengths. I am indebted to Sir Alfred Egerton and Dr. G. H. Young for this spectrum.

(e) Inner cone of premixed cyanogen/oxygen flame. Medium quartz spectrograph. The upper spectrum is of the flame of the carefully dried gases, taken with an exposure of only 2 secs. The lower spectrum is of moist gases (the O_2 being bubbled through water at 80°C.) with an exposure of 1 min. The latter shows CH, NH and OH bands and it is especially notable that the moisture suppresses the C_2 and CN radiation; even with the thirty-fold increase in exposure the C_2 bands are much weaker. I am indebted to Dr. G. Pannetier for this spectrum.

PLATE 4

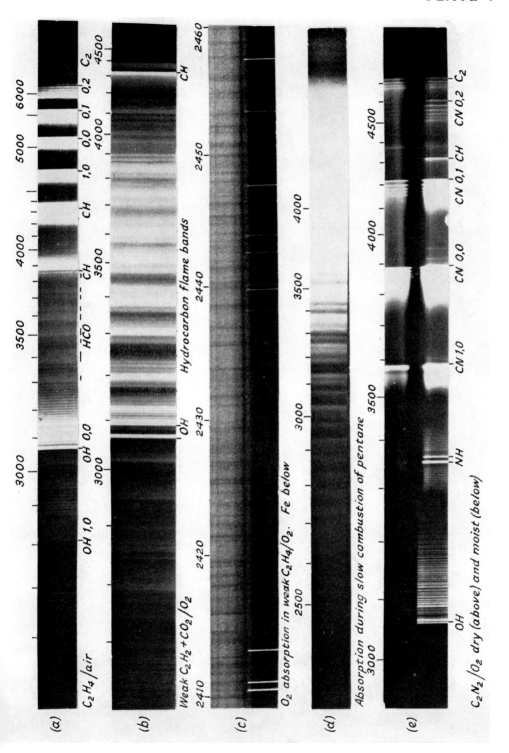

(a) C_2H_4/air

OH 1,0   OH 0,0   HĊO   CH   CH   1,0   0,0   0,1   0,2 C_2
3000   3500   4000   5000   6000

(b) Weak $C_2H_2+CO_2/O_2$

Hydrocarbon flame bands

OH   CH
2410   2420   2430   2440   2450   2460

(c) O_2 absorption in weak C_2H_4/O_2. Fe below

2500   3000   3500   4000

(d) Absorption during slow combustion of pentane

3000   3500   4000   4500

(e) C_2N_2/O_2 dry (above) and moist (below)

OH   NH   CN 1,0   CN 0,0   CN 0,1   CH   CN 0,2 C_2

DESCRIPTION OF PLATE 5

(a) Spectrum of low-pressure oxy-acetylene flame photographed with a quartz Littrow spectrograph crossed with a Fabry-Perot interferometer. For suitable lines it is possible to measure the line contour, and hence to estimate the Doppler broadening and effective translational temperature. For the 3900 band of CH most lines are double, but the $Q_{2\frac{1}{2}}$ line is single and suitable for measurement. For the 4315 band all lines are double due to spin, but with suitable choice of interferometer spacing the two components of lines in the R branch may either be resolved or exactly superposed, to enable measurement.

(b) Spectrum, on medium quartz instrument, of glow of products of discharge through water (H_2O) reacting with heavy acetylene (C_2D_2). The volume flows of water and acetylene were equal, but the spectrum shows mainly CH and OH bands.

(c) Glow of products of discharge through D_2O reacting with C_2H_2 similar to previous strip. This shows CD and OD. These plates indicate that in this atomic flame the H atoms in CH and OH originate from the water, not from the acetylene.

(d) Flame of methyl alcohol with O_2 containing a little $Fe(CO)_5$ at 12 torr pressure, on medium quartz spectrograph. The upper part of the spectrum is of the interconal gases and shows only weak OH and the resonance lines of Fe. The lower part of the spectrum corresponds to the reaction zone of the flame and shows strong OH and CH and strong abnormal electronic excitation of the full spectrum of Fe I.

(e) Spectrum of tip of " feather " of rich oxy-acetylene welding flame. Medium glass spectrograph. The enlargement was made with a cylindrical lens on paper of high contrast. This shows the 4050 Comet-head group of C_3 unusually free from overlapping by the weaker CH bands.

PLATE 5

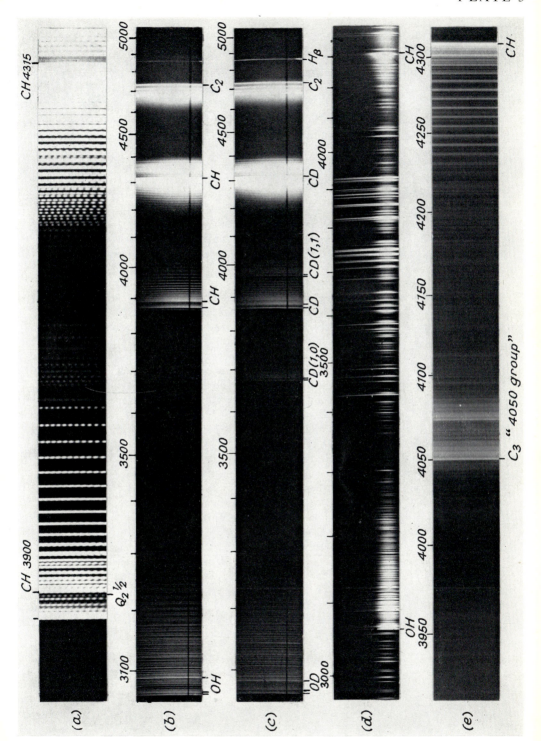

(a.)

(b)

(c)

(d)

(e)

DESCRIPTION OF PLATE 6

(a) Stoichiometric oxy-hydrogen flame at 7 torr pressure. Medium quartz spectrograph; exposure 2 hours. This shows the OH bands with normal rotational temperature.

(b) The same flame as above with 0·75 per cent. of acetylene added. The exposure time is reduced to 6 min. because of the much greater intensity of OH emission, and the abnormally high effective rotational temperature is shown by the longer " tail. "

(c) Reaction zone of stoichiometric low-pressure oxy-acetylene flame, containing Pb ethyl, photographed against the pole of a carbon arc as background. Quartz Littrow spectrograph. This shows the resonance line of Pb at 2833 Å in absorption, but other lines of Pb in emission. The brightnesss temperature of the arc far exceeds the equilibrium flame temperature, and this emission of Pb lines is clear proof of the lack of thermal equilibrium in the reaction zone.

(d) Low-pressure oxy-acetylene flame on small quartz spectrograph. The hair line across the slit serves as a fiduciary mark, and it is clear that the CH bands have their maximum intensity higher in the flame than bands of C_2 or OH.

(e) Spectrum of glow of fluorine impinging on surface of cold water. Quartz Raman spectrograph; Fe arc comparison below. This shows, besides some OH, the Schumann-Runge bands of O_2 the latter having unusually clear heads because of the low rotational temperature. I am indebted to Dr. R. A. Durie for this and the two following strips.

(f) Spectrum of diffusion flame of ethyl alcohol burning with fluorine. Quartz Raman spectrograph. The C_2 bands are very strong, but the CH is weak, with the main 4315 head no stronger than the (2, 0) band of C_2. (R. A. Durie.)

(g) The upper spectrum is of a flame of C_7F_{14} mixed with excess hydrogen burning in air. This shows unusual strength of the (1, 0) and (1, 1) bands of the 3900 Å system of CH. The lower spectrum, for comparison, shows a rather heavier exposure on an ethylene/air flame. Quartz Raman spectrograph (R. A. Durie).

(h) Spectrum of shock-wave excitation of acetylene in argon (1: 9) at 5 torr pressure. The driver gas was H_2 and the pressure ratio about 800: 1. This shows strong C_2, with a little CN, Na and other impurities but no CH.

PLATE 6

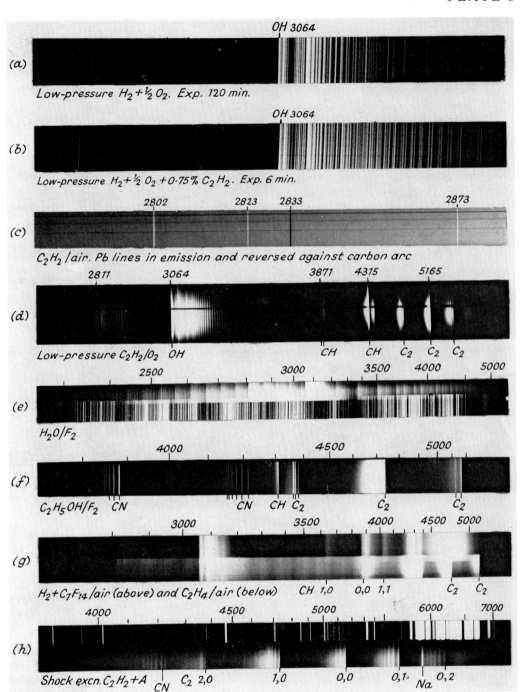

(a) Low-pressure $H_2 + \frac{1}{2} O_2$. Exp. 120 min. — OH 3064

(b) Low-pressure $H_2 + \frac{1}{2} O_2 + 0.75\% C_2H_2$. Exp. 6 min. — OH 3064

(c) C_2H_2/air. Pb lines in emission and reversed against carbon arc — 2802 2823 2833 2873

(d) Low-pressure C_2H_2/O_2 — 2811 3064 3871 4315 5165 — OH CH CH C_2 C_2 C_2

(e) H_2O/F_2 — 2500 3000 3500 4000 5000

(f) C_2H_5OH/F_2 — 4000 4500 5000 — CN CN CH C_2 C_2 C_2

(g) $H_2 + C_7F_{14}$/air (above) and C_2H_4/air (below) — 3000 3500 4000 4500 5000 — CH 1,0 0,0 1,1 C_2 C_2

(h) Shock excn. $C_2H_2 + A$ — 4000 4500 5000 6000 7000 — CN C_2 2,0 1,0 0,0 0,1 Na 0,2

DESCRIPTION OF PLATE 7

The carbon monoxide flame bands, due to CO_2, as emitted by the afterglow of a low-pressure discharge through carbon dioxide. Taken in the second order of a grating spectrograph with an exposure of 5 days and reproduced by courtesy of Prof. R. N. Dixon. This shows the vibronic structure in the region 3230 to 3510 Å. The quantum numbers are $(2v_1'' + v_2'')$ and n where n is a running number which takes the value l for the uppermost level of each "polyad". The alternating structure, Σ, Δ, Γ for even values of $(2v_1'' + v_2'')$ and Π, Φ, H for odd values is clear.

PLATE 7

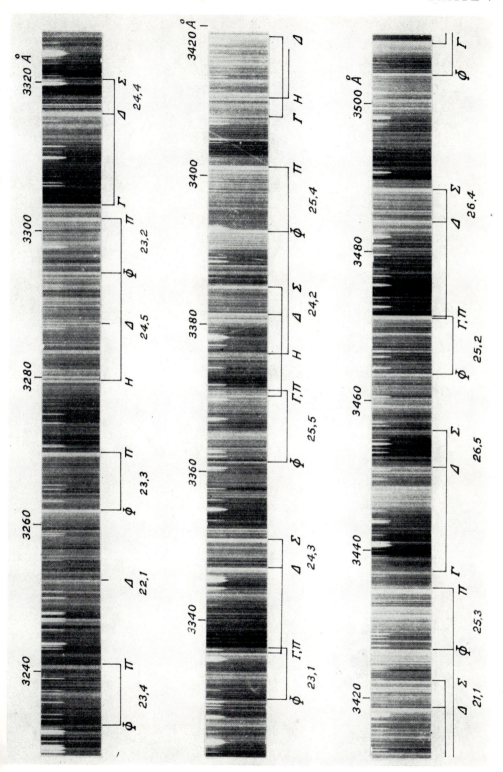

DESCRIPTION OF PLATE 8

Rotational structure of three bands of the A_1 progression of Vaidya's Hydrocarbon Flame Bands, due to HCO; reproduced by courtesy of Prof. R. N. Dixon. Photographed in an acetylene atomic oxygen flame with an Ebert Spectrograph. Lines marked with an asterisk* are due to OH.

PLATE 8

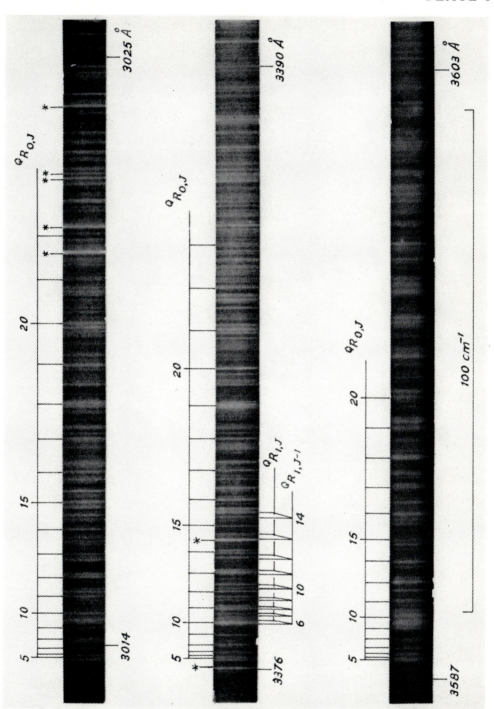

emission showed bands of C_2, CH and C_3. With a small amount of water vapour present the C_2 rotational temperature was only 4300 ± 250 K, but with more water (4.46%) the C_2 rotational temperature rose to 7130 K; this suggested the occurrence of two reactions leading to the formation of electronically excited C_2. Although atomic oxygen appears necessary for this flame with atomic hydrogen and acetylene, the spectrum is different from that of the $C_2 H_2 /O$ flame, since C_3 is present but OH is not, so that it seems that other active species e.g. H, take part in the flame, as well as atomic oxygen.

The dissociation products of a discharge through water vapour gave surprisingly bright flames with acetylene and some other hydrocarbons (Gaydon and Wolfhard, 1952); it was thought that the dissociation would lead to OH + H and that the active species would be the hydroxyl. However, the later work of Bayes and Jansson seems again to indicate that the initiating reaction must involve atomic oxygen; however, OH and perhaps H are probably also involved in subsequent reactions e.g. the CH emission may result from formation of C_2 and then the usual reaction $C_2 + OH = CO + CH^*$. These flames are greener than those with O, but the flame does show CH and OH in addition to the strong C_2. It is interesting to note that the flame of methyl alcohol reacting with the dissociation products from water gives OH bands with high rotational temperature, although the flames of methyl alcohol with O or with O_2 do not give the high rotational temperature.

It was also found by Gaydon and Wolfhard (1952) that various halogenated compounds — CCl_4, $CHCl_3$, C_2HCl_3, $CHBr_3$, C_2H_5I — gave flames with O, H and water products and these sometimes showed additional band systems such as CCl and even unidentified features. The flames with atomic hydrogen were in some cases much brighter than with hydrocarbons and could possibly be genuine, i.e. not dependent on atomic oxygen. Tarry polymers were often formed by these halogenated compounds.

Spectra Excited by Shock Waves

Experimental techniques have been discussed on page 58 and by Gaydon and Hurle (1963). The only hydrocarbon which can readily be excited to luminescence in a simple shock tube without dilution with an inert such as argon is acetylene. This gives strong continuous emission from carbon particles, with weak C_2 bands superposed. In the hotter shock waves through acetylene diluted with argon (see Plate 6 b) the C_2 bands are much stronger but there is little sign of the CH band at 4315 Å. Similar spectra are obtained from benzene in argon, but if moisture is present then CH appears immediately; once again we have evidence for $C_2 + OH = CO + CH^*$. Usually shock-tube spectra are similar to those from furnaces, giving the equilibrium radiation, but it is difficult to see how addition of a trace of water would increase the equilibrium CH concentration, so presumable decomposition processes in the shock front are involved in the C_2 and CH emission. Fairbairn made some temperature measurements by a modified spectrum-line reversal method during the shock pyrolysis of acetylene and cyanogen; the electronic excitation temperature near the shock front tended to be low, rising later towards the equilibrium value, and the strong emission near the front appeared to be due to a high concentration of C_2 during the initial pyrolysis rather than to a high effective temperature.

Shock waves can also be used to ignite combustible mixtures. Gaydon found that shock-ignited acetylene/oxygen explosions showed OH emission relatively more strongly and the CH less strongly than a stationary flame of a similar mixture. Conditions in the shock-ignition would, of course, be closer to those of a detonation and the temperature is higher. Gaydon and Hurle (1963) found that the chromium-line reversal temperature was very high in the detonation front of shock-ignited ethylene/oxygen. In acetylene/oxygen and acetylene + cyanogen/oxygen detonations, initiated by a shock, Fairbairn (1963) found a sharp emission spike and high C_2 excitation-temperature in the front, and compared the decay of this with the expected relaxation time for vibrationally excited CO molecules; he concluded that vibrationally excited CO might be formed during the reaction and that the strong emission of C_2 and high excitation temperature might be due to transfer of energy from the CO to other species.

Occurrence of C_2 and CH in some Special Flames

While the emission of C_2 and CH bands is a general feature of the spectra of flames of organic compounds, especially hydrocarbons burning with oxygen, there are certain flames in which either hydrogen or oxygen is absent and which emit C_2 bands and sometimes CH bands.

Flames of carbon suboxide, C_3O_2, and of cyanogen, C_2N_2, burning with oxygen emit strong C_2 bands, and the latter also gives the Violet and Red systems of the radical CN. Both these flames normally give CH bands as well and very careful drying of both fuel and supporting gas (air or O_2) is necessary to eliminate CH. In cyanogen flames the addition of a little moisture or hydrogen not only produces strong CH emission but also has a marked effect in reducing C_2 and CN emission (Pannetier and Gaydon, 1947); see Plate 4e. This is further support for the view that C_2, and probably also CN, react with H_2O, H_2 or more probably OH to produce excited CH.

Many organic substances, especially hydrocarbons, will burn directly with fluorine or chlorine trifluoride in the absence of oxygen. Durie (1952b), working in the author's laboratory, was the first to study the spectra of flames of this type. He used small diffusion flames which were shielded from external air by a surrounding flow of nitrogen. The flames were very bright and hydrocarbon/fluorine flames showed strong C_2 and CH bands; some CN and OH emission was also shown on his plates. When the burner became heated the flame turned luminous and smoky due to carbon formation; dilution with nitrogen prevented the carbon formation and increased the strength of the CN bands. These flames also showed the 4050 Å "Comet-head" group of C_3.

Later Skirrow and Wolfhard (1955) extended the work to flames supported by chlorine trifluoride. Using hydrocarbons or hydrocarbon/hydrogen mixtures they found strong emission from C_2, CN, C_3, CCl and CF but were unable to obtain CH when the gases were pure. On addition of quite small amounts of oxygen, however, the CH bands appeared strongly. Once again it seems that the CH is formed from C_2 by reaction with an oxygen containing molecule, presumably OH. Flames of methanol and ethanol burning with fluorine show strong OH and in this case both C_2 and CH are also emitted, although rather weakly. Rich flames of ethanol also show C_3 and give a little carbon formation, but methanol does not.

Flames can also be obtained with halogenated organic compounds (CCl_4, $CHCl_3$, CF_4, $CHBr_3$ etc.) reacting with alkali-metal vapours (K, Na). Experiments have been carried out (Miller and Palmer, 1963; Tewarson, Naegeli and Palmer, 1969) at low pressure in vessels heated to around 400°C, the organic vapours being carried in an inert gas, usually He. For fully halogenated compounds such as CCl_4 there is strong C_2 emission, many band systems being observed but the strongest emission arising from bands with $v' = 6$ and 7 of the Swan system, i.e. the "High Pressure" bands as they used to be known. These are attributed to the reaction $C + C + M = C_2^* + M$. In fuels containing hydrogen, especially chloroform and bromoform, both CH and C_2 bands are emitted. In this case we definitely have a process leading to CH emission which does not involve oxygen or OH. In these haloform flames the C_2 Swan bands show preferential excitation of bands with $v' = 2$. Palmer et al attributed this to reactions

$$CH + CH = C + CH_2$$

followed by

$$C + CH = C_2^* + H$$

However taking $D(CH) = 3.47$ e.V and using the value of $D(C_2) = 6.11 \pm 0.04$ e.V from the recently observed predissociation (Messerle and Krauss, 1967), the energy available seems more likely to excite the $C_2(A\,^2\Pi)$ to the $v' = 0$ or possibly $v' = 1$ rather than $v' = 2$, and this reaction cannot be considered as proven.

Halogenated compounds, and certain other substances, act as inhibitors to combustion, and also produce changes in the flame spectra. These inhibiting processes are most important in diffusion flames because of the use of such compounds in fire extinction. Ibiricu and Gaydon (1964) found that the changes produced were quite complex, but for their counter-flow diffusion flames there was a general tendency for the inhibitors (Cl_2, CCl_4, CH_3Br, $POCl_3$ etc.) to reduce OH emission, to increase C_2 emission and soot formation and to have rather little effect on CH emission. The results are rather complicated but are probably to be explained as removal of OH radicals by the reaction

$$OH + HX = H_2O + X$$

where X is the halogen. This removal of an important radical slows the combustion processes and gives better opportunity for competing pyrolysis of the fuel; the C_2 may be associated with hydrocarbon radicals being produced by pyrolysis rather than oxidation. As OH decreases and C_2 increases the $C_2 + OH = CO + CH^*$ reaction may occur with comparable frequency with and without inhibitor. The spectra of flames with added halogen will be discussed in more detail in Chapter XII.

Isotope Tracer Experiments

The use of isotopes to follow the reaction leading to the production of particular chemical species in flames has been referred to in Chapter III, page 62. Experiments with deuterium as tracer (Gaydon and Wolfhard, 1952; Broida and Gaydon, 1953b) have given some information about the processes of formation of excited OH, CH and HCO radicals in flames. At atmospheric pressure, flames of mixtures of deutero-acetylene with ordinary hydrogen, and of ordinary acetylene with deuterium, give the CH/CD ratio and the OH/OD ratio approximately the same as the overall H/D ratio in the original mixture, irrespective of which way round the mixture occurs. This is taken to indicate that both CH and OH bands arise from some secondary reactions rather than from direct breakdown of the acetylene; direct breakdown should tend to give CD and OD from $C_2D_2 + H_2$, and similarly CH and OH from $C_2H_2 + D_2$ flames.

In similar experiments with low-pressure flames, however, there did appear to be some strengthening of the isotope coming from the acetylene at the base of the thick reaction zone. This could be interpreted either as due to direct breakdown of acetylene to give this isotope at the base, with rapid interchange of H and D preventing observation of the effect in the upper part of the reaction zone, or more probably it could be explained by preferential combustion of acetylene at the base of the flame, thus giving CH and OH from the acetylene's hydrogen, with the hydrogen, which has a higher ignition temperature, only burning in the hotter upper part of the reaction zone.

The experiments with atomic flames were more conclusive. The flame of the products of a discharge through ordinary water vapour reacting with deutero-acetylene gave CH and OH bands in emission, with only a weak trace of OD and CD (see Plate 5b and c). Similarly, products of a discharge through D_2O reacting with C_2H_2 gave CD and OD bands. This result was independent of the relative flows of water vapour and acetylene over a wide range. It was also found that for the flames between the products of a discharge through D_2O reacting with C_2H_2, which gave CD and OD bands, the addition of either H_2 or H_2O in excess to the acetylene stream did not affect the spectrum. If, however, a little O_2 was added to the D_2O stream, then the CH and OH bands were strengthened. These results were consistent with the formation of CD and of excited OD from ground-state OD radicals produced in the discharge through D_2O, while in the presence of added oxygen the discharge produced atomic oxygen which gave an atomic oxygen/acetylene flame which produced some CH and OH.

These results with atomic flames indicate that the CH and OH excitation (i) are the result of chemical processes, (ii) are due to secondary reactions rather than to direct breakdown of the acetylene to CH, and (iii) that interchange between H and D atoms in the acetylene and hydrogen or water is not important under the conditions of atomic flames. The results are consistent with the reactions

$$C_2 + OH = CH^* + CO$$

and

$$CH + O_2 = OH^* + CO.$$

Since this work, it has been shown (see page 213) that the flames with the decomposition products of a discharge through water vapour involve atomic oxygen rather than OH for the initiating step. However, although OH may not react with the acetylene, there can be no doubt that rapid interchange of O, H, OH, H_2 and H_2O will maintain a fair proportion of OH in the discharge products and the above conclusions are not affected.

Some preliminary measurements of absorption spectra were made by Gaydon, Spokes and van Suchtelen (1960). In low-pressure flames of mixed deutero-acetylene and ordinary hydrogen of compositions $C_2D_2 : H_2 : O_2 = 1:2:4$ and $1:2:2.25$ the OD/OH absorption ratio appeared consistent with the equilibrium ratio $1:2$ and did not indicate that OH was formed preferentially from either the acetylene or hydrogen.

The hydrocarbon flame bands of HCO are emitted quite well by low-pressure flames and Broida and Gaydon (1953b) studied the $C_2D_2 + H_2/O_2$ flame and found that both HCO and DCO were emitted. This seems to disprove any simple peroxide formation and breakdown

$$C_2H_2 + O_2 = C_2H_2 \cdot O_2 = 2\,HCO$$

as the spectrum would then have shown only DCO. Formation of HCO via C_2 is rather unlikely as these bands normally occur in weak-mixture flames in which C_2 is very weak. Formation via ground state CH, possibly by

$$CH + HO_2 = HCO + OH$$

seems possible and would be consistent with chemical detection of highly oxidizing substances in samples from chilled flames which show the HCO bands. Reaction with hydrogen peroxide

$$CH + H_2O_2 = HCO + H_2O$$

is another possibility.

Isotope studies of the mechanism of excited C_2 formation in flames have been made in some interesting experiments by Ferguson (1955). He burnt a mixture of ordinary acetylene and acetylene enriched with C^{13} and studied the relative intensities of the C_2^{12}, C_2^{13} and $C^{12}C^{13}$ bands. He found almost, though not quite complete, randomization. This indicates that each C_2 is formed from carbon atoms which come from different acetylene molecules. It rules out successive stripping, such as

$$C_2H_2 + H = C_2H + H_2$$

$$C_2H + H = C_2 + H_2$$

or similar stripping reactions by OH or O. It would be consistent with C_2 formation from two free C atoms, or from two CH radicals to form $C_2 + H_2$, or perhaps more likely by $CH_2 + C = C_2 + H_2$ which is probably about 1.55 e.V exothermic. The light-yield experiments on mixed flames (see page 164) also show that the strength of C_2 emission is not simply proportional to the concentration of the hydrocarbon, but usually involves two or more hydrocarbon molecules or fragments; this led to the view that polymerization of the hydrocarbons might precede C_2 formation. It may be possible to reconcile this polymerization mechanism with the isotope results, but it is not easy to do so. Fairbairn (1962) carried out some similar experiments using C^{13} as tracer on the C_2 emission during pyrolysis of acetylene by shock heating acetylene/argon mixtures. Again he found that the C_2 did not come from a single acetylene molecule but the high randomization (>67%) indicated complex reactions.

Ferguson (1957) also studied soot formation using C^{13} as tracer, but with mass spectrometric analysis of the soot, and showed that in these circumstances there was again almost complete randomization of C^{12} and C^{13}. However, if labelled carbon monoxide was added to the flame it was found that this did not contribute to the soot formation.

Chapter IX

The Infra-red Region

The Amount of Radiation from Flames

This book deals mainly with the interpretation of the visible and near ultra-violet spectra of flames, but it should not be forgotten that the major part of the radiation lies in the infra-red. The visible and ultra-violet emission seldom exceeds 0.4 per cent of the energy released by the combustion processes, and for some flames, such as the hydrogen flame, it is almost negligible. The infra-red part of the radiation is, however, much greater, varying usually between 2 per cent and 20 per cent of the heat released by the combustion.

The actual amount of energy radiated by a flame varies considerably with gas flow conditions. Obviously, if the hot combustion products could be thermally isolated from cool surrounding air, they would continue to radiate until they had lost all their heat by this process. In practice most of the heat is lost by conduction and convection, and the conditions of mixing of the flame gases with the surrounding cool air, and also conduction losses to walls and burner tubes, are important. Any turbulence must tend to increase mixing and heat loss to the surrounding air and so reduce radiant heat transfer; the change from laminar to turbulent flow conditions is frequently abrupt (see *Flames*) and we may expect a sharp change in radiant efficiency at this point. In diffusion flames there may be a smooth decrease in the height to turbulence if the flow rate is increased, but in some "sensitive" flames there is again an abrupt change in flow conditions. Self-absorption of infra-red radiation in the flame gases is very important and the amount of radiation escaping from the flame depends on flame shape and size; there is a general tendency for the proportion of energy radiated away, i.e. the radiant efficiency, to decrease with increasing flame size.

Old measurements of total radiation from small flames, made by

Helmholtz, Callendar, David and others, were conveniently summarized by Hartley (1932–3). According to these results, the carbon monoxide flame radiates the greatest fraction of its combustion energy, although the actual amount radiated is very dependent on the degree of dryness of the gas. Helmholtz found that moist CO flames radiated about 2.4 times as much as H_2 flames, and that the amount of radiation from small flames of methane, ethane, coal-gas etc. could be calculated approximately from the number of CO_2 and water vapour molecules formed, by using the measurements for flames of CO and H_2. However, Haslam, Lovell and Hunneman (1925) obtained different results, finding that methane radiated 14.9 per cent but CO only 10.4 per cent of the available heat; this probably serves to emphasize the importance of experimental conditions such as burner size and flow rate.

Hartley gave curves for the amount of energy radiated by coal-gas flames for varying aeration, and reproduced similar curves by Callendar. For very high aeration the energy radiated was about 10 per cent of that set free by combustion; it rose to 18 per cent for normal full aeration, then fell slightly for richer mixtures, but then rose again to a second maximum when the flame became luminous due to soot formation. David studied the effect of pressure and found that the radiation fell with increasing pressure; this is probably due to increased self-absorption.

Radiant heat transfer is, of course, particularly important in industrial flames. For clear (i.e. soot-free) flames we shall see that the emission is mainly due to banded spectra of CO_2 and H_2O which show partly resolved rotational fine structure. The width of these emission bands depends on the rotational and vibrational temperatures, and the contour of individual lines is affected by both temperature and pressure broadening, so that the flame emissivity is a complex function of conditions. However, the overall emissivity can be calculated fairly well if the CO_2 and H_2O contents and temperature are known; there is appreciable mutual interference between CO_2 and H_2O because some of the bands overlap and change the self-absorption of radiation. Old empirical measurements of emissivity by Hottel and Mangelsdorf (1935) are usually used in these calculations and some of their curves have been redrawn in scientific units in *Flames* (Gaydon and Wolfhard, 1970); Leckner (1971) has recently proposed some minor corrections to the emissivity data for CO_2. However, most industrial flames are luminous due to the presence of soot particles. Although soot particles are not strictly black or grey, because they are often comparable in size with the wavelength of visible light, the spectrum is continuous and calculations of emissivity and heat transfer should be easier, but practical limitations of distribution of the particles within the flame and variations of flame temperature with position have to be considered. Books by Thring (1962) and Hottel and Sarofim (1967) and publications about the IJmuiden furnace work by the International Flame Research Foundation (e.g. Chedaille and Braud, 1972) deal with radiation from large-scale furnace flames.

In small Bunsen-type flames we are familiar with the blue-green inner cone or primary reaction zone, the non-luminous interconal gases and the outer cone which is a secondary diffusion flame. There is also a region of hot gases above this outer diffusion flame. Hartley's measurements indicated that about a seventh of the radiation came from the interconal gases and six-sevenths came from the outer diffusion flame and the hot gases above. Although the visible radiation is so much stronger from the inner cone than from the interconal gases, this does not appear to be so for the infra-red radiation. Even photographs of Bunsen flames in near infra-red light (Bonhoeffer and Eggert, 1939) do not show the well-defined inner cone. Measurements of temperature by brightness and emissivity methods (Silverman, 1949) do not indicate abnormally high values. Most observations thus appear to be consistent with normal thermal emission of infra-red radiation by hot product gases.

There are, however, some effects which appeared at one time to show a chemical rather than a purely thermal origin for some of the infra-red radiation. An interesting observation, first made by Helmholtz and confirmed by Haslam et al (1925) was that preheating of the gases of a small premixed flame caused the infra-red radiation to decrease, although heat was being added to the flame; it has now been shown by Gaydon and Guénebaut (see page 38) that this apparent effect is due to an increase in the burning velocity which produces a marked change in both the shape and size of the flame and that this accounts for the decrease in radiation. Guyomard (1951) also found that for diffusion flames preheating of the air and gas supplies up to 400°C also reduced the flame radiation. In these diffusion flames most of the radiation will be from soot particles and although no quantitative explanation has been given it seems quite likely that, in the hotter flames produced by the preheating, the soot radiation could be decreased; this could happen if the soot particles burnt away more quickly or if they were formed in a different position where the oxygen content was lower so that they were less heated by surface combustion. Experiments on the effect of moisture and other catalysts on the radiation from carbon monoxide explosion flames will be discussed in a separate section later in this chapter; the effects reported appear at face value to indicate chemiluminescent emission in the infra-red for carbon monoxide flames, but here again there may be another explanation.

Infra-red Spectra of Organic Flames

First observations on the spectra of simple flames in the infra-red were made by Julius in 1890 and by Paschen around 1893 and numerous papers have since appeared on the spectrum of the Bunsen and other flames. The strongest bands are around 4.4 and 2.8 μm and there are numerous weaker features including bands around 15, 6.7 and 2.0 μm. The main features in the near infra-red as seen with small dispersion are shown in Figure IX.1 (after Plyler, 1948).

The flame of carbon monoxide also shows the strongest band at 4.4 μm and some emission around 2.8 μm, while the hydrogen flame shows its main emission at 2.8 μm. Thus the 4.4 μm band has been identified with CO_2, which is known to show strong absorption in this region due to the fundamental asymmetrical vibration ν_3. The band at 2.8 μm is due to superposition of bands due to H_2O and CO_2; H_2O has its asymmetrical fundamental at 2.67 μm while CO_2 has two combination bands in this region, $\nu_3 + \nu_1$ at 2.69 and $2\nu_2 + \nu_3$ at 2.78 μm. The author pointed out in *Spectroscopy and Combustion Theory* in 1942 that the OH fundamental vibration also lay at 2.8 μm, and Plyler has since shown that OH emission does also contribute to the rotational structure on the long-wave side of the 2.8 μm band in Bunsen flames, and the OH band has been observed under high resolution in hotter flames.

The appearance and structure of these two main bands are greatly modified by absorption both in the cooler outer layers of the flame and by moisture and CO_2 in the surrounding atmosphere. The centres of the bands are weakened most by this, and the region of maximum intensity is displaced to longer wavelengths. The extent of the absorption depends on flame size and shape and also the subsequent path length in the atmosphere. For the 4.4 μm band the CO_2

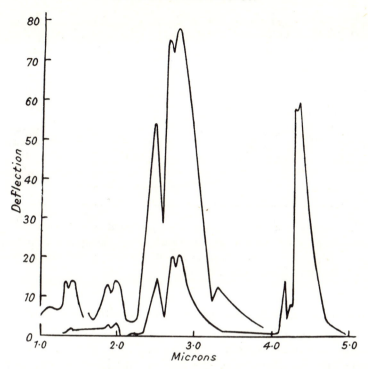

Fig. IX.1. Traces of the near infra-red spectrum of a Bunsen flame (natural gas/air)

absorption in the air may reduce emission around 4.25 μm, just on each side of the band origin, almost to zero, leaving the strongest emission between 4.3 and 4.6 μm. This is well shown in a trace redrawn from Plyler and Ball (1952) which shows a methane/oxygen flame as observed with a grating spectrometer involving a 4 m path through air (Figure IX.2); the heads of two bands, formed at high rotational quantum numbers in the R branch, are clear and show structure, and there is a little emission getting through near the band origin where absorption is less, but the main R branch emission between the origin and band heads is completely absorbed. In some cases the absorption may be strong enough to show resolved band structure of the isotopic molecule $C^{13}O_2^{16}$ at longer wavelengths around 4.4 μm superposed on the flame emission; this is well shown in traces reproduced by Plyler, Blain and Tidwell (1955).

The structure is further complicated by superposition of bands involving higher vibrational quantum numbers. For CO_2, as already

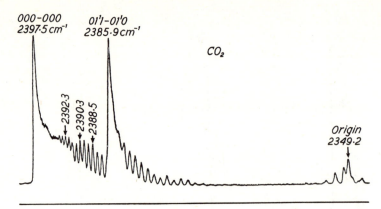

Fig. IX.2. Emission at the head of the 4.4 μm band from a methane/oxygen flame, observed through a 4 m path of air. Absorption by atmospheric CO_2 has completely obscured the main part of the R branches. Redrawn from Plyler and Ball.

pointed out (pages 86 and 130), there are three vibrational modes but because of the degeneracy in v_2 it is necessary to specify four quantum numbers, usually denoted v_1, v_2 ^{l} and v_3. Thus, as an example of this complexity the 4.4 μm has been found (Benedict, Herman and Silverman, 1951) to show structure and R-branch heads due to the following bands

v_1'	v_2'	l'	v_3'	v_1''	v_2''	l''	v_3''	Origin cm^{-1}	R head cm^{-1}	μm
0	0	0	1	0	0	0	0	2349.3	2396.7	4.17
0	1	1	1	0	1	1	0	2337.6	2385.3	4.19
1	0	0	1	1	0	0	0	2327.4	2377.1	4.20
0	2	0	1	0	2	0	0	2326.3	2374.1	4.21
0	2	2	1	0	2	2	0	2325.8	(2373.6	4.21)
0	0	0	2	0	0	0	1	2324.2	2372.4	4.215

The first two only of these heads are visible in Figure IX.2, but Benedict et al show five clear heads. Many of the vibrational transitions doubtless contribute to the structure of this band but the R heads, lying at slightly longer wavelengths are lost in stronger structure or masked by self-absorption. Indeed, the strongest part of this band between 4.3 and 4.6 μm is usually without resolvable rotational structure apart from absorption lines from cool CO_2 in the atmosphere.

In addition to the CO_2 emission bands at 2.8 and 4.4 μm, the bending frequency ν_2 (667 cm^{-1}) produces a band at about 15 μm. Early failures to observe this band in emission (Bailey and Lih, 1929; Garner and Johnson, 1927) led the author to speculate that the weakness of this band compared with that from ν_3 might indicate non-thermal emission. However, it is now clear that even black-body emission is relatively very weak at such long wavelengths; studies by Plyler (1948) showed that strong absorption weakens this band and that, when this is allowed for, the band occurs with about the expected strength, calculations indicating that this band should only be about 1/200 of the strength of the 4.4 μm band. The 15 μm band lies at the limit of transparency of NaCl, but Tagirov (1959) has shown traces of the emission spectrum of CO diffusion flames taken in this region through KBr.

The symmetrical frequency ν_1 of CO_2 at 1355 cm^{-1} is inactive in the infra-red. Band structure around 2.0 μm is due to CO_2 $2\nu_1 + \nu_3$ and $\nu_1 + 2\nu_2 + \nu_3$; many other bands of CO_2 which are known in absorption through relatively long path lengths may also contribute a little to flame emission in the near infra-red.

We have noted that the ν_3 fundamental of H_2O contributes to the strong emission band at 2.8 μm; the origin of the $(0,0,1 \rightarrow 0,0,0)$ transition is at 3756 cm^{-1} or 2.67 μm and the P branch lies under the CO_2 around 2.8 μm, but the R branch usually produces an intensity maximum around 2.5 μm and extends to 2.35. The ν_2 fundamental vibration of H_2O (1594 cm^{-1} or 6.27 μm) gives a band of moderate strength which usually has maximum intensity around 6.7 μm. This band shows partly resolved rotational fine structure and extends to about 8 μm. There are several combination bands due to H_2O. $\nu_2 + \nu_3$ produces emission at 1.9 μm and Benedict and Plyler (1954) listed eleven satellite vibrational transitions which contributed to this. The $\nu_1 + \nu_3$ combination gives emission around 1.4 μm, and again eleven vibrational transitions are known to be involved. These weaker bands of H_2O in the near infra-red have been studied under large dispersion and the rotational fine structure has been analysed; for the band at 1.9 μm nearly 1000 lines have been measured (Benedict, Bass and Plyler, 1954) and rotational structure of the transitions $(0,1,1 \rightarrow 0,0,0)$, $(0,2,1 \rightarrow 0,1,0)$ and $(0,3,1 \rightarrow 0,2,0)$ has been analysed. There is a band due to the transition $(1,1,1 \rightarrow 0,0,0)$ at 1.03 μm, and still higher vibrational overtone bands lead to weak emission right into the visible reagion (see page 107). Emissivity data for these vibration-rotation bands of water vapour have been published by Ferriso and Ludwig (1954), who used a small supersonic oxy-hydrogen flame giving temperatures up to 2250 K.

In the far infra-red, beyond 9 μm and extending to at least 23 μm Plyler (1948) found a lot of complex rotational structure which he attributed to the pure rotation spectrum of water vapour, with a contribution also from OH. The H_2O lines do not correspond well with absorption lines at low temperature because the emission from hot gas involves higher rotational energy levels. Ludwig, Ferriso, Malkmus and Boynton (1965) have shown emissivity curves for H_2O down to 22 μm, based on measurements with their supersonic oxy-hydrogen flame; there is a lot of fine line structure, but the emissivity between 14 and 22 μm is quite high, around 0.5 for this hot flame at 1 atm. (See Figure IX.3.)

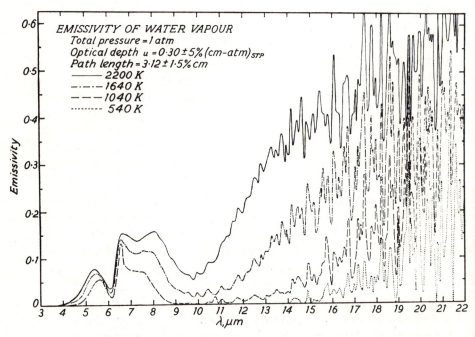

Fig. IX.3. Spectral emissivity of H_2O in the far infra-red, covering the pure rotation band and the 6.3 μm fundamental, as obtained from emission of a small oxy-hydrogen flame at the temperatures indicated. Redrawn from Ludwig et al.

The other stable species which is often present in flame gases and contributes to the infra-red emission is carbon monoxide. The CO fundamental is at 2143 cm^{-1} giving a band with origin at $4.67\,\mu$m. This is normally almost completely masked by the strong $4.4\,\mu$m band of CO_2, but in high resolution studies some lines of the P and R branches of the (1,0), (2,1) and (3,2) bands have been observed (Plyler, Blaine and Tidwell, 1955). At shorter wavelengths the overtone bands of the (2,0) sequence give clear R heads and the bands show resolved rotational structure of P and R branches (Plyler, Allen and Tidwell, 1958); the following are the band heads and origins,

v', v''	R head cm^{-1}	μm	origin cm^{-1}
2, 0	4360	2.294	4260
3, 1	4305	2.323	4205
4, 2	4251	2.352	4154
5, 3	4196	2.383	
6, 4	4142	2.414	

The homonuclear molecules N_2, O_2 and H_2 are of course, inactive in the infra-red and do not give either vibration-rotation or pure rotation bands.

In hot flames, such as the oxy-hydrogen and oxy-acetylene flames, the hydroxyl radical has an equilibrium concentration of several per cent, and vibration-rotation bands of OH can be observed under high resolution. The origin of the main (1,0) band lies at $2.80\,\mu$m and is completely masked by the strong $2.8\,\mu$m band of H_2O and CO_2, but lines of the double P branches of the (1,0) and (2,1) bands between 3.0 and $3.8\,\mu$m have been observed by Plyler and Ball (1952). The (2,0) band has its R head at $1.39\,\mu$m but is not usually a prominent feature because of overlapping H_2O bands. Additional vibration-rotation bands of OH extend weakly into the photographic infra-red, see page 365.

Thus, in summary, all the strong infra-red emission from ordinary organic flames comes from the two main combustion products, H_2O (strongest bands, pure rotation 9–24 μm, 2.67 and 6.7 with weaker bands 1.9, 1.4 and 1.03 μm) and CO_2 (strong bands 4.4, 2.8, 15 μm and weaker band 2.0). Small contributions from CO and OH are largely masked by the stronger CO_2 and H_2O emission, and all bands are modified by atmospheric absorption.

In a Bunsen flame of methane Plyler (1948) also observed a weak band at 3.32 μm (3000 cm^{-1}); this is about the C–H stretching frequency; Herzberg (1945) gives average values of 2960 cm^{-1} for –C–H, 3020 for =C–H and 3300 for ≡C–H. The emitter for this band at 3.32 could be methane, ethylene or formaldehyde; emission in a similar region also occurs in cool flames (see later).

In addition to these vibration-rotation bands, electronic transitions in C_2 and CN also give some bands in the near infra-red (Benedict and Plyler, 1954). For C_2 the (1,0) and (0,0) bands of Phillips's system at 1.01 and 1.20 μm occur, and in rich oxy-acetylene flames the (0,0) band of the Red System of CN at 1.09 μm is prominent; the (0,1) and (1,0) bands of this system are at 1.40 and 0.91 μm. Both the C_2 and CN bands are largely confined to the inner cone and differ in this respect from the other infra-red bands.

Relative Band Intensities and Temperature Measurements

For the main bands the wavelengths at which maximum flame emission is observed lie to appreciably longer waves than the corresponding absorption bands of cold CO_2 and H_2O. This is the combined effect of three basic causes. The higher rotational temperature causes the rotational structure to extend further in both directions, but because all vibration-rotation bands are weakly degraded towards longer wavelengths the extension of the P branch to longer wavelengths is greater than that of the R branch, which tends to close up towards a head on the short-wave side. The vibrational temperature is also higher and leads in the hot flame to a superposition of bands due to transitions between higher vibrational levels, as already indicated (page 226). The absorption in the atmosphere and in the cooler outer layers of the flames weakens the central region of the band and leaves the long-wave wing to appear most strongly; Daly and Sutherland (1949) have made some computations (*a*) on the expected emission only from hot CO_2, (*b*) on absorption by cool CO_2, and (*c*) on the emission seen through the cool gas, and these computations clearly indicate how the strongest emission is shifted to longer wavelengths.

The extent of the shift in the position of the band maximum obviously depends on flame temperature, but it is not easy to use it as a measure of temperature. Silverman (1949) made measurements of temperature by a brightness-emissivity method at various wave-lengths in the 4.4 μm CO_2 band; he used a glow-bar and light chopper to obtain the absorption and thus the emissivity. He found that near the band origin the measured temperature came quite low, corresponding to that of the outer layers of the flame, but that at longer wavelengths (>4.6 μm) a higher value for the temperature, corresponding to that in the core of the flame, was obtained. All the temperatures, in both carbon monoxide and hydrocarbon flames, seem to be around or a little below the adiabatic flame temperature, indicating that the CO_2 emission is thermal and not, like the ultra-violet emission, dominated by chemiluminescence. Tourin and Krakow (1965) and Krakow (1966) have shown that by making brightness and emissivity measurements at several wavelengths it is possible to determine the mean temperature in a number of zones of a complex flame; in particular, they made measurements at 4.555, 4.696 and 4.865 μm to determine the temperature in three zones of a carbon monoxide/oxygen flat flame.

The relative intensities of the two main bands at 4.4 μm (CO_2) and 2.8 μm ($H_2O + CO_2$) may be expected to vary with composition as well as temperature. In the simple Bunsen flame the 2.8 μm band is stronger, relative to the 4.4 μm band, in the inner cone than in the outer cone (Hartley, 1932–3); this is because the interconal gases around the inner cone contain a fair amount of water vapour and CO but little CO_2, which is only formed in the outer diffusion flame. The relative strength of the two bands depends on flame temperature, mixture strength and the nature of the fuel gas. Bailey and Lih (1929) recorded that the ratio of intensities of the 4.4 to the 2.8 μm band was 2.8 for a Bunsen flame of coal gas but 3.5 for a Meker burner. The carbon monoxide flame gave a ratio of 8.1, varying somewhat with the dryness of the gas; for methane the ratio was rather surprisingly given as 10.0. For flames of mixtures of CO and H_2, the ratio at first rose with increasing CO content, to reach 9.6 for 90 per cent CO, and then fell a little to 8.1 for pure CO. In hot flames supported by oxygen instead of air the total infra-red radiation is considerably stronger and the 2.8 μm band is strengthened more than that at 4.4 μm, and bands at shorter wavelengths, right up into the visible region, are particularly enhanced.

The spectra of the triatomic emitters CO_2 and H_2O are too complex for detailed study of the rotational intensity distributions, but for some of the diatomic emitters it has been possible to measure individual lines of the fine structure and to make estimates of the effective rotational and vibrational temperatures.

Some rather puzzling values have been obtained by Silverman (1954) and Plyler, Benedict and Silverman, (1952) who studied the (2,0) sequence of CO; these bands are in the 2.3 μm region and are so weak that they are unaffected by self-absorption. Using an essentially two-dimensional CO/O_2 flame, they found that the vibrational temperature near the reaction zone was below the calculated equilibrium flame temperature and only approached this temperature some 10 mm above the reaction zone. The effective rotational temperature, however, rose from a fairly low value in the reaction zone to a maximum which was about 500 K too high at 10 mm above the reaction zone, and then fell slowly to rather below the adiabatic flame temperature. Addition of diluents, especially CO_2, seemed to reduce the temperature anomalies. Table IX.1 gives the value they obtained.

TABLE IX.1

Effective rotational and vibrational temperatures 10 mm above the reaction zone, using the (2,0) sequence of CO, compared with the adiabatic flame temperatures.

Fuel mixture	T_{rot}	T_{vib}	T_{adiab}
$6\,CO + 1.5\,O_2$	3400 K	2550 K	2850 K
$6\,CO + 1.5\,O_2 + 1.3\,CO_2$	2500	2300	2625
$6\,CO + 1.5\,O_2 + 1.3\,N_2$	2750	2400	2680

It is probable that similar effects occur in other flames. Plyler et al noted that the rotational temperature for CO in the inner cone of an oxy-acetylene flame was 3600 K, but detailed studies of the variation with height were not reported. In the outer cones of the flames, well away from the reaction zone, there was equilibrium between the rotational and vibrational temperatures; this appears to show that the method was reliable; the accuracy was about 10 per cent. These results seem to indicate some lack of local thermodynamic equilibrium in these flames; it would not be difficult to understand that incomplete combustion could lead to low rotational temperatures, but it is not easy to see how to account for high rotational temperatures because equilibrium between rotation and vibration should be established in times much shorter than the radiative lifetime for infra-red bands. Further work is perhaps desirable.

For the rotation-vibration bands of OH, Plyler et al found the effective rotational temperature in oxy-acetylene flames was about 2400 to 2600 K, which was rather below the adiabatic temperature and several hundred degrees lower than for the ultra-violet bands of OH in the same flame; it is not quite clear whether this measurement referred to the reaction zone or more probably to the mean for the whole flame.

The infra-red radiation from steady flames thus appears to be mainly thermal, but there is still a suspicion that some non-equilibrium effects occur. The extent of any anomaly is certainly much smaller than for ultra-violet emission. This is because the longer radiative lifetime for transitions in the infra-red allows more time for collisions to establish equilibrium before the molecule radiates. The transition probability for radiation contains a term in ν^4 ; thus a band at 4 μm tends to have a radiative lifetime 10,000 times greater than for a transition giving visible emission at 4000 Å. For CO_2 at 4.4 μm the radiative lifetime is about 2 ms (Benedict and Plyler, 1954, quote 2.25 ms) while the vibrational relaxation time at flame temperatures appears to be less than 0.1 μs. Only for CO are both radiative and vibrational relaxation times significantly long; the radiative lifetime for the vibration-rotation band may be about 30 ms; the vibrational relaxation time is very sensitive to the presence of impurities but in the pure gas is about 45 μs at 2200 K (Gaydon and Hurle, 1963). This could explain the anomalies of effective vibrational temperature for CO in CO/O_2 flames, but not the rotational temperature anomalies, as rotational relaxation times of around 10^{-8} s are so very much shorter than the radiative lifetime.

Radiation from Explosion Flames

Although most of the radiation from stationary flames can be explained as thermal, there are many old observations which appear to show that for explosion flames, especially of CO, much of the infra-red emission is associated with the flame front and is chemical rather than thermal in origin. The author is not aware of any satisfactory explanation of this old work, but until the observations have either been disproved or explained it seems wrong to discard them.

Garner and colleagues made a long series of observations, between 1925 and 1935, on the effect of moisture, hydrogen and other catalysts on the emission from CO/O_2 explosion flames; most of the work was done in a metal tube of 1 inch diameter, closed at both ends and with ignition near one end, the radiation being viewed end-on through a fluorite window. The infra-red spectrum of these explosions was similar to that of a stationary flame, but the 2.8 μm band was relatively considerably stronger. When the gases were carefully dried there was a marked increase in the total radiation and a relative strengthening of the 4.4 μm band. The strength of this band was increased by a factor of 4 on careful drying, while the 2.8 μm band only strengthened by 65 per cent. It was shown that the reduction of radiation in the presence of moisture was not due to absorption by water vapour. Besides moisture and hydrogen, many other catalysts caused a reduction in infra-red radiation from CO explosion flames; ethyl nitrate and ethyl iodide were more effective than water, and Garner, Johnson and Saunders (1926) reported that the radiation could be reduced to less than 1 per cent of that from dried gases. Carbon tetrachloride on the other hand caused a slight increase in radiation; this was attributed to its "getting" residual traces of water.

In these experiments (e.g. Garner and Johnson, 1928) there was a close inverse correspondence between the amount of radiation and the flame speed*, the latter increasing rapidly with the addition of small amounts of catalyst, while the radiation fell. A few results are collected in Table IX.2.

The active catalysts all contain hydrogen and hydrogen itself has a strong effect on carbon monoxide/oxygen explosions. For carefully dried mixtures addition of hydrogen reduces the total radiation; there is a gradual fall in emission with increasing hydrogen concentration and then an abrupt fall which occurs at a critical pressure of hydrogen (Garner and Hall, 1930; Bawn and Garner, 1932). At atmospheric pressure this abrupt fall, or "step", in the radiation occurs with 0.03 per cent of hydrogen, and moves to greater hydrogen concentrations at lower pressures of the explosive mixture. Garner and colleagues discussed the reasons for these effects and attributed the gradual change of radiation with H_2 content to

*The "flame speed" is the rate of travel of the flame along the tube, not the true burning velocity of the mixture.

TABLE IX.2
Measurements of total radiation and flame speed for explosions of dry CO/O_2
with added catalysts

Catalyst	% Catalyst	Radiation (calories x 10^6)	Flame speed cm/sec
−	0.00	6.00	100
H_2O	0.23	2.76	780
H_2O	0.44	2.39	900
H_2O	1.95	1.63	>1000
$C_2H_5NO_3$	0.28	2.30	930
$C_2H_5NO_3$	0.89	1.56	2500?
C_2H_5I	0.32	1.67	1100
$CHCl_3$	0.47	2.30	560
CCl_4	0.35	6.34	79
CCl_4	0.72	6.60	44

deactivation of vibrationally excited CO_2 molecules (formed by the chemical reactions) by H_2O and the abrupt step to a change from a dry to a moist-type reaction mechanism.

Garner and Johnson (1928) studied the variation of radiation with time from dry CO/O_2 explosions. They used a galvanometer to record the emission and studied the propagating flame end-on, but were able to conclude that the strong radiation coincided with the duration of the flame and fell sharply to a small residual value when the flame became extinguished, this suggesting that emission was due to chemical processes in the flame front and not to thermal emission from the hot products. For moist gases, with higher flame speeds, the radiation outlasted the passage of the flame front and the thermal radiation was then more important; the stronger thermal radiation was associated with the higher temperature attained in explosions of moist gases.

Similar observations on anomalies in slower burning CO/air flames were made by David and colleagues. Pressure measurements showed that in $CO + H_2 +$ air mixtures the radiation followed the gas temperature, but with pure dry CO the maximum radiation occurred before the attainment of maximum pressure (David and Parkinson, 1933). Leah, Godrich and Jack (1950) compared the radiation from an expanding spherical flame in a dry CO/air mixture exploding in a long open-ended tube, with the radiation emitted during the cooling period, and made simultaneous measurements of temperature; the results showed clearly that the radiation came from the flame front, not from the hot gases. Again, addition of ¼ per cent of water vapour reduced the radiation to a quarter of its previous value but raised the temperature; the results also indicated that the reaction zone in dry CO/air explosion flames was extremely thick.

Bullock, Hornbeck and Silverman (1950) also made time-resolved studies of the rise and fall in radiation at 2.8 and 4.4 μm; they used a spherical bomb 15 cm diameter and employed a scanning spectro-meter at 125 scans/sec. For CO/O_2 mixtures containing a trace of H_2, the 4.4 μm band reached peak intensity in 0.02 sec and the 2.8 μm band in 0.03 sec. The tracings showed pulsations in light intensity, due to an "afterglow", with a period of around 0.2 sec; this time corresponded roughly with the time of travel of a flame across the vessel rather than with sonic vibrations. It appears to the author that the rapid fall of light intensity found in these studies is consistent with the other observations which indicate that in CO explosions the infra-red radiation comes from the flame front rather than from the hot products.

These results on CO explosion flames are difficult to explain. Present knowledge on vibrational relaxation times in CO_2 makes it difficult to accept the view of Garner that vibrational levels are highly populated by chemical processes and are responsible for the flame-front emission. For explosions in tubes the higher flame speed in the presence of moisture etc might increase heat transfer to the walls because of the higher gas velocity and greater turbulence, thus reducing the thermal radiation. This explanation can, however, hardly be applied to the observations made in large spherical vessels. In closed-vessel explosions the temperature is, of course, higher than in flames at constant pressure, and the temperature distribution in the vessel after explosion may be non-uniform; this is because the first part of the mixture to burn does work (and is thereby itself cooled) by compressing the rest of the mixture which is thus preheated and so attains a higher final temperature. There seems a need for more work on these CO explosion flames using modern fast recording techniques.

Cool Flames

The infra-red emission from the cool flames obtained in the slow combustion of organic compounds is much weaker than that from hot flames, but a detailed examination has been made (Donovan and Agnew, 1955; Agnew, Agnew and Wark, 1955) of both the emission and absorption of cool flames of ether in the infra-red. A modified flat-flame burner of the Egerton-Powling type was used and a two-stage flame was stabilized; emission from the first-stage cool flame, the region between the flames, the second-stage flame and also the region above this were studied; absorption was also measured below the first stage flame, as well as in the other positions.

The emission spectra were quite complex and showed a number of bands, some unidentified. The usual CO_2 bands near 2.8 and 4.4 μm were quite weak in and above the first-stage cool flame, but were rather stronger in the second-stage flame. The main features were in the regions 3.2—3.6, 5.5—6.0, 6.9—8.0 and near 8.96 μm; the relative intensities of these main bands, and the detailed structure within each,

TABLE IX.3

Peaks in the infra-red emission from first and second-stage cool flames of rich di-ethyl ether with oxygen

λ peak μm	First stage	Second stage	Possible emitter	λ peak μm	First stage	Second stage	Possible emitter
13.96	vw	vw		6.89	vw	m	C_2H_4
13.74	vw	w	C_2H_2	5.85	s	s	C=O group
13.48	—	vw		5.735	vs	vs	HCHO
13.32	vw	w		5.67	m	s	C=O group
13.07	—	vw		5.59	—	m	
12.43	vw	m		4.85	vw	s	C_2H_4, CO
12.30	vw	w					or CH_3OH
12.11	—	w		4.7	s	vs⎫	
11.79	—	w		4.2	m	s⎬	CO_2
11.03	vw	w		3.53	s	m	
10.57	w	m	C_2H_4	3.47	s	s	HCHO
9.99	—	vw	C_2H_4	3.38	m	s	C_2H_4*
9.50	vw	vw		3.25	—	m	C_2H_4*
8.96	s	s	ether	3.0⎫			
7.81	w	m	HCHO*	2.5⎬	—	s	H_2O, CO_2
7.35	s	m	*	2.32	—	vw	
7.02	w	w		1.88	—	w	H_2O

*Possible alternative identifications for these bands might be methane which has fundamentals with origins at 7.67 and 3.32 and might, owing to self absorption, show intensity maxima on each side of the origin positions.

change to some extent through the first and second-stage flames. It is difficult quickly to summarize these valuable results, but Table IX.3 gives the peaks, with estimates of intensity (w = weak, m = moderate, s = strong, v = very) in the first-stage and second-stage flames. These estimates by the author are based on the paper by Agnew, Agnew and Wark; Dr Agnew has informed the author that in

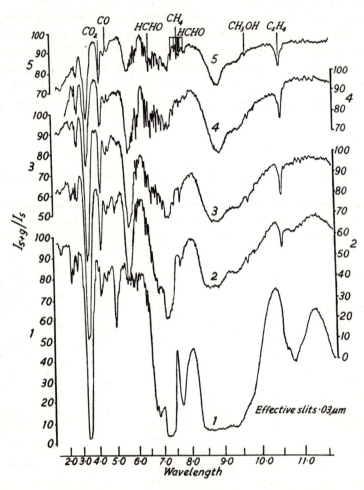

Fig. IX.4. Absorption spectra taken during the first and second stage cool flame of rich diethyl ether-air. Positions 1 to 5 are respectively below the first-stage flame, in the first-stage flame, between the two flames, in the second-stage flame and above the second-stage. From Donovan and Agnew.

this paper the captions to Figures 3 and 4 were accidentally interchanged, and this has been allowed for in this revised edition, but led to some difficulty in the first edition of this book.

The absorption spectra were very similar to those in emission. Indeed it appears clear that the emission is thermal in origin, and Donovan and Agnew (1955) were able to estimate the temperature in the various regions from the strength of emission and absorption bands; they obtained about 750 K in the first-stage flame, 840 K between and 1000 K in the second-stage flame. Figure IX.4 shows the absorption spectra at the five positions. Below the first-stage flame (position 1) the ether bands 10.6, 7.6 and 5.0 are clear; these disappear quickly in the first-stage flame (position 2). Other ether bands at 3.4, 7.3 and 9 μm appear to persist, but this is probably because these bands are characteristic of group frequencies, e.g. for the C—H bond, and are emitted by products as well. The strongest new feature in absorption is the band at 5.65 μm which appears quickly; this is characteristic of the C=O group and could be due to a variety of absorbers, especially aldehydes such as acetaldehyde CH_3CHO. This spectroscopic work does not give a complete analysis of the intermediate products during these slow combustion processes, but formaldehyde, ethylene and methane are certainly formed, and probably acetaldehyde and methyl alcohol; CO and CO_2 appear later.

A similar study (Donovan, 1957) has been made of the two-stage cool flame of acetaldehyde. In this case it was necessary to dilute with 10 per cent N_2 to obtain a stable flame. The emission spectrum (Figure IX.5) is very similar to that of the ether flame. Bands of methane at 7.67 and of formaldehyde at 6.65 and 7.8 μm are prominent, while the bands of acetaldehyde itself at around 5.7, 7.3 and 9.2 μm persist for some distance.

In these two-stage flames a number of the weaker emission bands remain unidentified. The ultra-violet absorption, which has been discussed previously (page 180) shows a band around 2600 Å which has been attributed to unsaturated ketones and β-dicarbonyls. The infra-red emission of these complex molecules would be difficult to identify, but may contribute to the emission. The products at a later stage of the reaction may also contain acids which only give far ultra-violet absorption but could contribute in the infra-red as well. Although interesting, this work has not added greatly to knowledge of cool-flame processes.

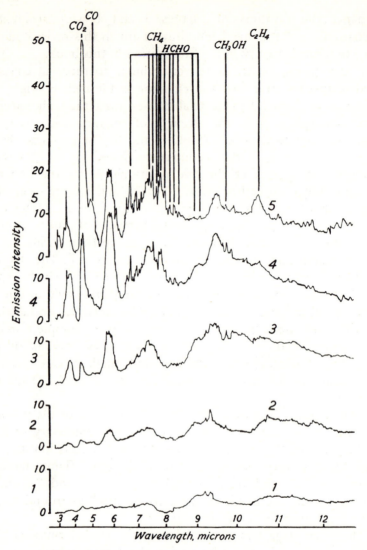

Fig. IX.5. Emission spectra of a rich acetaldehyde-air flame (+10% N$_2$). From Donovan. Flame positions as Fig. IX.4.

Flames with Nitrous Oxide

The infra-red emission of flames supported by nitrous oxide was studied by Bailey and Lih (1929). With carbon monoxide or coal gas the spectrum did not differ significantly from that of the corresponding flames in air, a strong band being observed at 4.50 μm and a less intense one at 2.88 μm. With hydrogen premixed with nitrous oxide, the spectrum was also similar to that of the hydrogen/air flame, but with hydrogen burning as a diffusion flame in an atmosphere of nitrous oxide the spectrum was quite different, there being a strong band at 4.75 μm and bands of moderate intensity at 4.00 and 5.49 μm, in addition to the water-vapour band at 2.7 μm and other weak bands due to this molecule. The bands at 4.75 and 4.0 are probably due to N_2O, which shows absorption near these wavelengths. The band at 5.49 was tentatively identified with water vapour by Bailey and Lih, but the agreement is unconvincing and it seems probable that this band was really due to NO which has its fundamental band at 5.3 μm.

Flames with Halogens

It was found, in early work by Julius, that the flame of chlorine burning with hydrogen gave a strong infra-red emission band due to HCl. The maximum intensity in emission from the flame was at 3.68 μm, whereas absorption by cold HCl gas lay at 3.46 μm. The difference between the wavelengths in emission and absorption was attributed by Kondratiev (1933) to the emission being due to transitions between higher vibrational levels, such as $(3 \rightarrow 2)$ at 3.72 μm, $(4 \rightarrow 3)$ at 3.86 μm and $(5 \rightarrow 4)$ at 4.02 μm, and he calculated that a very high proportion (90 per cent) of the energy released by the reaction went initially into vibrational energy of the newly formed HCl molecules.

Although there is indeed now evidence that the energy is initially released largely in the vibrational form, in flames at atmospheric pressure the vibrational relaxation time is so much shorter than the radiative lifetime that thermal emission should mask that from any chemical process. Thus in later work Newman (1952) studied the diffusion flame of Cl_2 in H_2 and found that most of the structure in the fundamental band was due to the main $(1 \rightarrow 0)$ transition, the rotational lines of which formed an R head at 3181.0 cm^{-1}. He also observed overtone bands, the $(2 \rightarrow 0)$ forming a head at 5833 and the $(3 \rightarrow 1)$ at 5620 cm^{-1}. The effective vibrational temperature was about 2500 K, which was rather below the adiabatic flame temperature and did not support the view of there being excess vibrational energy under the conditions of this flame. As with the CO_2 bands in organic flames, there would undoubtedly be strong self-absorption in the main fundamental band which would displace the position of the band maximum. Newman also attributed the apparent high rotational temperature (4500 K) to this effect; the weaker lines at higher rotational quantum number gave a rotational temperature of 2500 K, in agreement with the vibrational temperature.

However, in low-pressure flames the vibrational anomaly could well show up. Recent work to develop the HCl laser has shown that reaction of atomic hydrogen with molecular chlorine

$$H + Cl_2 = HCl + Cl$$

at low pressure (around 0.02 torr) leads to high populations in upper vibrational levels, especially $v = 3$, and produces a population inversion (Anlauf et al, 1967). High vibrational excitation leading to population inversion in certain rotational levels of the vibrational states $v = 1$ and $v = 2$ was also obtained by Kasper and Pimentel (1965) in the flash initiated reaction between hydrogen and chlorine at between 3 and 10 torr.

The infra-red emission from a premixed bromine/hydrogen flame burning above a platinum mesh shows bands of HBr around 2800 cm^{-1} (3.6 μm), and a detailed study (Bullock et al 1952) of the structure of the emission from a region about 7 mm above the flame front leads to effective rotational and vibrational temperatures of around 1400 K, consistent with the products being in equilibrium. Here again Anlauf et al have shown that low-pressure reaction of atomic hydrogen with bromine leads to high vibrational excitation of HBr and population inversions.

The diffusion flame of fluorine burning in hydrogen (Benedict, Bullock, Silverman and Grosse, 1953) shows the fundamental and first and second overtone bands of HF; the $(1 \rightarrow 0)$ and $(2 \rightarrow 1)$ form heads at 4412.3 and 4212.7 cm^{-1} (2.27 and 2.38 μm), the $(2 \rightarrow 0)$, $(3 \rightarrow 1)$ and $(4 \rightarrow 2)$ bands have heads at 7881, 7635 and 7288 cm^{-1}, while the weak $(3 \rightarrow 0)$ group has a head at 11536 cm^{-1}. Measurements indicated a temperature of about 4000 K for this very hot flame, but owing to lack of resolution and the usual troubles from self-absorption it was not certain whether equipartition of energy was established. Laser-type emission has been observed by Deutsch (1967) who used pulsed electric discharges to start reactions in flowing gas mixtures, such as $CF_4 + H_2$, and obtained stimulated emission lines of the $(1 \rightarrow 0)$, $(2 \rightarrow 1)$ and $(3 \rightarrow 2)$ bands of HF between 2.7 and 3.3 μm; he also excited corresponding lines of DF between 3.6 and 4.0 μm. This work was, of course, carried out at low pressure, 0.15 to 1.2 torr. Jonathan et al (1970) have shown that reaction of atomic hydrogen with F_2 leads to formation of HF particularly in the vibrational states $v = 5$ and 6, the $v = 5$ showing the greatest population inversion.

It thus seems a general feature of these halogen-hydrogen reactions that much of the energy released by the chemical reaction initially resides as vibrational energy in the newly formed halogen hydride molecules, and in low-pressure conditions this leads to strong infra-red emission and even laser action due to population inversion. In flames at atmospheric pressure, however, collision deactivation of the vibrationally excited molecules quickly establishes equilibrium and the infra-red emission is not obviously stronger than that due to thermal excitation.

Chapter X

Flame Structure and Reaction Processes

In the preceding chapters I have concentrated on observations relating to the spectra of flames and related sources. Now the time has come to talk of many things, in fact to "stick my neck out" and give my opinions on the interpretation of the observations and their bearing on theories of chemical reaction processes in flames and on the detailed structure of the reaction zone. As usual, the arrangement of the chapter is similar to that in the previous edition, but ideas have matured considerably and this has led to much detailed revision.

As pointed out in Chapter I, no single tool can be an open sesame to all combustion problems; major developments in mass spectrometry and gas chromatography have recently added to our knowledge of the intermediate species which occur in flames and slow combustion, but in a book on Flame Spectroscopy it is natural for me to weight the spectroscopic data rather highly. The visible and ultra-violet light from flames represents a very small fraction of the energy released, and this radiation comes from a very limited selection of electronically excited molecules, but despite this the emission of light is the most distinctive single property of flame and must surely be capable of giving much information about what is happening. Astronomers do not hesitate to use the spectrum of the extremely faint light from the stars to make pronouncements about their composition, temperature, structure and motion; a flame is, by comparison, a convenient and simple source in which many of the variables can be controlled at will, and its spectrum should be capable of giving quite detailed information on the state of the flame gases.

Reactions in Cool Flames

There is ample evidence for the formation of intermediate oxidation products during the slow combustion accompanying the cool flames of hydrocarbons and other organic compounds. Absorption spectra reveal bands of formaldehyde, acetaldehyde and propionaldehyde. The far ultra-violet extinction is probably due to organic acids and peroxides, and we have seen that the diffuse absorption band near 2600 Å indicates the presence of unsaturated ketones and β-dicarbonyls prior to the passage of the cool flame. In emission from cool flames the ultra-violet bands of formaldehyde are the main characteristic of the phenomenon, while in the infra-red, bands of formaldehyde and unsaturated hydrocarbons like ethylene and acetylene are found. Chemical sampling reveals a variety of aldehydes, peroxides, acids, peracids and alcohols.

The literature on this subject is vast and still growing. Most senior workers in the combustion field — Bone, Burgoyne, Cullis, Egerton, Hinshelwood, Laffitte, Lewis, Neumann, Newitt, Norrish, Pease, Prettre, Townend, Ubbelohde and Walsh — have been interested in cool flames and have made significant contributions. Lewis and von Elbe (1961) take ninety-six closely written pages to deal with slow combustion and cool flames. Obviously it is very difficult to summarize concisely all this work here.

Cool flames are associated with a discontinuity, or probably an actual reversal, in the usual rise of reaction rate with temperature. The cool flame and allied processes provide a mechanism for oxidation at much lower temperatures than would normally be required, and the apparent absence of this mechanism at higher temperatures suggests that one of the intermediary compounds may be thermally unstable or be removed rapidly by some competing process at high temperature. Another pointer to the type of mechanism is the observation that in many cases a succession of cool flames may be seen to pass through a mixture held in a furnace at apparently constant temperature.

Two general types of mechanism have been postulated to explain the periodicity of cool-flame phenomena, (i) a system of chain reactions with interlinked chain-branching and chain-terminating steps, and (ii) a self-accelerating thermal decomposition of some unstable product after this product has reached a critical concentration.

A formal system of chain reactions was proposed by Frank-Kamenetsky and developed by Walsh (1947) to account for a succession of cool flames. The initial fuel A is assumed to react with a radical X to give a product, B, and two more radicals, also denoted X, although not necessarily identical chemically. The reaction is thus autocatalytic in the formation of the radicals X and the oxidation of A. However, another intermediate Y is also formed, which may also react with X to give a stable product, also denoted B, and two more molecules of general type Y. The main steps are thus

$$A + X = B + 2X \tag{1}$$

$$X + Y = B + 2Y \tag{2}$$

and

$$A + Y = B \tag{3}$$

Thus we have an initial increase in [X], the concentration of X, but the removal of X by Y becomes more and more rapid until reaction is nearly stopped, when production of Y will in turn fall and removal of Y by (3) will allow a further build-up of X to commence. Thus the reaction rate may fluctuate, with the varying concentrations of X and Y out of phase. If we denote the rate constants for the above reactions by k_1, k_2 and k_3, then

$$d[X]/dt = k_1 [A] [X] - k_2 [X] [Y]$$

and

$$d[Y]/dt = k_2 [X] [Y] - k_3 [A] [Y].$$

Thus $d[Y]/dt$ becomes positive and Y increases at the expense of X when [X] reaches the critical value $k_3 [A]/k_2$. Similarly [Y] has a critical value $k_1 [A]/k_2$ above which $d[X]/dt$ becomes negative and so leads to a fall in [X] to below the critical value $k_3 [A]/k_2$. The reaction rate, measured by the rate of fuel consumption, will be

$$d[A]/dt = [A] (k_1 [X] + k_3 [Y])$$

and may fluctuate as [X] and [Y] reach their successive maxima. Thus Fish (1966) assumes that the criteria for the first, third, fifth etc. cool flames will be that [X] exceeds a critical value, while for the second, fourth, etc., cool flames [Y] exceeds its critical value. This kinetic explanation requires that successive odd-numbered cool flames decrease in intensity as fuel is consumed, and successive even-numbered ones also do so, but the relative intensities of odd to even numbered flames is not determined. For his work on 2-methyl-pentane Fish says that the pressure fluctuations indicate that the strength of the cool flames is in the order 1st > 3rd > 2nd; thus he favours this type of kinetic mechanism.

There has been some mathematical discussion of the conditions under which this type of process can produce oscillations. Gray and Yang (1969) initially maintained that it could not, but later (Gray, 1970) conceded that under some conditions, allowing for consumption of reactants, it might. Perche, Pérez and Lucquin (1970) made some computations on a simulated system and these showed that oscillations could occur in a non-equilibrium system such as a cool flame.

This mechanism is purely formal, and in practice there is difficulty in the chemical identification of the radicals and intermediates X and Y. Aldehydes and peroxides are detectable by sampling, and some of these are known to accelerate and others to inhibit the reaction; thus under some conditions formaldehyde may prolong the induction period, while acetaldehyde and peroxides (especially alkyl hydro-peroxides) may shorten it. Walsh tentatively identified X and Y as peroxides and formaldehyde respectively, and later Walsh (1963), Cullis et al (1966) and Barat et al (1972) stressed the importance of alkyl peroxy radicals ROO· and their isomerization in the chain-branching process. Déchaux, Flament and Lucquin (1971) also favour this basic scheme and discuss the interplay of two mechanisms involving the formation and reactions of peroxy radicals $RO_2 \cdot$ and $RCO_3 \cdot$, and the competition between chain propagation and radical isomerization.

The second general mechanism suggested to account for multiple cool flames involves decomposition of a reactant by a mild thermal explosion. Knox and Norrish (1954) pointed out that under conditions of slow reaction self-heating will normally occur when the rate of heat production exceeds the rate of heat removal by the surroundings. However, for many hydrocarbons undergoing slow oxidation, the reaction rate has a negative temperature coefficient at around 400°C. At this temperature the aldehyde which is formed in the early stages of the self-heating process may be rapidly destroyed, and, because of competing reactions, may not be replenished. The rate of reaction and temperature then fall rapidly until thermal stability is re-established when the reaction can accelerate again. An alternative to aldehyde decomposition is peroxide decomposition; Bardwell and Hinshelwood (1951) concluded, from kinetic studies, that the cool flame was due to a mild explosive decomposition of peroxides rather than to their reaction with either oxygen or formaldehyde. This would require the build-up of a certain critical concentration of peroxide prior to the passage of the cool flame and

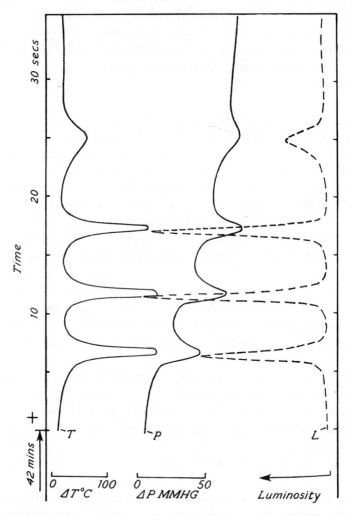

Fig. X.1. Records of temperature (*T*), pressure (*P*) and light emission (*L*) as a function of time for the passage of four cool flames through a 1:1 methyl ethyl ketone: oxygen mixture at 250 torr and 590 K. (From Barnard and Watts.)

a sharp decrease in it during the passage of the flame; the chemical evidence is, however, rather in conflict with this, since the peroxide concentration appears to rise, not fall, during the cool flame period. A possible way out of this difficulty would be to associate the cool-flame luminescence with reactions leading to peroxide formation rather than with its explosive thermal decomposition. It is interesting to note the spectroscopic evidence, from the diffuse absorption at 2600 Å (see page 180), for the fall in concentration of β-dicarbonyls etc. during the passage of a cool flame.

Cool flames are usually accompanied by a small temperature rise; this may be quite trivial or up to 200 K. Barnard and Watts (1969) have used silica-coated thermocouples to study temperature changes during the passage of cool flames through ketones, and have also made measurement of pressure changes and intensity of light emission. Their results for methyl-ethyl ketone are shown in Figure X.1. The light emission appears to precede the temperature rise by at least half a second, which is more than the response time of the thermocouple (0.2 sec). In some of these observations the third cool flame appears definitely brighter than the first, which would be inconsistent with the purely kinetic mechanism. In this work a considerable build-up of organic peroxides was observed in the cool flame period, followed by a sharp fall as the temperature reached its maximum. Barnard and Watts conclude that these observations support the thermal mechanism put forward by Knox and Norrish.

With temperature rises of the magnitude indicated in Figure X.1, gas movements due to buoyancy effects may occur. Melvin (1969) has shown interferograms which indicate an asymmetrical temperature distribution during the passage of the cool flame; he notes the existence of appreciable convective movements and considers a possible mechanism involving successive ignitions near the top of the vessel and gas movements. Two-stage cool flames in steady flow systems, as observed by Topps and Townend, could not be explained in this way; gas movements do not seem, to the author to be a likely cause of cool flame oscillations, although they may well modify them.

Obviously the last word on cool-flame processes has not yet been said. It might perhaps be possible to bring the kinetic and thermal mechanisms into one by assuming that temperature changes also have important effects on the rates of reactions (1), (2) and (3), and Fish (1966) has compared the implications of the various theories. Perche,

Pérez and Lucquin (1971) have shown that periodicities in reaction rate can be accounted for if the chain termination step has a greater activation energy than the chain branching step.

The subject of cool flames is not just academically interesting. There is a close relationship between the two-stage ignition process and knock in internal combustion engines. The mechanism of peroxide formation and the point of attack on the hydrocarbon are important; straight-chain hydrocarbons are more vulnerable than branched-chain molecules like iso-octane and so have their two-stage ignition region at lower pressures and knock more easily (see also page 289).

The Formaldehyde Excitation Process

Cool flames normally occur at relatively low temperatures when thermal emission from formaldehyde molecules could not possibly account for ultra-violet light emission. This must be due to chemiluminescence, and we should like definitely to identify the specific reaction.

It was shown by Topps and Townend (1946) that the light yield is very low. For the relatively bright second-stage cool flame of acetaldehyde only about one light quantum is emitted for every 10^6 acetaldehyde molecules consumed. This very low light yield may be partly accounted for by collision deactivation of the excited formaldehyde molecules; however, the weak quenching of fluorescence of formaldehyde shows that the efficiency of collision deactivation is not high. About 125 formaldehyde molecules may be detected chemically in the products for each light quantum emitted. It thus seems that the formation of the excited formaldehyde molecules is not a part of the main chain reactions of the oxidation process, but involves some secondary process. Walsh (1947) concluded that the emission must result from radical-radical reactions; this would account for the low light yield, and also these radical-radical reactions are those most likely to be sufficiently exothermic to produce the electronic excitation, which requires about 340 kJ/mole. The formaldehyde light emission from cool flames is so very characteristic of the phenomenon that the author is reluctant to dismiss it as an incidental side reaction. The excited molecules are not formed in the direct chain reactions but the radical-radical reaction responsible for the emission may well be one of the major chain terminating steps and the strength of the emission should depend on the square of the radical concentration and thus be some measure of the reaction rate.

Among the many specific reactions producing excited formaldehyde, Walsh (1947) suggested

$$RCH_2O + OH = ROH + HCHO* \qquad (4)$$

while Lewis and von Elbe (1961) favour

$$CH_3 + HO_2 = H_2O + HCHO* \qquad (5)$$

or

$$CH_3 + CHO_3 = H_2O + CO + HCHO* \qquad (6)$$

Damköhler and Eggersglüss (1942) proposed the more general reaction

$$RCH_2O + R' = RR' + HCHO* \qquad (7)$$

where R' is either an alkyl radical or an H atom and RCH_2O is especially likely to be methoxy, CH_3O. Bailey and Norrish (1952) made a thorough study of cool-flame processes in hexane and favoured this reaction (7) and also discussed the reaction of acetaldehyde and atomic oxygen

$$CH_3CHO + O = HCHO + HCHO* \qquad (8)$$

This reaction (8) does not fulfil the radical-radical condition but would be sufficiently exothermic. The author is not aware of any direct studies of the atomic flame of acetaldehyde with free O atoms which could confirm or disprove (8), but with all flames studied the normal and atomic flames give a similar spectrum, and as the normal hot flame of acetaldehyde does not show formaldehyde bands the reaction (8) seems unlikely. Another reaction of this type is that between methanol and atomic oxygen

$$CH_3OH + O = H_2O + HCHO* \qquad (9)$$

It is known that both the normal and the atomic flames of methanol do show formaldehyde emission; reaction (9) probably is responsible for the emission from methanol flames, but seems unlikely to be a general reaction in cool flames; methanol itself does not possess a cool flame.

Reactions (5) and (6) both involve methyl radicals and if they occurred should lead to particularly strong formaldehyde emission from methane flames. Methane itself does not possess a two-stage combustion region and cool flames only occur under special conditions. The hot flame of methane certainly contains CH_3 radicals and might be expected to contain some HO_2 as well but does not show formaldehyde emission. Despite the advocacy of these reactions by Lewis and von Elbe they appear unlikely.

This leaves (4) and (7) as the most probable reactions. These are very similar because they both involve reaction of an alkoxy radical with another radical to form excited formaldehyde and a stable product molecule. It is difficult to say whether the second radical is OH, as in (4), or an alkyl radical or H, as in (7), or perhaps even some other radical. All these reactions would be sufficiently exothermic. The work of Gaydon, Moore and Simonson (1955) on the formaldehyde emission from the auto-ignition of methane is most readily explained by reaction of methoxy with OH radicals. For other cool flames it must suffice to accept a general reaction of an alkoxy radical, RCH_2O, with another unspecified radical.

In addition to the cool flames obtained during oxidation processes, Gray and colleagues (Gray, 1954; Gray, Hall and Wolfhard, 1955; Gray and Yoffe, 1949) have found that some alkyl nitrites and nitrates give a glow on thermal decomposition. In the case of methyl nitrite this glow is still visible when it is diluted a million times with argon. The primary step is believed to be decomposition to methoxy,

$$CH_3ONO = CH_3O + NO$$

These glows apparently give the formaldehyde emission, and their existence supports the assignment of cool-flame emission to a reaction of methoxy, or more generally the radical RCH_2O. For the glow of the highly diluted methyl nitrite it is not easy to see where the second free radical would come from; reaction of two methoxy radicals

$$CH_3O + CH_3O = CH_3OH + HCHO^*$$

seems the most likely; taking Gray's value for the heat of formation, the reaction would be exothermic by about 76 kcal/mole (318 kJ), which is nearly enough to form the formaldehyde in the electronically excited state.

Reactions in Diffusion Flames

Studies of absorption and emission spectra of flat diffusion flames, already reviewed, have given a clear general picture of processes in this type of flame. Apart from the small region at the base of the flame, the hydrocarbon and oxygen are separated by a wedge of hot combustion products and the hydrocarbon is pyrolysed to carbonaceous soot and hydrogen in a region where the concentration of molecular oxygen is negligible. The OH concentration is greatest rather on the oxygen side of the region of highest temperature (see *Flames*, page 144), but these radicals do penetrate further to the fuel side than molecular oxygen. Measurements of H_2O concentrations were not made in these flames, but some observations by Cole and Minkoff (1956) indicated that in rather similar flames the infra-red emission from H_2O was strongest near the luminous carbon zone. The hydrocarbon concentration is also likely to be highest near the carbon zone, or even slightly to the fuel side of it. Thus the pyrolysis of the hydrocarbon occurs in a region practically free from molecular oxygen, but where hydrogen atoms probably have an appreciable concentration and where a few OH radicals may still penetrate. The soot particles are also in a region of low, probably negligible, oxygen concentration and are consumed mainly by reaction with water vapour to form CO and H_2. The oxygen, as it diffuses towards the carbon, is consumed by reaction with hydrogen. Carbon monoxide will also react to form CO_2 on the oxygen side.

The detailed processes of pyrolysis and soot formation have been the subject of much research and discussion; work up to about 1969 has been reviewed in *Flames* (Gaydon and Wolfhard, 1970). In forming soot we start with small hydrocarbon or other organic molecules and end with particles containing many thousands of atoms and a much higher carbon/hydrogen ratio. Thus we must have hydrogen removal (dehydrogenation) and polymerisation or condensation to larger aggregates. Many routes appear possible, and in *Flames* five of these were considered in some detail (i) via acetylene and polyacetylenes (C_4H_4, C_4H_2, C_6H_2, C_8H_2 etc.), (ii) polymerization to large molecules with low vapour pressure which condense to droplets and graphitize from within, (iii) ring closure to form first simple aromatics, then larger aromatics like circumanthracene and finally graphite, (iv) nucleation on small nuclei, possibly C_2, to form a small particle, with further growth by surface decomposition of hydrocarbons at the particle surface, and (v) nucleation by positive ions, such as $C_n H_m^+$, which are known to be formed in flames (Howard, 1969). There is probably no single route by which all soot is formed; in diffusion flames the process is relatively slow and there is sufficient time for polymerization as a first step, so that mechanisms (ii) and (iii) may be relatively more important. The recent studies of the absorption spectrum of the pre-carbon zone of diffusion flames by Laud and Gaydon (1971) show that the pyrolysis "continuum" is not really continuous but diffusely banded; this may be taken as evidence of a process involving larger molecules rather than nucleation on C_2 or ions, but might be consistent with mechanisms (i), (ii) or (iii).

The pyrolysis to form polymers and unsaturated compounds will probably be catalysed by radicals diffusing from the flame, so that it occurs more rapidly and under conditions of lower temperature. The products may thus differ from those of uncatalysed pyrolysis at higher temperature. Polymerization probably involves chain reactions and it has been known for a very long while (Storch, 1934) that a trace of oxygen catalyses the polymerization of ethylene; Volman (1951) has shown that butadiene is rapidly polymerized at 150°C by CH_3 radicals obtained from butyl peroxide. Free hydrogen atoms will certainly be present in diffusion flames and will diffuse fastest into the fuel side, but observations on "atomic flames" of moist hydrogen reacting with hydrocarbons do not provide any evidence of polymerization, but it may be that polymerization chains could be started by hydrogen atoms at rather higher temperature. Partly halogenated compounds do polymerize rapidly to oily substances in the presence of hydrogen atoms.

Near the base of diffusion flames, on the oxygen side of the carbon zone, there is some C_2 and CH emission. At one time the view was expressed that the position of these bands indicated that C_2 could not be an intermediate in the carbon-formation process, but that the C_2 might be formed from the carbon particles. Certainly it is most improbable that C_2 precedes soot formation in diffusion flames, but we must remember that in these flames there are gradients of both composition and temperature; we know that for the other main emitters, OH and O_2, the radiation depends on the state of local equilibrium determined by temperature and concentration, and not upon the processes of formation. The C_2 and CH radiation becomes very weak higher in the flame, even though hot carbon particles are still present. This emission seems to be associated with a small region near the base of the flame where the wedge of combustion products is still very thin so that there is some interdiffusion of oxygen and unpyrolysed fuel.

In diffusion flames with air the carbon luminosity is less strong and the C_2 and CH emission zone extends higher. The thickness of the flame is also less. These differences are no doubt due to lower temperature; fuel molecules are able to penetrate nearer to the oxygen zone before they are pyrolysed; the dilution with nitrogen will also reduce the number of collisions between fuel molecules and between radicals and fuel molecules, so delaying the pyrolysis. In these flames with air, and still more in flames with additional nitrogen as diluent, the green region, where the C_2 and CH emission occurs, also shows some of the anomalies such as high electronic excitation and predissociation of CH which are usually associated with premixed flames. Low-pressure diffusion flames, even with pure oxygen, closely resemble pre-mixed flames; the more rapid inter-diffusion at low pressure again enables unpyrolysed fuel to come into contact with oxygen.

Reactions in Premixed Flames

The emission spectra of organic flames show as strongest features the bands of two unexpected radicals, CH and C_2, which do not fit into the main chain reactions of any accepted reaction mechanism. Intermediate oxidation products, such as aldehydes, are conspicuous by their absence in either emission or absorption. Even in combustion initiated by flash photolysis, when very long absorbing paths can be used, these bands are still not obtained.

Early theories of hydrocarbon oxidation were mostly based on observations of slow combustion reactions and on the ignition process. The late W. A. Bone visualized step-wise hydroxylation to form alcohols, aldehydes (with elimination of H_2O) and acids. Later a peroxidation mechanism was developed by H. L. Callendar, Sir Alfred Egerton and others; according to this, reaction was initiated when an energy-rich hydrocarbon molecule combined momentarily with oxygen to form a peroxide which quickly broke down to aldehyde and water, the aldehyde molecule carrying excess energy which was passed on to continue the combustion processes. These early theories led to a lot of controversy. It now seems that for slow combustion and the initiation of explosive reaction the peroxide mechanism is nearer the truth, as discussed on page 247, but for actual propagation of established flames neither is relevant.

In flames the propagation involves heat transfer to the unburnt gas mixture and diffusion of free radicals, especially OH and H, from the burnt gases into this preheated region where chain reactions are thereby started. These main chain reactions involve many radicals and species, and many side chains also participate. The simplest case to consider is the flame of methane. The detection of CH_3 radicals by absorption spectroscopy has already been mentioned (page 176) and recently Harvey and Jessen (1973) have found strong absorption by this radical well into the preheating zone of low-pressure flames. In early studies with a mass spectrometer Eltenton (1947) detected CH_3, acetylene and some intermediate oxidation products including formaldehyde and the radicals HCO and CH_3O. The CH_3 became more prominent in hotter flames but the oxidation products like formaldehyde and HCO had a marked negative temperature co-efficient being much weaker when additional heat was supplied to the flame.

The main reaction chain for the oxidation of methane in flames appears to be

$$CH_4 + OH = CH_3 + H_2O$$
$$CH_3 + O_2 = HCHO + OH$$

Fristrom and Westenberg (1965) in *Flame Structure* page 345, have listed these and fourteen other reactions which are important in premixed methane/oxygen flames,

$$CH_4 + H = CH_3 + H_2 \qquad O + H_2 = OH + H$$
$$CH_4 + O = CH_3 + OH \qquad O + H_2O = OH + OH$$
$$HCHO + OH = HCO + H_2O \qquad H + H_2O = H_2 + OH$$
$$HCO + OH = CO + H_2O \qquad H + H + M = H_2 + M$$
$$CH_3 + O = CO + ? \qquad O + O + M = O_2 + M$$
$$CO + OH = CO_2 + H \qquad H + O + M = OH + M$$
$$H + O_2 = OH + O \qquad H + OH + M = H_2O + M$$

The last eight of these are, of course, those used in discussing hydrogen flames (Chapter V). The low concentration of formaldehyde in hot flames suggests that its removal, either by reaction with OH or by thermal decomposition, is rapid. HCO is detectable by its emission spectrum (Vaidya's hydrocarbon flame bands), especially in the cooler flames, but is known to have low thermal stability.

In hydrogen flames the free atoms and radicals only recombine by the relatively slow three-body reactions which are the last four listed above; the evidence for persistence of excess free radicals in hydrogen flames has already been discussed (page 122). In hydrocarbon flames there is less evidence for persistence of abnormal concentrations of free radicals. This may be attributed to fairly rapid removal of OH in bimolecular reactions with formaldehyde and HCO, as indicated above. Plots of the temperature course through the reaction zone often show that the curve of heat-release rate against distance through the reaction zone is not symmetrical but has a marked shelf on the burnt-gas side (for discussion see Gaydon and Wolfhard 1970, *Flames* page 89). The main heat release is associated with the formation of carbon monoxide and water vapour, and the later slower reactions of CO to form CO_2 (see Chapter VI) are responsible for this shelf on the curve. There is now quite a lot of data for the reaction-rate constants for many of the reactions listed, but even for methane, the simplest hydrocarbon, it is still not possible to give a complete quantitative treatment of the combustion.

Larger molecules and unsaturated molecules like ethylene and acetylene, which are detectable by mass spectroscopy, may be produced by radical polymerization chains, such as

$$CH_3 + CH_4 = C_2H_5 + H_2$$

and chain termination steps such as

$$C_2H_5 + H = C_2H_4 + H_2$$
$$C_2H_5 + OH = C_2H_4 + H_2O$$

or

$$CH_3 + CH_3 = C_2H_4 + H_2 .$$

These reactions, and side chains involving other radicals such as CH_2 and C_2H, appear to be important in producing the spectroscopically observable species CH and C_2 and also in soot formation. They are, of course, relatively important under fuel-rich conditions when the polymerization can compete with oxidation to H_2O and CO_2. These processes are discussed in the following sections.

Carbon Formation in Premixed Flames

Conditions in premixed flames are significantly different from those in diffusion flames, because whereas in the latter pyrolysis of fuel takes place in a region where the oxygen concentration is negligible, in the premixed flames oxygen is always present. Many premixed flames show luminosity due to hot carbon particles at mixture strengths for which carbon formation would not be expected. When the total number of oxygen atoms exceeds the total number of carbon atoms in a hot gas mixture, then under equilibrium conditions, practically all the carbon should be present as CO (and CO_2) and there should be no free solid carbon. However, carbon luminosity frequently occurs when oxygen is present in considerable excess of the quantity which should suppress it. (Gaydon and Wolfhard, 1970). Quantitative measurements on the onset of luminosity, reported in a most valuable paper by Street and Thomas (1955), show that for paraffins (ethane, propane, n-pentane and n-octane) with air luminosity starts at an oxygen/carbon ratio of 2.1 or 2.2 instead of the expected 1.0. Obviously under some conditions the processes of dehydrogenation and aggregation are able to compete successfully with those of oxidation. For acetylene/air, however, luminosity only starts at an oxygen/carbon ratio of 1.20, which is nearer the expected mixture strength.

Temperature is probably an important influence, and it is interesting to compare the temperatures of some flames with oxygen and with air at the point at which solid carbon should just be formed under equilibrium conditions, i.e. O/C = 1.0.

Mixture	Calc. temp., K	Mixture	Calc. temp., K
$C_2H_2 + O_2$	3325	$C_2H_2 + O_2 + 4N_2$	2200
$C_2H_4 + O_2$	2300	$C_2H_4 + O_2 + 4N_2$	1350
$C_2H_6 + O_2$	1150	$C_2H_6 + O_2 + 4N_2$	800
$C_6H_{14} + 3O_2$	1450	$C_6H_{14} + 3O_2 + 12N_2$	1020

Many of these mixtures are too rich and have too low a temperature to propagate a flame, i.e. are beyond the rich limit. Yet in practice we know that premixed flames of these fuels can give carbon. The special position of acetylene with its very hot flame of rich mixtures is clear.

The spectra of rich premixed flames show strong C_2 and CH emission. Acetylene flames show these bands particularly strongly, and in the "feather" above the tip of the reaction zone a band of C_3, as well as C_2, can be detected in emission and absorption, and there is also some other structure of uncertain origin around 3900 Å (see page 177); concentrations of C_2, but not C_3, are well above equilibrium in this reaction zone. The possibility that soot might be formed by polymerization of C_2 radicals has been considered (e.g. Smith, 1940*b*) but semi-quantitative estimates of the amount of C_2 in flames (Gaydon and Wolfhard, 1950), observations on the relative positions of C_2 and soot in flames and some work on flash photolysis make this theory untenable. There is indeed, usually close correspondence between carbon formation and the strength of C_2 radiation from flames, but it is not invariably so; thus sulphur trioxide may greatly increase soot formation without strengthening C_2 (Gaydon and Whittingham, 1947), while chlorine can strengthen C_2 without increasing the tendency to soot formation. In very rich flames on a flat-flame burner Egerton and Thabet (1952) noted that the carbon luminosity occurred well above the bluish flame front, being separated from it by a dark zone.

In rich flames of simple paraffins chemical sampling with probes shows the formation of unsaturated compounds, especially ethylene and acetylene. For the acetylene flame itself, sampling with a mass spectrometer shows the formation of a number of carbon rich species, including the radical C_2H and several polyacetylenes (Bonne, Homann and Wagner, 1965). Curves for variation of concentration of some of these species with height in a low-pressure flame are redrawn in Figure X.2.

The possible soot-forming mechanism via acetylene has already been mentioned (page 253). However, although acetylene may be an important intermediary, it has been stressed by Street and Thomas that acetylene itself is less likely to give carbon in excess above equilibrium amounts than most other fuels; it is therefore unlikely that all carbon is formed via acetylene.

Once a hot carbon particle is formed it is probable that surface reaction will lead to the growth of the particle by thermal decomposition of most organic fuels. The nature and process of formation of the initial particle, i.e. the carbon nucleus, then assumes special importance. We know that C_2 radicals are present in premixed flames and might serve as nuclei. Fairbairn and Gaydon

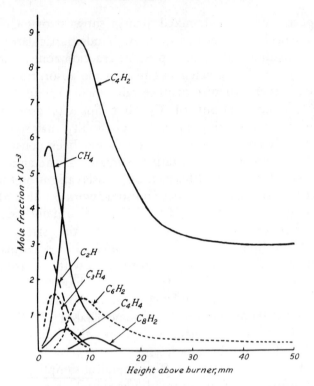

Fig. X.2. Concentrations of carbon-hydrogen species as observed by mass-spectrometric sampling of an oxy-acetylene flame ($C_2H_2/O_2 = 0.95$) at 20 torr. Redrawn from Bonne, Homann and Wagner (1965).

(1955) studied the afterglow of carbon monoxide in a heavy-current discharge tube at around 3½ torr and found that solid carbon was deposited when the afterglow gases were led into acetylene or benzene vapour, but not when led into other hydrocarbons. The afterglow gases showed C_2 bands, especially the High Pressure bands due to members of the Swan system with $v' = 6$. This gave some experimental confirmation that C_2 could initiate the formation of solid carbon and that acetylene might be especially important in the initial growth of the particles. The mechanism of formation of C_2 is discussed in the next section.

Another possible mechanism for initiating the formation of a carbon particle was suggested by Goldfinger (1955). This involved reaction of CH radicals

$$CH + CH = C + CH_2$$

and

$$C_x + CH = C_{x+1} + H$$

This mechanism is believed to be important in steller atmospheres but in flames the concentrations of CH are too low for it to be responsible for the formation of large particles; it is just possible it could play a part in formation of nuclei.

In a premixed flame there will be competition between the soot-forming processes and those of oxidation of both fuel and of the soot particles themselves. We have already stressed that luminous carbon particles and even free soot, as smoke, can be obtained under conditions for which carbon should not be present under equilibrium conditions, so that the carbon-forming reactions must be able to race the oxidation reactions.

In discussing carbon formation in diffusion flames the routes via large polymers were favoured, but in premixed flames there will be less opportunity, because of the shorter time, higher temperature and competing oxidation, for these to be formed. The author visualizes the main processes in a rich premixed flame as follows:

(1) Diffusion of radicals from the flame front into the mixture, giving chain reactions leading to the formation of unsaturated compounds, including acetylene.

(2) Reactions of unsaturated compounds to give carbon nuclei, perhaps including C_2.

(3a) Reactions of carbon nuclei with acetylene and perhaps fuel molecules to give larger nuclei.

(3b) Competing oxidation of carbon radicals and small nuclei.

(4a) Growth of carbon nuclei to carbon particles by surface reaction with original hydrocarbons and intermediates.

(4b) Parallel oxidation of hydrogen, formed by reactions (3a) and (4a), to form water vapour.

(5) Reaction of carbon particles with water vapour to form H_2 and CO, removing carbon luminosity and establishing equilibrium.

Reaction (1) will occur in the preheating zone, (2) in the flame front, (3) in the dark space above it, (4) at the base of the luminous zone and (5) at the top of the luminous zone.

The extent of carbon formation will obviously depend on the amount of oxygen available for competing reactions and the ease with which a particular fuel is oxidized or dehydrogenated; thus normal paraffins tend to oxidize more readily than iso-paraffins (page 250) while the latter give more soot. The processes will also depend on temperature; Burgoyne and Neale (1953) found that carbon was not formed in some very rich mixtures which were still able to propagate a flame; this was probably because the temperature was too low for either reactions (2) or (4a) to take place. In conical flames the regions where carbon formation occurs may be influenced by selective diffusion processes of the type responsible for the formation of cellular and polyhedral flames (see Gaydon and Wolfhard, 1970).

Concentrations of Free Radicals

Since the dominant features of the emission spectra of flames are bands of diatomic free radicals, OH, CH and C_2, it is important to have some knowledge of the expected equilibrium concentrations to compare with experimental values. Free atoms are also present and data for the concentrations of these are also useful.

For a diatomic molecule AB it is possible, if the heat of dissociation is known, to derive the equilibrium constant K_p relating the partial pressures of AB, A and B

$$K_p = P(A) \cdot P(B)/P(AB)$$

using spectroscopic data for the moment of inertia and vibrational frequency of AB and the statistical weights of A, B and AB. A convenient formula is given in *Dissociation Energies* (Gaydon, 1968;

TABLE X.1

Equilibrium constants, K_p, in atmospheres, for the dissociation of CO, C_2 and CH

Temp., K K_p	2000	2300	2600	2900	3200	3500
$\dfrac{P(C) \cdot P(O)}{P(CO)}$	7.5×10^{-22}	3.9×10^{-18}	2.8×10^{-15}	5.2×10^{-13}	3.7×10^{-11}	1.3×10^{-9}
$P(C)^2/P(C_2)$	3.9×10^{-10}	4.7×10^{-8}	1.9×10^{-6}	3.5×10^{-5}	3.7×10^{-4}	2.7×10^{-3}
$\dfrac{P(C) \cdot P(H)}{P(CH)}$	2.8×10^{-4}	4.5×10^{-3}	3.9×10^{-2}	0.21	0.86	2.8

see page 152). Constants for equilibria between H_2, H, O_2, O and OH are well known, being required for the calculation of flame temperatures, and values are tabulated in *Flames*. For the dissociation of CO, C_2 and CH to free atoms, some values, taken from Gurvich et al (1962) are listed in Table X.1. These use $D(CO) = 255.79$ kcal/mole (1070kJ), $D(C_2) = 143.17$ kcal (600kJ)* and $D(CH) = 80$ kcal (335kJ).

Thus if we know the partial pressures of CO and atomic oxygen in a flame we can derive the equilibrium partial pressure of carbon atoms; from this we then get that of the C_2 radicals, and also by combining $P(C)$ with the partial pressure of H atoms in the flame we can get $P(CH)$. In flames of mixtures so rich that there is insufficient oxygen to convert all carbon to CO, solid carbon should be present and then the partial pressures of atomic carbon, C_2 and also C_3 will be determined by the vapour pressure of carbon at the gas temperature. Some values, abridged from Gurvich et al, are given in Table X.2. The values for $P(C_3)$ are based on a dissociation energy,

TABLE X.2

Partial pressures (in atm.) of atomic carbon, C_2 and C_3 in equilibrium with hot solid carbon

Temp., K	2000	2300	2600	2900	3200	3500
$P(C)$	3.4×10^{-11}	9.3×10^{-9}	6.9×10^{-7}	2.1×10^{-5}	3.3×10^{-4}	3.2×10^{-3}
$P(C_2)$	3.0×10^{-12}	1.8×10^{-9}	2.5×10^{-7}	1.2×10^{-5}	2.8×10^{-4}	3.8×10^{-3}
$P(C_3)$	3.8×10^{-11}	1.8×10^{-8}	1.8×10^{-6}	7.2×10^{-5}	1.4×10^{-3}	1.6×10^{-2}

to free C atoms, of 320.755 kcal/mole. This value is perhaps in some doubt, and all these partial pressures are very sensitive to the dissociation energies used in the calculations. Values are now more reliable than previously, but although partial pressures are usually quoted to several significant figures, for these carbon species at high temperature it is difficult to be sure of even the first significant figure!

For all but the hottest flames, equilibrium concentrations of the carbon radicals, C, C_2, C_3 and CH, are trivial. Even for hot flames, if

---

*A more recent value for $D(C_2)$ based on observation of a predissociation, by Messerle and Krauss (1967), is 6.11 ± 0.04 e.V. (141 kcal, 591 kJ); this would give a slightly higher equilibrium constant. The value of $D(C_2)$ used in the previous edition of this book (113 kcal) is definitely wrong.

TABLE X.3

Concentrations of free radicals and atoms in acetylene flames, expressed as partial pressures in atmospheres. For oxy-acetylene flames values for the temperature, CO, O, H and OH are taken from Gay et al (1961) but with values for C, C_2 C_3 and CH recalculated from data by Gurvich et al (1962). The acetylene-air flame is the author's calculation, again using Gurvich et al

Mixture	Temp., K	P(CO)	P(H)	P(O)	P(OH)	P(C)	P(C$_2$)	P(C$_3$)	P(CH)
C_2H_2 + 2.5 O_2	3345	0.334	0.077	0.119	0.097	6.1×10^{-10}	3.7×10^{-16}	1.0×10^{-21}	3.0×10^{-11}
C_2H_2 + 2.0 O_2	3398	0.407	0.108	0.100	0.087	1.7×10^{-9}	1.9×10^{-15}	8.9×10^{-21}	9.4×10^{-11}
C_2H_2 + 1.5 O_2	3435	0.498	0.154	0.057	0.059	5.4×10^{-9}	1.6×10^{-14}	1.8×10^{-19}	3.8×10^{-10}
C_2H_2 + 1.1 O_2	3383	0.584	0.185	0.0081	0.013	2.5×10^{-8}	4.7×10^{-13}	3.8×10^{-17}	2.5×10^{-9}
C_2H_2 + 1.0 O_2	3325	0.607	0.179	5.0×10^{-6}	1.1×10^{-5}	2.1×10^{-5}	3.9×10^{-7}	5.4×10^{-7}	2.6×10^{-6}
C_2H_2 + 0.9 O_2	3266	0.591	0.159	1.0×10^{-7}	2.5×10^{-7}	5.5×10^{-4}	5.2×10^{-4}	2.5×10^{-3}	7.7×10^{-5}
C_2H_2 + 0.8 O_2	3233	0.567	0.154	8×10^{-8}	2.4×10^{-7}	4.2×10^{-4}	3.8×10^{-4}	1.8×10^{-3}	6.6×10^{-5}
C_2H_2 + 12½ air	2525	0.039	0.0019	0.0022	0.0095	1.1×10^{-14}	1.9×10^{-20}	3.5×10^{-29}	9.0×10^{-16}

there is sufficient oxygen to convert carbon to CO, i.e. O/C > 1.0, the partial pressures of carbon species are still very low, but for rich flames, O/C < 1.0, they may become appreciable; thus the C_2 and C_3 radiation from the "feather" above the tip of the rich oxy-acetylene flames is apparently equilibrium thermal radiation. Acetylene gives the hottest of the ordinary hydrocarbon flames because of the heat released when the acetylene is decomposed. Table X.3 gives values for equilibrium partial pressures of radicals in acetylene flames, partly based on the work of Gay et al (1961) and partly recalculated by the author.

Early theoretical calculations indicated that larger carbon molecules such as C_7 and C_9 should also be present, and these were considered by Gay et al, but experimental studies (Drowart et al, 1959) and later calculations (Strickler and Pitzer, 1964) show no molecules larger than C_5, even at 4000 K; it is rather difficult to bring particles into these equilibrium calculations, but presumably large particles would grow at the expense of smaller ones or of large molecules, leaving only species up to C_5 and a solid surface which would have a vapour pressure of C, C_2 etc.

For the reaction zone, we have seen (page 175) that the measurements of absorption indicate ground-state C_2 and CH concentrations far above those to be expected, or found, in the burnt gas above. There is no evidence, however, for excess C_3. The excited states of C_2 and CH, responsible for the flame emission, must have concentrations even more above equilibrium, as shown by the high excitation temperatures (page 186).

The partial pressures of OH, H and O may be quite high (of the order ten percent) in hot flames, and although departures from equilibrium probably occur in the reaction zone, these are more noticeable in flames of lower temperature when adiabatic equilibrium values are much lower. In hydrogen flames mass-spectrometric studies and observation of alkali-metal emission spectra (page 122) show excess concentrations of OH, H or O in the reaction zone and a persistence of these high values for some distance up the flame, due to slow recombination of these species by three-body collisions. In organic flames the concentrations of OH, H and O also appear to be high in the reaction zone, and that of excited OH, $OH(A\ ^2\Sigma^+)$, is outstandingly high, but the recombination of the excess free radicals above the reaction zone is much more rapid than in hydrogen flames, afterburning effects in organic flames being due mainly to slow burning of CO. There is evidence of diffusion of OH, and in some cases of CH_3 as well, into the unburnt-gas region, leading to relatively high concentrations of these important radicals in the preheating zone.

Reactions Forming Excited Species

We have seen that the emission by certain radicals, especially OH, CH and C_2, from the reaction zone of premixed flames is often very much stronger than would be the case in thermal and chemical equilibrium. This could be due to three possible causes, or a combination of them, (i) chemical disequilibrium, with excess concentrations of the radicals, (ii) thermal disequilibrium, with failure of the Maxwell-Boltzmann distribution law, leading to an abnormally high proportion of electronically excited species, and (iii) true chemiluminescence in which the radicals are formed in electronically excited states as the result of the chemical reactions. There is no doubt that (i) and (ii) are operative to some extent, but it is the object of this section to discuss specific reactions which could lead to the formation of radicals or molecules in electronically excited states.

The OH Radical.   There appear to be at least five ways in which OH emission may occur, (i) the weak thermal radiation from the interconal gases and burnt gases, (ii) the weak anomalous excitation to vibrational levels $v' = 2$ and 3 in some hydrogen flames, (iii) the weak abnormal excitation in the reaction zone of relatively low-temperature hydrogen flames, (iv) the very strong excitation which occurs in the reaction zone of hydro-carbon flames which also leads to very high effective rotational temperatures, and (v) the strong excitation which occurs in the reaction zone of some other flames (methyl alcohol, formaldehyde etc.,) which does not, however, lead to abnormal rotational temperatures.

The thermal excitation (i) does not require any specific reaction mechanism, but we have noted (page 109) that it may be relatively inefficient in hydrogen flames, so that there is appreciable radiation depletion, especially in flames at low pressure; if the reversible predissociation/preassociation affecting $v' = 2$ and 3 occurs it may help to reduce the radiation depletion and so strengthen bands with $v' = 2$ and 3 even in chemical equilibrium.

The anomalous excitation to $v' = 2$ and 3 in hydrogen flames (ii) has also been discussed, page 110. It is due to an inverse predissocation; the intermediate state involved is not quite certain; it was originally thought to be a $^2\Sigma^-$, but is now believed to be $^4\Sigma^-$,

$$O + H \rightarrow OH(^4\Sigma^-) \rightarrow OH(A\,^2\Sigma^+).$$

This occurs when there is an excess population of free atoms, and the intensity tends to be proportional to the square of the free-atom population. In hot flames this is masked by stronger thermal emission.

The weak abnormal emission from the reaction zone (iii) of low-temperature flames is again associated with an excess population of free atoms and radicals and in this case (see page 103) is due to

$$H + OH + OH = H_2 O + OH^*$$

and tends to vary as the cube of the population of free atoms, since fast bimolecular reactions maintain a pseudo-equilibrium between concentrations of H, O and OH.

The strong excitation in the inner cones of hydrocarbon flames (iv) leading to high rotational temperatures is due to the reaction

$$CH + O_2 = CO + OH^*$$

This reaction is exothermic to the extent of about 159 kcal/mole (765 kJ) and gives sufficient energy for both electronic and rotational excitation. One may see classically that the reaction, if written in the form

$$
\begin{array}{ccc}
\text{C–H} & & \text{C} \quad \text{H} \\
+ & \longrightarrow & \parallel + \mid \\
\text{O=O} & & \text{O} \quad \text{O}
\end{array}
$$

involves the formation of the strong C=O bond and rupture of the weak C–H bond and may lead to the H atom being repelled so that it rotates rapidly about the O atom, giving high rotational energy to the newly formed OH.

This reaction was suggested by the observation that high rotational temperatures for OH were only obtained from flames in which CH emission was also strong. It has been supported by the deuterium tracer work (page 217) and the further observation that atomic flames using pure O by titration did not give strong OH emission but that addition of O_2 to these flames did give OH strongly and with high rotational temperature (page 210).

The strong OH emission (v) from flames of methanol, form-aldehyde, formic acid etc. in which the rotational temperature is in approximate equilibrium with the translational temperature, has not been assigned to any specific chemiluminescent reaction. It is probably due to the general level of high electronic excitation in the reaction zone of flames, which has been measured for various added metals (page 182); its cause is not really known, but will be discussed later in this chapter. See also Becker et al (1969).

The CH Radical.   The readiness with which flames of moist cyanogen (Pannetier and Gaydon, 1947) or moist carbon suboxide give CH bands suggests they involve a reaction of C_2. The absence of CH bands from flames supported by oxygen-free chlorine trifluoride, and the tracer work with deuterium, both show that CH is not normally formed by direct breakdown of the organic molecule, but by some side reaction. The studies on ClF_3 flames indicate that this side reaction involves oxygen, and the tracer work with atomic flames shows that H_2 and H_2O are not active molecules. The initial evidence for the reaction

$$C_2 + OH = CO + CH^*$$

is thus very strong. It has been confirmed by quantitative measure-ments of C_2 and OH concentration and CH emission (Bulewicz, Padley and Smith, 1970); this process does not, however, account for the formation of ground-state CH.

This C_2 + OH reaction is strongly exothermic and in addition to forming CH in either the $A\,^2\Delta$ or $B\,^2\Sigma$ state leads to molecules with high translational energy, as shown by the Doppler broadening (page 188). The effective rotational temperature is not, however, very high, although, at least with methane, it does appear to be determined by the excitation process rather than by the flame temperature (page 199).

In low-pressure flames CH emission usually comes above that of C_2, but detailed study of rich flames shows that there is also relatively strong CH at the very base of the flame, suggesting that there is also another mechanism not dependent on C_2. In some flames, e.g. of formaldehyde, there is some CH emission but no C_2, and in pre-ignition glows, even of rich mixtures, CH occurs without C_2. The reaction in pre-ignition glows may be the same as that at the very base of normal flames. These flames usually show HCO as well, and in chilled flames the HCO is strong, CH is present and C_2 usually absent. These chilled flames are also associated with peroxide formation and it is likely that the breakdown of a peroxide produces the CH; it may be produced in the ground state and then excited thermally. Peeters et al (1971) studied CH emission from hydrogen flames with added traces of hydrocarbons and concluded that the CH was excited by radical recombinations, such as $CH + OH + H = CH^* + H_2O$. It is under this condition that the predissociation in the 3900Å band of CH is suppressed. Concentrations even of ground state CH are apparently above equilibrium for that for complete combustion. Since the formation of ions (see later) and possibly atomic carbon, depends on reactions of ground-state CH, we are still in need of more information about CH formation, even though the $C_2 + OH$ reaction accounts for the strong emission.

The C_2 Radical. The very strong C_2 emission poses an interesting problem. Most simple reactions which one can postulate are very endothermic and would have difficulty in producing even unexcited C_2, yet the high rotational, vibrational and electronic excitation temperatures suggest a very exothermic end step, with chemiluminescent formation of C_2 in the excited state.

The rather obvious simple reaction

$$CH + CH = C_2 + H_2$$

is indeed exothermic (by 84 kcal/mole or 350 kJ) but all indications are that CH is formed from C_2, not the reverse. Quantitative measurements of CH concentration (Bleekrode and Nieuwpoort, 1965; Bulewicz, Padley and Smith, 1970) definitely eliminate this reaction.

The reaction

$$C + CH = C_2 + H$$

is fairly exothermic (61kcal or 255 kJ), but also depends on CH; the atomic carbon is most likely to be formed by $CH + H = C + H_2$ and so the C_2 emission would again tend to depend on the square of the CH concentration; Fairbairn (1962b) has discussed reactions of this type forming atomic carbon. A similar reaction

$$CH_2 + C = C_2 + H_2$$

has been suggested by Peeters et al (1971); however, it is only exothermic by about 38 kcal (160 kJ) and could not form the C_2 in the excited state. These mechanisms do, however, have the advantage that the C atoms in C_2 are drawn from different hydrocarbon molecules, and would therefore show the observed randomization of the carbon isotopes (page 219).

It has been suggested that successive stripping of H from the hydrocarbon by reaction with H atoms or OH radicals might lead to C_2 formation. Thus for acetylene

$$C_2H_2 + H \text{ (or OH)} = C_2H + H_2 \text{ (or } H_2O)$$
$$C_2H + H \qquad\qquad = C_2 + H_2$$

The C_2H radical is indeed detected by mass spectrometry, but the end reaction is endothermic, and also this process would not give randomization of the carbon atoms since they both come from the one acetylene molecule.

Studies of C_2 light yield from low-pressure flames of CO or H_2 with added traces of hydrocarbon (page 163) showed C_2 emission varying as a high power (square to sixth) of the hydrocarbon concentration, indicating its formation from several hydrocarbon molecules or fragments from them. This led Gaydon and Wolfhard (1950) to suggest that C_2 formation occurred via a polymerization or condensation step in which large unsaturated molecules were formed. For some fuels, like methane, acetylene might be formed as an intermediate step. The decomposition of acetylene to solid carbon and hydrogen is, of course, very exothermic, and Gaydon and Wolfhard suggested that larger molecules might decompose exo-thermically to C_2. There is now further evidence that these larger unsaturated molecules like polyacetylenes are present; they can be detected by mass spectrometry. Thus one of the most abundant, C_8H_2, might decompose to C_6 (graphite nucleus), C_2 and H_2. This theory of "big molecules that go bang in the night" has not received much support. One objection has been that polymerization/condensation should be favoured by high pressure, whereas the C_2 light yield is higher in low-pressure flames. However, the poly-merization probably occurs by a radical chain mechanism which need not be pressure dependent, and of course collision quenching of excited C_2 molecules at high pressure will tend to reduce light yield at the higher pressures. Experimentally, the observations of C_8H_2 by mass spectrometry were also made in low-pressure flames. It is not quite clear how this mechanism would reconcile with the isotope tracer results, but it seems possible that the C atoms in the final C_2 could come from different original molecules.

Bayes and Jansson (1964), who studied atomic flames, have proposed a modified polymerization mechanism, involving initial formation from acetylene and atomic oxygen of a complex $C_2 H_2 O$, but there is little evidence of exactly what subsequent reactions would occur.

In flames containing oxygen it would also be possible for unsaturated polymers to react very exothermically with O_2. Thus if diacetylene is formed

$$C_4 H_2 + O_2 = C_4 H_2 O_2 = C_2 + H_2 + 2\, CO$$

would give sufficient energy. This has the advantage that in flames both species are likely to be present; again it might be possible to account for the randomization of carbon isotopes, although it is not obvious that this would happen. This mechanism would not account for C_2 emission from oxygen-free flames supported by fluorine or in the decomposition of pure acetylene by shock heating (Fairbairn, 1962b).

There is a general similarity in those flame conditions which give high electronic excitation and those which lead to high chemi-ionization, and Knewstubb and Sugden (1958) suggested that electronically excited C_2 might be formed by recombination of an electron with an ionized radical, and gave as example

$$H_2 C-C{\equiv}CH^+ + e^- = CH_3 + C_2^*.$$

$C_3 H_3^+$ is known to be one of the most important positive ions in the reaction zone of organic flames. However, there are several objections to this. The $C_3 H_3^+$ is believed to be formed from $CH(A\,^2\Delta) + C_2 H_2 = C_3 H_3^+ + e^-$, and thus the C_2 would again be formed the wrong way round, from CH^*. Kinbara and Noda (1971) have shown from time-resolved flash photolysis that C_2 emission precedes ion formation, which is definitely the wrong way round. Bulewicz (1967) noted that addition of electron acceptors, such as cyanogen and chlorine, to flames reduced the electron concentration but did not reduce the C_2 or CH emission; Cl_2 reduced $[e^-]$ to a half but increased $[C_2^*]$ by a quarter.

Fairbairn (1962a) in considering C_2 formation in shock-heated gases and the randomization of the C isotopes, considered reactions

of atomic carbon. $C + C + M = C_2 + M$ is, of course, very exothermic, but brings us back to the old problem of formation of
atomic carbon, which is most likely to involve CH. It would be most
helpful if the concentration of atomic carbon in the reaction zone of
flames could be measured, but this appears to be still an unsolved
problem.

Thus no entirely satisfactory scheme for producing excited C_2 has
been devised. A basic difficulty is the mutual involvement of C_2, CH
and OH in the formation of each other. The author favours the
mechanism of polymerization/condensation to largish unsaturated
molecules which decompose exothermically to C_2; the evidence for
the actual presence of $C_8 H_2$ and even larger polyacetylenes has been
obtained since the theory was originally put forward. CH emission
would then come from the C_2, but a separate mechanism to form
ground state CH to give the OH emission and also the ionization
would be necessary. The most promising alternative mechanisms
involve atomic carbon.

The HCO Radical. The hydrocarbon flame bands are not such a
strong feature of flame spectra as the bands of the diatomic emitters,
but they are relatively strong in flames of very weak mixtures, in
chilled flames and in pre-ignition glows, in all of which thermal
excitation would be slight because of the rather low temperature. It
is not certain whether the radicals are produced in the electronically
excited state (i.e. chemiluminescence), but some exothermic reactions may be written down. The two most likely exothermic reactions
(see page 219) are

$$CH + HO_2 = HCO + OH + 142 \text{ kcal } (594 \text{ kJ}) \text{ mole}^{-1}$$

or

$$CH + H_2O_2 = HCO + H_2O + 174 \text{ kcal } (725 \text{ kJ}) \text{ mole}^{-1}$$

The deuterium tracer work seems to rule out direct breakdown of a
peroxide, which would not, in any case, be sufficiently exothermic.
The above reactions would, however, agree with observations that
peroxidic substances, in this case H_2O_2 or HO_2, are associated
chemically with HCO emission in chilled flames. Reactions of CH_3
with O_2 or of formaldehyde with a radical also seem to be excluded
by the tracer observations.

Excitation of CO. Excitation of the Fourth Positive and other
systems of CO has been discussed on pages 156 and 212. The main

electronic excitation is now usually accepted to be

$$C_2O + O = CO(A\,^1\Pi) + CO(x\,^1\Sigma).$$

The reaction

$$CH^* + O = CO(A\,^1\Pi) + H$$

is sufficiently exothermic but could not occur in the $C_3O_2 + O$ flame.

Blades (1967) suggested

$$C_2 + O_2 = CO(A\,^1\Pi) + CO(x\,^1\Sigma).$$

This was not accepted by Bayes because C_2 emission was not observed from his $C_3O_2 + O$ flame. However, the mechanism only requires ground state C_2; it is true that if a lot of C_2 were present some excitation by energy transfer could be expected, but this is a qualitative argument depending on collision cross-sections. The formation of CO^* from C_2 has the advantage that in hydrocarbon flames the order with respect to hydrocarbon concentration is greater than unity, whereas if C_2O were formed directly from C_2H_2 it should have an order unity with respect to C_2H_2.

Vibrational excitation of CO (see page 211) is attributed to

$$O + C_2H_2 = CH_2 + CO_{v=14}$$

and

$$O + CH(x\,^2\Pi) = H + CO_{v=33}$$

Ionization and Electron Temperatures

Ionization is abnormally high in and above the reaction zone in organic flames, ion concentrations usually ranging from 10^9 to 10^{12} ions/ml, several orders higher than for equilibrium at the flame temperature, as calculated by the Saha equation. Indeed this is the basis of detecting organic compounds in gas chromatography, the increased ionization in a small hydrogen/air flame serving to record the arrival of the various organic species. There has been a lot of discussion about the reactions producing the ions (see Gaydon and Wolfhard, 1970; Lawton and Weinberg, 1969; Boothman et al, 1969; Bulewicz, 1967). The two most important appear to be

$$CH + O = CHO^+ + e^-$$

and

$$CH + C_2H_2 = C_3H_3^+ + e^-$$

Both CHO^+ and $C_3H_3^+$ have been detected by mass spectroscopy in the reaction zone of organic flames, although rapid change transfer makes it difficult to determine the first species to be formed. H_3O^+ is normally the most abundant ion, probably being formed by proton transfer

$$CHO^+ + H_2O = CO + H_3O^+.$$

The CH + O mechanism is fairly well established, but it is likely that the CH + C_2H_2 also occurs, at least under rich conditions. Both reactions would be slightly endothermic with the CH in the ground state but slightly exothermic with the CH in an excited state such as CH($A\ ^2\Delta$). In the CH + O reaction it is probably that the CH is in the ground state (e.g. Peeters et al 1971, Bulewicz, 1967), although Kinbara and Noda (1971) favoured the excited state. The CH + C_2H_2 reaction is probably less important but is usually believed to involve excited CH.

There is also evidence from probe measurements (Calcote, 1963, Kinbara, Nakamura and Ikegami, 1959) that electron temperatures (i.e. the kinetic energy of free electrons) are abnormally high in flames. Von Engel and Cozens (1964) suggested that the high electron temperatures and also the high ionization might be due to electrons gaining large amounts of energy in inelastic collisions with electronically or vibrationally excited molecules which had been formed by the combustion processes. Owing to the big mass difference between the fast electrons and ordinary molecules, elastic collisions would be inefficient in reducing the electron temperature, but inelastic collisions could cause further ionization.

There is still lack of agreement whether these very high electron temperatures are real. Early work was mainly done with single Langmuir probes and Silla and Dougherty (1972) have shown that although single probes can give reliable temperatures they do not do so near the condition of electron saturation. Some double-probe measurements (Travers and Williams, 1965) have failed to confirm high electron temperatures, while others, especially for carbon monoxide flames* (Bradley and Matthews, 1967) show temperatures up to 30 000 K.

A full discussion of these ionization processes and of electron temperature measurements is beyond the scope of this chapter. We

---

*These CO flames contain small additions of methane.

may summarize by saying that pure hydrogen flames and pure carbon monoxide flames do not show much abnormal ionization, but hydrocarbon flames do. Addition of methane or other hydrocarbon in quite small amounts produces high ionization in both hydrogen and carbon monoxide flames, but the high electron temperature is only observed with the CO + hydrocarbon, not with H_2 + hydrocarbon. There is some evidence that the high electron temperatures, if real, are associated with the presence of vibrationally excited CO_2 molecules (Bell and Bradley, 1970). The electrons will be produced from the added methane by the reaction $CH + O = CHO^+ + e^-$, and the electrons so formed will be maintained at high temperature by collision with the vibrationally "hot" CO_2 formed by the combustion of CO. However, further experimental work to establish the reality of the high electron temperatures is still necessary.

The High Electronic Excitation in the Flame Front

In many flames the electronic excitation temperature of lines and bands in the ultra-violet far exceeds the adiabatic flame temperature. This was shown quantitatively by spectrum-line reversal measurements on low-pressure flames (page 184), and the effect also occurs at atmospheric pressure, but the thinness of the reaction zone then makes measurement difficult. The anomaly was not observed in H_2/O_2 CO/O_2 flames at low pressure and it was at first thought that it might, like the high ionization, be due to some specific reaction occurring in organic flames. However, it was found that some inorganic flames, such as NH_3/O_2, H_2/N_2O and CO/N_2O also show the effect at least to some extent, whereas CS_2/O_2 and some organic flames, such as formaldehyde/oxygen and hydrocarbons/nitric oxide, do not. Although oxy-hydrogen flames do not show abnormal excitation of added metals, it has been found that hydrogen flames at lower temperature (when there is less masking by thermal emission) do show anomalous excitation of Pb and other metals (page 117).

One possibility which has been considered is that the effect is associated with an excess enthalpy wave, discussed in relation to minimum ignition energies by Lewis and von Elbe (1961). Briefly the gases ahead of a flame front are preheated by conduction so that they might retain their chemical energy and acquire additional thermal energy; if the chemical energy could then be released quickly it would be possible to attain local temperatures in excess of the normal adiabatic value. However, it is now realized that forward diffusion of combustion products (H_2O etc) dilutes the mixture and prevents any major wave of excess enthalpy. In any case preheating to a few hundred degrees could not produce the very high temperatures which would be required to account for strong excitation of ultra-violet lines.

The high electronic excitation appears to be a general phenomenon which occurs when there is a failure of the Maxwell—Boltzmann energy distribution, due to the rate of chemical reaction producing energy-rich species exceeding the rate of the processes by which the energy is redistributed among its various possible forms. A failure of the distribution law is most likely when the chemical activation energy E_{ac} is low and the temperature is high; the failure is likely to become important when $E_{ac}/kT \geqslant 5$. The tendency will be for translational temperatures to be below the equilibrium values, with excess of vibrational and rotational energy of the newly formed molecules. Under some conditions an equilibrium distribution may be maintained within each degree of freedom, but not between the various degrees of freedom. When this pseudo Maxwell—Boltzmann distribution occurs we may have separate effective translational, rotational, vibrational and electronic temperatures. Shuler (1955) stressed that the Maxwell—Boltzmann law is a probability law and involves exponential terms; even in the absence of equilibrium there will be some form of probability of distribution, and this may involve exponential terms and so show some formal similarity to the Maxwell—Boltzmann expression.

The exchange between translational energy and electronic excitation tends to be very slow, as shown by the extremely small effective cross sections for quenching of resonance fluorescence of various metal atoms by rare gases. Exchange between vibrational energy and electronic excitation is, however, in many cases very rapid, and shock-tube experiments (e.g. Gaydon and Hurle, 1963) have shown that spectrum-line reversal temperatures for Na and Cr in N_2, CO and probably other molecular gases follow the vibrational temperature during the period of vibrational relaxation. In pure gases vibrational relaxation may be very slow, requiring many thousands of collisions; at atmospheric pressure pure N_2 has a relaxation time of about 1 millisecond at 1000 K, 350 μs at 1500 K and 140 μs at

2000 K, while pure CO takes $600\,\mu s$ at 1450 K, $100\,\mu s$ at 2200 K and $35\,\mu s$ at 3000 K. Fairbairn (1963) has shown that in shock-initiated detonations the electronic excitation of carbon radicals relaxes at the rate expected for vibrational relaxation of the CO molecules. In flames this relaxation is much faster than in pure gases, due to the presence of moisture and other molecules. Direct evidence of strong vibrational excitation of CO in atomic flames by Clough, Schwartz and Thrush (1970) has been noted (page 211), and it is likely that in all fast reactions a considerable part of the energy will initially be released as vibrational energy of the newly formed molecules. For organic flames all reactions forming CO tend to be very exothermic, because of the high dissociation energy (11.1 e.V) of this molecule, and vibrationally excited CO molecules are probably the main source of the abnormal electronic excitation in organic flames. For some other flames, such as ammonia/oxygen and hydrogen/nitrous oxide the N_2 molecule (dissociation energy 9.76 e.V) may play a similar role.

We have seen that while some flames show very strongly the abnormal excitation, other flames hardly show the effect at all. The differences are no doubt due to differences in the speed of the chemical reactions and to the form the energy takes after its initial release. The effect of speed is apparent in the difference between premixed and diffusion flames. When highly exothermic bimolecular reactions occur, vibrationally excited products will be formed and lead to strong excitation in the flame. When the main exothermic reactions involve three-body collisions, then they will occur less frequently, there will be a smaller departure from the Max-well–Boltzmann distribution of energy, and the high excitation effects will be less prominent. Thus in the hydrogen flame, although the propagation occurs by bimolecular reactions these are roughly thermo-neutral (see Chapter V); the really exothermic reactions are all three-body collisions leading to recombination of O, H and OH; this accounts for the absence of strong electronic excitation in this flame. One would not expect the high excitation to be as marked in the slower flames with air as in the faster flames with oxygen, but actually the flame of hydrogen with air appears to show the effect more strongly. This must be because the normal thermal radiation is so very much stronger in the hot flame with oxygen that it masks the non-thermal effects, but in the cooler flame with air the thermal radiation is small, so that although the non-thermal processes themselves are probably also weaker, they are easier to observe because of the absence of masking by thermal radiation. In these hydrogen flames the relatively high electronic excitation in the reaction zone is not due to conversion of excess vibrational energy but to excitation by the three-body collisions of the type of, for example, $H + OH + Pb = H_2O + Pb^*$ (page 117).

Effects on Flame Structure of Lags in Equipartition of Energy

The processes of conversion of energy from one form to another and the delays in these processes have been discussed in the book on Shock Tubes (Gaydon and Hurle, 1963) and in a review by Callear (1965). In gases at atmospheric pressure and flame temperature a molecule makes about 2×10^9 collisions/sec. Equipartition of translational energy of molecules will require only a few collisions, and data from ultra-sonic dispersion and light reflection at shock fronts show that interconversion of rotational and translational energy only requires from 5 to 20 collisions. Interchange between translational and vibrational energy is rather slower, as discussed in the previous section; for pure gases it could cause significant lags, but in ordinary flames in which water is a product even vibrational relaxation is quite quick, probably around a microsecond. Only for dry carbon monoxide and cyanogen flames are vibrational lags likely to extend beyond the reaction zone; the low vibrational temperature for CN in cyanogen flames, for some 2 mm above the reaction zone, has been noted (page 205). Interconversion of electronic excitation energy and translational energy is likely to be very inefficient, but may vary rather widely with the nature of the colliding species. However, energy transfer between electronic and vibrational forms is often very easy and the electronic excitation tends to be dominated by the vibrational energy, i.e. the effective vibrational temperature. Excited electronic states may be depopulated by emission of radiation, radiative lifetimes for allowed transitions in the visible and ultra-violet being of the order 10^{-8} to 10^{-6} sec. However, in flame gases at atmospheric pressure the efficient electronic-vibrational energy transfer keeps the excitation and vibrational temperatures practically the same and radiation depletion effects are very small, temperatures above the reaction zone by spectrum-line reversal being within a few degrees of the equilibrium values.

For the fastest flames, such as oxy-acetylene, the burning velocity is around 10 m/s and, at 1 atm., the luminous reaction zone is about 1/50 mm. thick. Thus the time of passage through the zone is around 2 μs, during which time a molecule makes about 400 collisions. Translational and rotational disequilibrium cannot therefore persist beyond the reaction zone. Some slight persistence of departure from vibrational and equipartition may, however, occur just above the reaction zone, and high electronic excitation will follow this too. For most flames, with thicker reaction zones and lower burning velocities, the effects of these physical disequilibria will be smaller still, and the only departures from equilibrium of importance will be chemical, due to three-body recombination processes of free atoms and OH radicals, to the rather slow combustion of carbon monoxide and to freezing of some chemical equilibria such as that between NO, N_2 and O_2.

Fig. X. 3 will help us to visualize the main features of the structure of the reaction zone of a premixed flame. The regions are denoted as U, the unburnt cool mixture, P the preheating zone, L the luminous reaction zone, A the afterburning zone, and finally B the burnt gases in equilibrium. In the preheating zone the mixture is heated by conduction and the temperature in this zone can be determined by

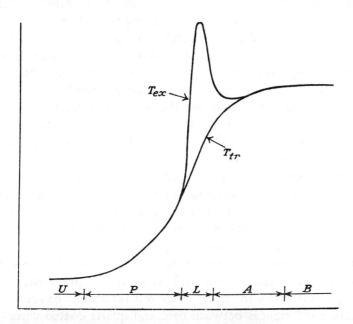

Fig. X.3. Structure of the reaction zone of a premixed flame.

refractive-index methods (Burgoyne and Weinberg, 1953; Weinberg, 1963; see also *Flames*); the amount of chemical reaction in this region is small, although forward diffusion of products (H_2O, H_2) and active species (H, OH) may modify the chemical composition slightly; in some very slow burning mixtures, especially for example rich flames of ether, there may be some cool-flame type reactions here, producing aldehydes and similar intermediate oxidation products. In the luminous zone the main chemical reactions are proceeding and there is rapid energy release; radicals such as OH, CH, C_2 and HCO are formed, in some cases in excited electronic states (i.e. chemiluminescence); newly formed molecules may hold excess vibrational energy and abnormal electronic excitation may occur both of naturally occurring molecules and added metal atoms; chemi-ionization processes also occur here. In this region L there is the greatest chance of departures from the Maxwell–Boltzmann distribution of energies, so that there is no single gas temperature; the translational temperature T_{tr} may tend to be rather lower than that expected for the energy released, while the electronic excitation temperature T_{ex} tends to be very high; for those species excited by chemiluminescence. such as OH, CH and C_2 we may have abnormal rotational, vibrational and even translational temperatures given by the emission spectrum. In the afterburning zone the gases will have reached fairly good physical equilibrium, so that the temperature again has a real meaning, but owing to incomplete chemical equilibrium, especially slow recombination of free atoms and radicals, and also of ions, the energy release will not be complete and so the temperature will be below the adiabatic value for the flame; only for very fast flames will delayed vibrational relaxation persist a little into this region. The duration and extent of the chemical disequilibrium, and hence the difference between the observed and adiabatic temperatures, will be least for hot fast-burning mixtures and greatest with slow-burning mixtures. For rich hydrogen/air flames afterburning is due to slow removal of H atoms and OH radicals (see Chapter V); in weak mixtures O and OH tend to persist. In hydrocarbon flames the afterburning effects are usually less noticeable although in weak mixtures the last traces of CO are slow to burn. In flames of carbon monoxide there is also persistence of free O atoms, as shown by the nitric oxide test for atomic oxygen and by temperature differences between bare and quartz-coated wires (page 141). This afterburning was attributed by the late Prof. W. T. David and colleagues at Leeds to a latent energy in the newly formed CO_2 molecules; however, metastable electronically excited molecules or slow vibrational relaxation should also be observed in fast hot flames, if the missing energy took this form, and it now seems clear that the departures from equilibrium are chemical, with atomic oxygen being the most likely species to persist.

This qualitative physical picture of the structure of the flame front appears to be consistent with most observations on stationary flames. There are certain observations on the infra-red emission from CO/O_2 explosion flames (see page 233) which do not fit into this picture based on stationary flames, and there is still a need to make the treatment more quantitative, using measurements of the various effective temperatures and combining them with reaction-rate data.

Chapter XI

Explosions, Engines and Industrial Flames

Explosions in Closed Vessels

The spectra of explosion flames of hydrocarbons mixed with air or oxygen were the subject of a number of older studies (e.g. Lauer, 1933; Bone and Outridge, 1936; E. C. W. Smith, 1940b). With weak mixtures the usual bands of OH, CH, C_2 and HCO are obtained and there is little obvious difference from the spectra of the corresponding stationary flames. Such differences as do occur are due to the higher temperature, because in closed vessels the specific heat of the gases is less — the constant-volume instead of the constant-pressure value — and there is less dissociation because of the higher final pressure. Thus the CH bands at 3143 Å ($c^2\Sigma - x^2\Pi$) and 3628 Å (the (1,0) of $B^2\Sigma - x^2\Pi$) are relatively strengthened in hydrocarbon explosion flames, and in CO/O_2 explosions the Schumann—Runge bands of O_2 occur. The (0,0) band of OH may appear relatively weaker, probably due to stronger self-absorption at the higher pressure and OH concentration. Another effect, also associated with the higher temperature, is a strengthening of bands involving nitrogen, such as CN; in carbon monoxide explosions at high pressure Bone and Newitt (1927) noted self-absorption by nitrogen peroxide.

Explosions of rich mixtures of hydrocarbons show strong continuous emission due to hot carbonaceous particles. There is probably a real increase in carbon formation associated with the higher final pressure, but the stronger thermal emission from the particles because of the higher temperature also causes their emission to dominate that from chemiluminescent processes in the flame front.

With faster-burning mixtures, around the stoichiometric point, the blue continuous spectrum associated with burning carbon monoxide becomes very strong and tends to mask other features. Photographs of flame travel show that immediately after the flame reaches the end of the vessel there is a strong reillumination of the hot gases. This increase in luminosity is usually referred to as afterburning and is believed, as in the engine work referred to later, to give the CO-flame spectrum. It is probably largely a manifestation of the pressure and temperature rise due to the combustion and the pressure waves in the vessel, but is also associated with some delay in completion of the combustion of carbon monoxide. The effect is strikingly shown in some early photographs (Bone, Fraser and Winter, 1927; Bone, Fraser and Witt, 1927). This afterburning is usually of short duration, but is bright enough to dominate photographs of the spectrum of the complete explosion flame. In explosions of carefully dried CO/O_2 mixtures it may persist much longer and W. T. David recorded afterburning persisting visually for fourteen seconds in large vessels (see also page 140).

Time variations of the spectra of explosions of H_2, CH_4 and CO with O_2 in long tubes were studied by Minkoff, Everett and Broida (1955) using a photomultiplier and oscillograph. The OH emission came mainly from the burnt gases rather than from the flame front, with a normal rotational temperature, but with some excess excitation of the (2,1) band due to recombination of excess free atoms. C_2 emission showed an initial sharp peak, indicating its presence mainly in the flame front. The CO-flame continuum was of rather longer duration and its intensity depended on pressure, varying about as the square of the pressure; this was attributed to the higher pressures suppressing dissociation and so raising the temperature. In all these explosion flames continuous (or more probably unresolved band) absorption was observed below 2800 Å; this was probably due to hot oxygen. No ozone absorption was observed, and evidence for hydrogen peroxide was inconclusive. With methane, bands attributed to formic acid were observed in absorption on occasions. Despite the additional information obtained from this type of time-resolved study, the complications from irregular flame movements and pressure changes make interpretation of the results difficult, and for fundamental research stationary flames are easier to control.

In explosion flames at high initial pressure (around 30 atm.) spectra taken by Bone, Newitt and colleagues have been examined by the author and found to show very strong CO-flame type continuum and OH bands. With explosion flames travelling away from the spectrograph, so that the light passed through the burnt gases, bands of NO_2 were often present in absorption and in some cases the OH was reversed.

Detonations

In explosions of fast-burning mixtures in long narrow tubes, the flame front accelerates rapidly until the pressure pulse ahead of it develops into a shock wave which is strong enough to cause ignition behind it. At this stage the form of the combustion processes change and the explosion flame propagates at very high speed (of the order 3000 m/sec). This very violent type of explosion is known as *detonation.* The basic processes leading up to the flame acceleration are not understood in complete detail but involve a big increase in flame area due to gas movements associated with the expansion behind the flame front, turbulence, and the setting up of shock waves which heat the gas and increase the burning velocity. A simplified account of the initiation and structure of detonation waves is given by Gaydon and Hurle (1963).

Experimental work on the spectra of detonating explosions is difficult both because of the dangers of handling these very destructive explosions and because the spectra of dust and material scrubbed from the walls (Na, Ca, CaOH, Cu) often masks the gas emission. Wagner (1957, 1961) has succeeded in obtaining some valuable spectra of detonations of gases initially at atmospheric pressure, while a bursting-diaphragm shock tube has been used in a number of investigations (Fairbairn and Gaydon, 1957; Lyon and Kydd, 1961; Miyama and Kydd, 1961; Miyama, 1962; Fairbairn, 1963) to study emission and absorption spectra of detonating gases initially at pressures around 10 torr.

Oxy-hydrogen detonations at 1 atm. show OH bands and a good deal of continuous emission. The appearance of the OH bands is rather different from that in an oxy-hydrogen flame because of the high temperature and pressure (around 18 atm. for gases initially at 1 atm.), these leading to high rotational and vibrational temperatures and strong self-absorption of the (0,0) band. Thus the (3,3) and (3,2) bands at 3254 and 2945 Å are relatively strong in detonations. Pressure broadening of individual lines is also much greater and obvious in Wagner's spectra. The continuum is probably due to radical recombination processes such as $H + OH = H_2O + h\nu$ (see page 106). When oxygen is in excess, thermal emission of the Schumann—Runge bands of O_2 in the ultra-violet also occurs.

In oxy-hydrogen detonations at reduced pressure the spectrum is more similar to that of flames. OH absorption has been used to study induction times for shock ignition, of the type occurring in detonations, and it has been found (Miyama and Takeyama, 1964) that below 1100 K the reaction $HO_2 + H_2 = H_2O_2 + H$ is rate determining, but at higher temperature, as in true detonations, the simpler reaction $H + O_2 = OH + O$ is rate determining. In the shock ignition of hydrogen/oxygen/argon mixtures (Schott, 1960) there is an overshoot in OH concentration close to the shock front, similar to that in flames and due to slow three-body processes being necessary to remove excess radicals.

Detonations of carbon monoxide/oxygen mixtures show very strong continuous emission. This is probably of similar origin to that in the CO flame; it extends from the ultra-violet, through the blue and weakly to longer wavelengths. Wagner's spectra show continuum to at least 6500Å , but thermal emission from dust particles and calcium impurities may contribute. In these high-temperature detonations the banded features, best seen in low-temperature flames, are not apparent.

Hydrocarbon/oxygen detonations, initially at 1 atm., show strong continuous emission throughout the visible and near ultra-violet. The OH is usually just visible, although for methane detonations it is a fairly strong feature. Both CO-flame emission and that from hot carbonaceous particles seem to contribute to the continuum, but we do not know quantitatively the proportions. In weak mixtures the continuum does appear to be stronger at the short-wave end and is probably mainly of the CO-flame type, while for rich mixtures the red end is strongest and soot emission is no doubt the main cause. It is known (Fraser, 1959) that soot may be formed from many hydrocarbons over almost the complete mixture-strength range, so particle emission may contribute to the continuum even for weak mixtures. Wagner (1957) noted that with oxygen excess there was emission of the Schumann—Runge bands of O_2 and the 4315 Å band of CH; this CH band is not very obvious in the published spectrograms. C_2 is not obvious in any of these spectra.

At reduced pressure Fairbairn and Gaydon found that detonations in shock tubes do show C_2 and CH bands, although much less strongly than the corresponding steady flames. Lyon and Kydd were able to detect absorption of the CH band at 3143 Å, noting that this absorption corresponded to at least 33 times the equilibrium concentration and that it preceded the OH absorption, which is the reverse of the normal sequence for flame emission. Miyama found that the OH concentration quickly reached the equilibrium value and that the rotation and vibration temperature of OH was normal. For shock-initiated detonations in both C_2H_2/O_2 and C_2N_2/O_2 Fairbairn (1963) found a high excitation temperature (by a brightness and emissivity method) for C_2 and CN bands and attributed this to excess vibrational energy in newly formed CO (or N_2).

The differences between the spectra of hydrocarbon flames, which show very strong C_2 and CH, and the corresponding detonations, showing rather weak CH and little or no C_2, may be partly due to the higher temperature and pressure in the detonation, but are also likely to be due to the different propagation mechanism. In flames the propagation involves heat transfer by thermal conduction and also diffusion, from the burnt gases, of free radicals, especially OH and H, so that initial reactions occur in a region of moderate temperature and rather high radical concentration; we know that "atomic" flames containing free radicals show strong C_2 emission. In detonations there is no time for normal heat transfer or diffusion, the strong shock front immediately producing a very high translational temperature, and the early stages of the combustion in this suddenly heated gas may resemble more nearly that in preignition glows, which are known (page 171) to show strong OH, moderate CH but hardly any C_2 emission.

The Internal Combustion Engine: Flame Fronts and Afterburning

The spectra obtained from ordinary internal combustion engines running on petrol (gasoline) with spark ignition are, of course, similar to those of premixed hydrocarbon flames. Among the earlier investigations are those of Clark and co-workers (Clark and Thee, 1926; Clark and Henne, 1927; Clark and Smith, 1930) and of Withrow and Rassweiler (1931) and Rassweiler and Withrow (1932). The latter used a stroboscopic method which enabled them to study the spectrum at any stage during the engine cycle. Ordinary photographs revealed a strong reillumination of the gases during the pressure rise accompanying the burning of the last part of the charge. Withrow and Rassweiler referred to this as the "afterglow", but it is more often described as the "afterburning", a term which will be used here as "afterglow" is in general use for continued luminescence of low-pressure gases in a discharge tube. The spectrum of the flame front was similar to the inner cone of a premixed flame; both petrol and benzene showed bands of OH, CH and C_2. When the engine was run on ethyl alcohol the spectrum showed mainly OH and CH with a trace of the stronger Swan bands of C_2. Methyl alcohol gave strong OH and a trace of CH but no C_2.

The afterburning, for all the fuels studied by Withrow and Rassweiler, showed only CO-flame type emission and OH, without CH or C_2. This clearly indicated that the light emission in the later stages was due not to continued burning of hydrocarbons but to delayed combustion of carbon monoxide or carbon monoxide and hydrogen. Withrow and Rassweiler suggested that the equilibria

$$H_2 + CO_2 \rightleftharpoons CO + H_2O$$

and

$$2CO_2 \rightleftharpoons 2CO + O_2$$

were influenced by pressure and temperature changes produced by the burning of the last part of the charge. A similar afterburning was observed by Withrow and Rassweiler when hydrogen was used as fuel in the engine, the spectrum then showing mainly OH bands.

Studies of infra-red emission were made by Steele (1931, 1935) using a stroboscopic method and filters to isolate the main 4.4 and 2.8 μm bands. He found that the radiation continued to increase for some $20°$ of crank angle after the passage of the flame front; he was inclined to attribute this to metastable excited CO_2 and H_2O molecules, but it seems really to be part of the afterburning phenomenon already discussed and due to pressure and temperature changes associated with the burning of the last part of the charge and perhaps also to delay in attaining full chemical equilibrium and to pockets of gas continuing to burn in regions of incomplete mixing in the combustion chamber. Smith and Starkman (1971) have shown that pressure changes and gas movements lead to considerable local variations in temperature, the gas in the centre of the cylinder often being 200 K hotter than the mean. Studies of OH bands showed that the concentration of this radical was up to an order of magnitude greater than the equilibrium value. The excess concentration was, however, less in engines than in flames because of the higher pressure under engine conditions.

Knock

The maximum compression ratio which can safely be used in an internal combustion engine, and thus the efficiency of the engine, is normally limited by the phenomenon known as "knock". Above a certain compression ratio a more violent type of explosion occurs; this was often referred to in the early literature as "detonation" although it is not a true detonation of the type discussed earlier in this chapter (page 284). Ball (1955) has shown that in engines, cool flames occur, these travelling at about a quarter of the normal flame speed, but that when knocking there is a second much faster flame front. The velocity of this is subsonic, so that it is not a true detonation, but it is strong enough to produce shock waves, which constitute the knock.

The spectroscopic studies of Withrow and Rassweiler (1931) and Rassweiler and Withrow (1932) included engines under knocking conditions. As already stated the flame front under normal conditions showed the usual bands due to C_2, CH and OH. For knocking explosions the spectrum sometimes showed a strong continuum in the red, which was apparently black-body radiation from soot formed by cracking of the lubricating oil at the higher temperatures attained when knock occurred. The bands of C_2 and CH were less strong in the "detonating" zone, although these bands were still emitted with normal intensity in the "non-detonating" zone of the knocking explosions. This indicated that the difference between knocking and non-knocking explosions was confined to the nature of the burning of the last part of the charge. The reduction of the strength of the C_2 and CH bands is well shown in the published spectrograms by Withrow and Rassweiler. They have pointed out that the effect could not be due to changes in density or physical movement of the gases or to temperature affecting the thermal excitation of radicals but that it must be due to a change in the chemical reactions which either produced the C_2 and CH radicals or consumed them. They suggested that the C_2 and CH radicals might be removed by reaction with OH radicals which were formed by thermal processes ahead of the flame front. However, on the evidence of later work, it seems more likely that the change is due to an alteration in the basic method of flame propagation; under non-knocking conditions the flame front is propagated by the usual radical diffusion processes and shows C_2 and CH, but in the "detonating" zone of the knocking explosions there is thermal auto-ignition due to the higher compression ratio producing higher temperature. Under this auto-ignition condition spectra tend to be more similar to that of the pre-ignition glow (page 173) or shock-ignited explosions (page 286) where CH and especially C_2 are much weaker.

Studies of absorption spectra (Rassweiler and Withrow, 1932; Withrow and Rassweiler, 1933, 1934) showed that when the engine was knocking, the gases immediately ahead of the flame front showed strong absorption. This consisted of a continuum in the ultra-violet with superposed formaldehyde bands; these bands always occurred when the engine was knocking or in a state of incipient knock, and were observed for all fuels studied. The gases in the flame front also showed OH, but all absorption quickly disappeared after passage of the flame front. When either isopropyl nitrite or diethyl peroxide were added to produce knock the formaldehyde bands appeared strongly, and conversely the addition of aniline to suppress knock was found to reduce the formaldehyde absorption. Affleck and Fish (1967) have made a gas-chromatographic analysis of products from cool flames at reduced pressure (0.1 atm.) and from a motored engine at high pressure (20 atm.) and shown that they are very similar indeed.

These valuable observations support other evidence that knock is due to preignition of the mixture ahead of the flame front and that it occurs through a two-stage mechanism involving reactions of a cool-flame type which produce formaldehyde. Methane and benzene, which do not possess normal cool-flame regions, are much less prone to knock than other hydrocarbon fuels.

It is now well known (following the work of Midgley and Boyd, 1922) that various substances, especially certain organo-metallic compounds like lead tetra-ethyl and iron carbonyl, reduce knock and lead tetra-ethyl has been used for many years as a dope in petrol. Thee (1929) found that addition of lead ethyl caused emission lines of atomic lead and bands, now assigned to PbO, to appear. Withrow and Rassweiler found that the PbO and Pb emission occurred in the flame front, and that the addition of the lead to prevent knock also restored the strength of C_2 and CH emission and stopped the continuum from hot carbonaceous particles. In absorption, however, they found that formaldehyde bands still appeared even when the knock was suppressed by the lead ethyl.

The flash photolysis technique has been used by Erhard and Norrish (1956, 1960) and Callear and Norrish (1960) to study knock and the effect of inhibitors. For oxy-acetylene explosions sensitized with amyl nitrite the usual induction period between the flash and the explosion was between 20 and 150 μs and this was increased by a factor of 2 or 3 on adding lead ethyl. During the induction period PbO bands appeared in absorption and these were replaced by Pb lines at the onset of ignition. It was concluded that the lead ethyl was not decomposed by the flash but that it reacted with oxygenated compounds (e.g. peroxides) during the cool-flame type first-stage combustion to form PbO and delay the second-stage ignition; the cool flame itself was not delayed. There was no sign of continuous absorption due to separation of a solid phase, and Norrish and colleagues concluded that the anti-knock effect was due to a homogeneous gas-phase reaction of PbO with oxygenated inter-mediates, such as peroxides, and free radicals. For some other anti-knocks (Sn, Mn and Fe) Callear and Norrish did detect some continuous absorption due to formation of particles, but still concluded that the basic anti-knock action was homogeneous gas phase.

A number of other investigators (Walsh, 1963; Hoare, Walsh and Li, 1971; Zimpel and Graiff, 1967; Salooja, 1967) have attributed the anti-knock action of lead to a heterogeneous process by decomposition of HO_2 radicals and other peroxides on the surface of lead oxide particles. This seems to be the most commonly accepted view at present, but the evidence of Norrish and colleagues that under the flash-photolysis conditions solid particles are not present seems very strong. Their spectra clearly show that PbO molecules are present, that continuous absorption does not occur and that the induction period is increased. When solid lead oxide is present in a hot gas, then some PbO vapour will necessarily be present as well, but the reverse is not always so.

Compression-Ignition (Diesel) Engines

In the ordinary compression-ignition engine of the Diesel type, the spectrum is dominated by continuous emission from hot carbon particles. Lyn (1957, 1963) found that this continuum extended into the infra-red and even the strong CO_2 bands at 2.8 and 4.4 μm, which occurred in emission and self-absorption, produced only slight irregularities in the continuum. Lyn distinguishes three phases during the burning. The first is of low luminosity and gives bluish light; it appears to be associated with the ignition in a region of premixed vapour and air. The second main highly luminous phase involves turbulent diffusion flames surrounding the burning droplets which were ignited during the first phase. The less well-defined third phase is also luminous with a continuous spectrum and may be due to delayed combustion of carbon monoxide and residual carbonaceous particles. During the main second phase the emissivity is quite high and the intensity distribution in the continuum was found by Lyn to be close to that of a black body.

Although this type of engine is normally used with a distillate fuel oil and spray injection, it may be employed to study combustion processes in more volatile fuels or gases. The compression-ignition system is similar to that of an internal combustion engine running under knocking conditions. By use of an auxiliary motor an engine may be kept running at conditions under which it does not develop sufficient power of its own (i.e. a "motored engine"). In this type of work blue flames are often found to precede the normal yellow flames of full ignition; they may be examined stroboscopically or by limiting the compression ratio so that normal ignition does not quite occur. These blue flames appear to be of cool-flame type, although occurring at higher temperature because of the shorter time. They are of importance because of their close relation to knock, and support the view that this occurs (for fuels other than methane and benzene) via a low-temperature two-stage ignition. Spectroscopic studies (Broida, Levedahl and Howard, 1951; Downs, Street and Wheeler, 1953) of these blue flames show emission of Emeléus's cool-flame bands of formaldehyde. Gaydon, Moore and Simonson (1955) found that even methane will give a cool flame under these motored-engine conditions, the spectrum showing formaldehyde bands superposed on a CO-flame type continuum; later Gaydon and Moore (1955) showed that the formaldehyde emission from methane only occurred when oxidation of carbon monoxide was also taking place.

A curious feature of these experiments with motored engines is the strength of the radiation from hot carbonaceous particles, even from flames of extremely weak mixtures. Contamination from lubricating oil is a possible cause, but it is likely that under these conditions of high temperature and high pressure the soot forming processes are able to compete successfully with oxidation. In the absence of a spark to start a propagating flame, combustion processes will be of the thermally initiated type, and in compression-ignition experiments on weak methane/air the HCO bands were found to be relatively strong.

Engines with Continuous Combustion

Gas turbines and rocket motors derive their thrust from the expansion in continuously running flames, usually at well above ambient pressure. The spectra of these flames depend on mixing conditions as well as on the nature of the fuel and oxidant. With gas turbines the flame is usually supplied with an oil spray (e.g. of kerosene). It has been shown in *Flames* (Gaydon and Wolfhard, 1970) that very fine oil droplets may vaporize in the flame front and burn like a premixed flame, giving a Bunsen-type spectrum, but larger droplets burn with a diffusion-type flame, giving emission from carbonaceous particles. Most gas turbines show the yellow flames in which the continuum from hot particles is dominant, although in a few cases combustion chambers have been designed for vapour injection or for burning very fine sprays, and flames in these are closer to the Bunsen type and show OH, CH and C_2 emission. The production of this type of flame in the primary combustion zone depends not only on spray size and fuel but also on local mixture strength, which depends on air-flow conditions.

With rocket motors the nature of the fuel is important. Thus hydrocarbon fuels tend to show hot carbon, whereas methyl alcohol shows mainly OH and CO-flame emission, with some CH. Where the fuel or oxidant contains combined nitrogen, bands of CN, NO and NH may appear.

In gas-turbine reheat systems additional fuel is sprayed into the hot combustion products downstream from the turbine to gain extra thrust. A flame holder, usually in the form of a V-gutter, is used to stabilize the flame. The spontaneous ignition of this fuel spray was studied by Mullins (1949); under some conditions a blue glow, showing the cool-flame bands of formaldehyde, was again obtained, and ethyl nitrate was found to strengthen this. When full ignition occurred the spectrum was similar to that of a weak premixed flame, showing strong OH and CH and some HCO bands.

Exhaust Flames

Powerful engines, such as aero engines, often show luminous exhaust gases. Spectroscopic study of these can often help empirically in determining the type of combustion which is occurring in the exhaust, and may sometimes help in diagnosing the cause.

The normal bright blue exhaust flames of aero engines under take-off conditions are due to burning of excess carbon monoxide in the surrounding air, and show the usual CO-flame spectrum, with OH. Bands and lines of inorganic compounds also occur if doped fuels are used; lines of Pb and Na and bands of PbO and Br_2 have been observed.

When C_2 and CH bands occur in exhaust flames it indicates primary combustion of hydrocarbons or related fuels, thus showing that pockets of unburnt fuel are passing through the engine.

In some cases pale grey or blue-grey exhaust flames may show Vaidya's hydrocarbon flame bands of HCO, and work (Gaydon, 1942a) on chilled flames suggests that the cause may be chilling in the engine, either by the primary flame being quenched by cold surfaces or by sudden entrainment of cold air.

Some reddish exhaust flames show a continuous spectrum which may be attributed to hot carbonaceous particles, and may indicate carbon formation either through bad mixing, through cracked lubricating oil or through an uneven spray size.

Furnace Flames

Large industrial flames, such as those used in power generation or for melting metals, usually show strong continuous emission from hot particles. These flames may burn natural gas, an oil spray or pulverized coal, but in all cases the flame appears to be of the diffusion type, where the fuel is first broken down to carbon and the mixing processes are more important than the chemical reactions. Spectroscopically they are of little interest, although the quantity and distribution of radiation is of major importance, for the efficiency of heat transfer depends on it. Some description of such furnace flames is given in the valuable book by Thring (1962) and quantitative observations on emission from furnace flames burning various fuels have been made on the IJmuiden furnace under the auspices of the International Flame Research Foundation; summarized results have been published since 1951 in a series of papers in the Journal of the Institute of Fuel. The emissivity of large furnace flames is usually quite high, often above 0.5.

Burning coal usually gives luminous (carbon) flames, but in the early stages of combustion the more volatile coals emit vapours, containing methane and other organic compounds, which normally burn with diffusion-type flames, but the base of such flames, especially if turbulent, may show some C_2 and CH emission. After the volatiles have been driven off the coke-like residue burns with a bluish flame surrounding it; this shows the carbon monoxide flame spectrum, and frequently also bands of metallic impurities. Bands of CuCl in the violet are for some reason particularly frequent, and diffuse bands of boron oxide (BO_2) are sometimes found in the green.

The spectrum taken looking into a large furnace is, of course, mainly continuous, approaching that of the ideal black-body source, but the author has observed that in practice some discrete features often persist. Thus in a blast furnace the sodium D lines are still visible. Bands of S_2 occur in flames above this type of furnace. In the flame above a Bessemer converter during the blow, bands of manganese oxide (MnO), together with lines of Mn, Na and Fe are prominent, and a few other atomic lines, such as those of Mo, K and Ca are also present.

Solid Propellants

When strands of double-base solid propellants, usually involving nitroglycerin, nitrocellulose and minor constituents, are burnt the main emission shows a continuous spectrum with additional broad lines and bands of impurities (Na, K, Ca, CaOH). Bands of OH, CN, NH, NO and perhaps C_2 have been reported (Rössler, 1954; Rekers and Villars, 1954, 1956) There may be a flame close to the burning surface, which is separated by a dark space from a second flame whose spectrum is mainly continuous (probably CO-flame type). There is no NO_2 in the dark space, but absorption indicates that NO_2 is formed in the surrounding air. In the infra-red the CO_2 bands at 2.8 and 4.4 μm have been reported.

Flame propagation in nitro-compounds is discussed in *Flames*, and spectroscopic work on flames supported by oxides of nitrogen is dealt with in the following chapter. Perchloric acid flames, related to the use of ammonium perchlorate propellants, are also discussed.

Chapter XII

Flames Containing Nitrogen, Halogens, Sulphur and Inorganic Substances

Although the main applications of flame spectroscopy to combustion studies are concerned with organic fuels, there are a number of examples where the spectra of other flames are of real interest. Flames of special high-energy fuels for propulsion systems have been studied in recent years. The effects of various additives which promote or inhibit combustion of organic fuels have also been studied spectroscopically, and recently there has been emphasis on pollution problems such as the mechanism of formation of oxides of nitrogen and sulphur. The interest tends to shift, from time to time, from one type of flame to another, and in this chapter I have tried to give some basic information on the spectra of a considerable variety of unusual flames.

Flames of Ammonia, Hydrazine etc.

The premixed oxy-ammonia flame is greenish yellow in colour and has a well defined inner cone. Above the inner cone there is a yellowish luminosity which is especially strong with weak mixtures. This yellowish luminosity shows the continuous spectrum characteristic of the recombination reaction $NO + O = NO_2 + h\nu$ (see page 106). The inner cone has been found (Fowler and Badami, 1931) to show several strong band systems. The NH bands with characteristic Q maxima at 3360 and 3370 Å are the strongest feature in the ultra-violet, but OH bands and the γ-system of NO are also present. The visible region shows the complex structure of numerous lines known as the α-band of ammonia, due to NH_2 ; this has been described in discussing the hydrogen/nitrous oxide flame, page 123. Kaskan and Hughes (1973) have extended the spectroscopic study of premixed ammonia flames by using a multiple-reflection system to follow absorption as well, and have been able to follow concentrations of 12 species; the hydrogen/oxygen species do not appear to be far from equilibrium, but concentrations of NH and N are abnormal.

The diffusion flame of ammonia burning with oxygen was one of the first to be studied on a flat-flame burner. This flame showed a yellow zone on the ammonia side, a dark central zone and a bluish zone on the oxygen side. The yellow zone emitted the ammonia α band (NH_2), with NH bands on the oxygen-edge of this zone. The dark central zone showed only OH emission, and the blue zone showed a continuous spectrum with superposed Schumann–Runge bands of O_2. Wolfhard and Parker (1949, 1952) also observed the OH, O_2, NH and NH_3 bands in absorption, but not the NH_2. The strengths of the emission and absorption bands were plotted semi-quantitatively across the flame and served to indicate the concentrations of the various species. The emission from OH, NH and O_2 was thermal, and it was shown that the ammonia broke down thermally, via NH_2 and NH in a region where the O_2 concentration was negligible. The products, mainly hydrogen, then reacted in the central zone with oxygen to form water and OH. Their diagram showing the structure of the diffusion flame is reproduced in Figure XII.1.

The spectra of diffusion flames of ammonia burning with nitrous oxide, nitric oxide, nitrogen peroxide and fuming nitric acid have been studied by Farber and Darnell (1954) and Hall and Wolfhard (1956); bands of OH, NH and NH_2 were again observed.

Hydrazine, N_2H_4, has some potential as a high-energy fuel and its combustion has been discussed in *Flames*, and spectroscopic studies have been made by Hall and Wolfhard (1956). It possesses a pure decomposition flame which is unusual because the decomposition products contain ammonia and so this flame is hotter than an adiabatic flame forming nitrogen and hydrogen. It has a thin yellow-brown reaction zone, above which there is a pale yellow-brown afterglow. Both show NH_2 and weak NH emission. The premixed flame of hydrazine with oxygen shows NH_2, NH, OH and weak NO γ bands and appears to involve a two-stage process, the decomposition preceding the oxidation. The hydrazine/nitric oxide flame also shows NH_2, NH and OH but with the OH relatively stronger.

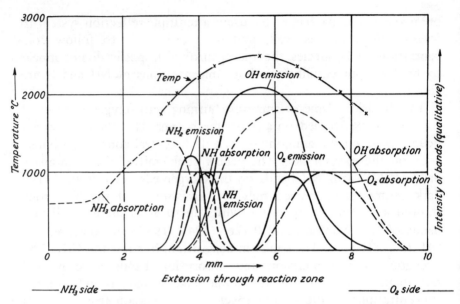

Fig. XII.1. Intensities of emission and absorption bands in the flat diffusion flame of ammonia with oxygen.

In the explosive thermal decomposition of hydrazoic acid N_3H (Pannetier, 1951; Pannetier and Gaydon, 1951) the main emission is from NH, for which weak bands of the $^1\Pi - ^1\Delta$ system at 3240 Å also occur; the NH_2 bands are weak.

Flames of Organic Nitrogen Compounds

Although flames of hydrocarbons etc. supported by air show little evidence of band systems involving nitrogen, apart from weak CN Violet in some of the hottest flames, fuels which contain combined nitrogen show the CN Violet system very strongly and usually NH and NO γ bands as well. Thus Vaidya (1935) found these systems in flames of nitrobenzene, pyridine, aniline etc, and the author has obtained them from flames containing ethyl nitrate and amyl nitrite. The bands of the CN Red system are strong in the infra-red, but do not usually extend with appreciable intensity into the visible. These flames still show the normal features of organic flames, OH, CH, C_2, HCO.

A fairly full discussion of the combustion of nitrates and nitrites has been given in *Flames*, and will not be repeated here. For rich mixtures two-stage combustion often occurs, the first stage showing cool-flame (CH_2O) emission. Some of these compounds form decomposition flames, which may show CH_2O in the first stage emission and an orange second-stage flame which shows only a continuum which is strongest in the red and of uncertain origin but possibly associated with NO_2; there are no other band systems, not even OH; for details of these decomposition flames see Needham and Powling (1955) and Gray, Hall and Wolfhard (1955).

Some features of the cyanogen flame have already been discussed (page 203). The diffusion flame in air is a rosy colour and shows CN Violet and CN Red systems; it does not show C_2 nor does it normally form solid carbon. The premixed flame with air is brighter and gives weak C_2 in addition to the CN. The premixed flame of cyanogen with oxygen is phenomenally hot, with a maximum temperature, for $C_2N_2 + O_2 = N_2 + 2CO$, of around 4835 K (Thomas, Gaydon and Brewer, 1952). The spectrum shows very strong C_2, CN, NO γ and NO β systems. When moisture is present the flame shows emission from CH and NH and the C_2 and CN are weaker.

Low-pressure flames of cyanogen/oxygen/nitrogen mixtures have recently been examined in detail by Bulewicz, Padley and Smith (1973), using a multiple reflection system to study the absorption spectrum. For a flame with an adiabatic temperature of 2700 K the CN concentration in the reaction zone was far above equilibrium, but C_2 was undetectable and therefore below equilibrium. For the CN ground state, $CN(x^2\Sigma)$, the rotational and vibrational temperatures were normal (i.e. not above the adiabatic flame temperature) but in emission the rotational and vibrational temperatures of $CN(B^2\Sigma)$ and $C_2(A^3\Pi)$ were between 5500 and 6500 K. When small quantities of hydrogen were added to this flame the emission of CH, NH and OH all reached maximum intensity for fairly low hydrogen concentrations. CH was not detectable in absorption, though.

Another extremely hot flame is that of carbon subnitride C_4N_2, which should have a maximum temperature of around 5260 K at 1 atm., but the author is not aware of any details of its spectrum.

Flames Supported by Oxides of Nitrogen

The spectra of flames of hydrogen supported by oxides of nitrogen have been discussed at the end of Chapter V. Flames of carbon monoxide with oxides of nitrogen all show the normal CO-flame

spectrum, together with some continuum from the NO + O association reaction.

Most organic fuels burn readily with nitrous oxide, N_2O, giving a
single reaction zone. Hydrocarbon flames show CN Violet, NH and
NO γ as well as the usual OH, CH and C_2. In the methane/N_2O flame
the ammonia α-band of NH_2 is also quite strong, but C_2 bands are
very weak. Some other fuels, such as formic acid, formaldehyde and
methyl alcohol, show quite strong NH_2, and in these flames C_2 is
absent and CH weak. In low-pressure flames (Gaydon and Wolfhard,
1949a) it was found that the ammonia α-band was emitted from the
base of the rather complex reaction zone, with C_2 and CN also fairly
low, but OH and NH extending higher. For the ethylene oxide flame
the NH_2 emission was so much below that of the other radicals that
it could be distinguished as a pale green fore-zone. In these flames
with N_2O the OH emission from the reaction zone is very strong but
it does not show such high effective rotational temperature as it does
for flames with O_2.

The strength of the various band systems of nitrogen-containing
molecules in these flames with N_2O indicates that the N_2O does not
just break down to N_2 and oxygen, but reacts directly with the fuel
molecule. Detailed reaction mechanisms have not, however, been
established. The formation of NH_2 at an early stage in these flames is
perhaps unexpected. Since this radical also emits quite strongly from
H_2/N_2O flames, it suggests that H_2 may be an early product of the
reactions; this might link up with views on hydrocarbon/oxygen
flames for which some initial dehydrogenation rather than direct
reaction seems to occur.

Flames of organic fuels with nitric oxide, NO, are much more
difficult to maintain than flames with O_2 or N_2O, despite the high
adiabatic temperatures of flames with NO. They have a very low
burning velocity and large quenching distance, but Wolfhard and
Parker (1953, 1955) succeeded in maintaining such flames and also
examined their structure by emission and absorption spectroscopy.
Nitric oxide itself was made to give a pure decomposition flame
provided the final temperature exceeded about 3070 K, and the
spectrum of this decomposition flame showed Schumann—Runge
bands of O_2 with some continuous background and, if moist, OH.

The methane/nitric oxide flame showed strong CH, CN, NH and
NO bands, with very weak C_2 and NH_2. The OH was present but was
not enhanced in the reaction zone compared with above the zone, as

it is with flames with O_2 and N_2O. Other hydrocarbons showed similar spectra, but with stronger C_2 and without NH_2. For rich mixtures there was some persistence of CN above the reaction zone. In lean mixtures the reaction zone became double, the lower part giving the bands already noted, while the upper part, which was due to the decomposition of the excess NO, showed the Schumann– Runge bands and much stronger OH, due to the higher temperature, and some continuum. The flame with acetylene was slightly different, because of its higher temperature, and did show some enhancement of OH in the reaction zone; the NO appeared to decompose in this zone instead of persisting and decomposing higher in the flame. In some cooler flames, such as NH_3 /NO, excess NO was undecomposed and could be detected by absorption of the γ bands above the flame.

Flames supported by nitrogen peroxide, NO_2, always show at least two reaction zones, the lower giving a continuous (NO + O) spectrum due to the decomposition of the NO_2 to NO and oxygen, and the second being similar to that of flames maintained by NO, showing band spectra of CN, NH etc, but only weak OH. With excess NO_2 a third reaction zone may develop where the excess NO is decomposed. In absorption, bands of NO_2, NO, NH and OH are detectable and serve to identify the character of the various layers of the reaction zone. Wolfhard and Parker also made some temperature measurements and studied abnormal electronic excitation in these flames.

Flat diffusion flames of hydrocarbons burning with NO or NO_2 are similar in general appearance to flames with O_2, indicating that the NO or NO_2 are decomposed thermally in the preheating zone before reaction. In the thin greenish zone at the base of the flame where C_2 and CH emission occurs there is some CN and NH emission; this is attributed to slight premixing at the base where the separating wedge of hot products is too thin to prevent it. The diffusion flames with NO_2 also show a yellowish zone on the NO_2 side where the NO_2 decomposition is occurring, this giving the NO + O continuum.

Formation of Oxides of Nitrogen

The formation of oxides of nitrogen in flames has become of considerable importance because of atmospheric pollution. Open flames of organic fuels do not produce much, and the absence of band spectra of nitrogen compounds in flames with air has already been noted. In hotter flames, such as those enriched by oxygen, or in engines where the flame temperature is raised by the pre-compression, appreciable quantities of nitrogen oxides may be formed. Attempts to reduce pollution by peroxides and unburnt carbon monoxide from engines by using higher temperatures and longer residence times to encourage complete combustion are indeed often negatived by the increased formation of nitrogen oxides.

Equilibrium concentrations of NO in slightly lean flames are quite high – of the order 0.2 per cent for flames at 1800 K, rising to one per cent at 2500 K and over two per cent at 3000 K. In practice this equilibrium is, fortunately, rarely approached. The rate of NO formation is limited by the very high dissociation energy of N_2, which is 9.76 e.V (225 kcal/mole or 940 kJ). The main reaction is

$$O + N_2 = NO + N \qquad (1)$$

which is followed by

$$N + O_2 = NO + O \qquad (2)$$

From shock-tube measurements, using far ultra-violet absorption to follow species concentrations, Wray and Teare (1962) have found a rate constant for reaction (1)

$$k_1 = 7 \times 10^{13} \exp(-75\,500/RT) \text{ cm}^3 \text{ mole}^{-1} \text{ sec}^{-1}$$

with $R = 1.986$ cal. deg^{-1} mole^{-1}. Reaction (2) has a much lower activation energy

$$k_2 = 13.3.10^9 \, T \exp(-7080/RT)$$

and so in the early stage of NO formation

$$d[NO]/dt = 2 \, k_1 \, [N_2] \, [O]$$

In organic flames the concentration of atomic oxygen, [O], is quickly equilibrated with that of molecular oxygen, [O_2] and the rate of formation of nitric oxide in post-flame gases can be calculated. Shahed and Newhall (1971) used absorption in the (0,0) γ band of NO at 2260 Å to follow nitric oxide formation in closed-vessel explosions of propane/air and of hydrogen/air mixtures at initial pressure of around 3 atm. (45 psia) and found satisfactory agreement with the calculated concentrations. Thompson, Brown and Beér (1972) also studied nitrogen oxide formation in oscillatory combustion of methane/air, using mass spectrometric sampling, and found that the amount formed was unaffected by the oscillations and could be calculated from the atomic oxygen concentration; they discussed the kinetics in some detail and conclude that the above mechanism is satisfactory, although it may be necessary in some cases to allow for atomic oxygen concentrations above equilibrium.

However, some studies on ethylene/air flames by Fenimore (1971) using proble sampling, showed that while most of the nitric oxide formation could be accounted for by the post-flame reaction $O + N_2 = NO + N$, as above, there was also a small amount of NO formed quickly in the reaction zone, and he suggested that this might be due to either

$$CH + N_2 = HCN + N$$

or

$$C_2 + N_2 = 2 CN$$

Both these reactions are only very slightly endothermic (about 2 kcal/mole) and CH and C_2 are known spectroscopically to be present in the reaction zone. The CN band at 3883 Å is weakly emitted by a stoichiometric ethylene/air flame. The CN or HCN would no doubt react quickly with O_2 to form NO.

Trace quantities of nitrogen oxides may be monitored by the chemiluminescent reaction with ozone, which is dealt with in the next section.

Ozone

The explosive decomposition flame of ozone is bluish and the spectrum has been found (Fairbairn and Gaydon, 1954) to show bands of the Schumann–Runge system of O_2; the strongest were the (0,13), (0,14) and (0,15) with heads at 3232, 3370 and 3516 Å. A few other bands of the $v' = 0$ and $v' = 1$ progressions were also observed. Emission was not strong and appears to be thermal rather than chemiluminescent in origin. The atmospheric band at 7594 Å was not detected. The thermal decomposition glow of dilute ozone in oxygen was very faint, the spectrum showing a continuum in the visible; it was probably due to reaction of atomic oxygen with traces of oxides of nitrogen.

In the flash photolysis of ozone kept isothermal with an excess of inert diluent, McGrath and Norrish (1957, 1960) found absorption bands of O_2 from vibrational levels right up to $v'' = 17$, with a maximum population in levels around $v'' = 13$ (53.5 kcal/mole). This vibrational excitation was attributed to

$$O + O_3 = O_2^* + O_2 .$$

The O_2 was, however, rotationally cold. In the explosive decomposition of ozone initiated by flash photolysis, the absorption from vibrationally excited O_2 was not observed, this being explained by the rapid quenching of the excited O_2, formed in the initial photolysis, by subsequent collisions with more ozone. Molecular oxygen and water vapour also rapidly quench the excited O_2 molecules. In the presence of moisture OH absorption was also detected, the formation of this being attributed to the reaction

$$O(^1D) + H_2O = 2\ OH.$$

The photolysis of ozone by red and by ultra-violet light has also been discussed at length by Norrish and Wayne (1965); as well as $O(^1D)$ and vibrationally activated O_2, excited molecules of oxygen, $O_2\,(^1\Delta)$ and $O_2\,(^1\Sigma_g^+)$, are involved in the various chain reactions.

Ozone gives chemiluminescent reactions with a number of substances. That with ethylene is used to monitor trace amounts of ozone (Nederbragt et al, 1965; Warren and Babcock 1970) and can apparently measure ozone concentrations down to two parts in a hundred million. The spectrum has been described as a broad continuum centred around 4350 Å and it could well be the usual cool flame spectrum (CH_2O).

There is also luminescence from ozone reacting with Rhodamine B absorbed on activated silica (Regener, 1964) and from ozone reacting with barium vapour and aluminium vapour (Gole and Zare, 1972).

The reaction between ozone and NO or NO_2 gives a red glow (Greaves and Garvin, 1959; Clough and Thrush, 1967) the spectrum of which shows unresolved bands on a continuous background in the region 5900 to 10 850 Å. This is due to the reaction

$$NO + O_3 = NO_2^* + O_2 \quad (\Delta H = -47.4\ \text{kcal/mole})$$

and the faintly banded NO_2 emission is a red-shifted modification of the afterglow reaction $NO + O = NO_2 + h\nu$. This red glow can be used to monitor either ozone or oxides of nitrogen (Fontijn, Sabadell and Ronco, 1970) and is sensitive down to one part in 10^9.

Halogens in Oxidizing Flames

The spectra of organic flames with air and added halogens or halogenated compounds have been studied mainly because of interest in the mechanism of flame inhibition by organic halides (see page 216), but several new band spectra have been observed for the first time in flames containing halogens.

When chlorine, bromine or iodine are introduced into a flame of hydrogen burning in oxygen, bands of ClO, BrO or IO are observed. The bands of IO and BrO were first reported by Vaidya (1937, 1938) in flames of methyl iodide and ethyl bromide, and all three systems have since been further studied and analysed (Coleman and Gaydon, 1947; Coleman, Gaydon and Vaidya, 1948; Pannetier and Gaydon, 1948). These systems, which are listed in the Appendix, are all very similar, consisting of long progressions of bands degraded to the red.

These flames also show a continuous spectrum as background to the band systems, and similar continua are emitted more strongly by flames of halogens burning with hydrogen (Ludlam et al, 1929) and also when halogens are heated to $1000°C$ in a quartz tube. These continua are due to association of normal and excited halogen atoms, e.g.

$$Br(^2P_{1/2}) + Br(^2P_{1\,1/2}) = Br_2 + h\nu$$

This is the reverse of the process responsible for the main visible absorption in the halogens, and it has been shown from examination of the potential energy curves of Br_2 that the continuum in flames has the expected energy distribution.

Vaidya (1941) also found that a flame containing methyl chloride, chloroform or carbon tetrachloride showed a complex band with head at 2796 Å known to be due to CCl (see Appendix). Corresponding bands of CF have been obtained from CF_4 in an oxy-acetylene flame (Mann, Broida and Squires, 1954) and from flames of methane etc. supported by chlorine trifluoride (Skirrow and Wolfhard, 1955). With methyl bromide in a flame Coleman and Gaydon (1947) found a strong diffuse band at 2900 Å in the inner cone, below the region which showed BrO. It was at first thought that this band was the corresponding band system of CBr, but Prof. K. Wieland pointed out its production in discharges through pure Br_2, indicating that the emitter was Br_2. The curious enhancement of this band in the inner cone of the flame is not properly understood but may be related to the abnormally high electronic excitation temperature in the reaction zone of organic flames. With organic iodides a similar strong diffuse band of I_2 at 3425 Å is observed, and chlorides give a band of Cl_2 at 2580 Å.

The bands of the diatomic carbon halides (CF, CCl) and the halogen bands (I_2, Br_2, Cl_2) are emitted from the inner cones, but the oxides, ClO, BrO and IO occur just above the inner cone. Thus with the flame of methyl bromide the inner cone is blue green and the BrO bands are emitted from a bluish region just above it; the outer cone of the flame is coloured orange and shows bands of Br_2 in the red but not the 2900 Å band. The ClO, BrO and IO bands which occur in flame emission do not appear to have been reported from any other source in emission, but are now known in absorption following flash-photolysis of various compounds. The occurrence in flames, especially of hydrogen, may be due to a persistence of an excess concentration of oxygen atoms shifting the chemical equilibrium in favour of the oxides.

Studies of the spectra of diffusion flames with added halogen-containing inhibitors have been described, page 216, and it has been shown that the inhibition is mainly due to removal of hydroxyl radicals. These inhibitors and flame extinguishers are of most practical importance for diffusion flames, but they do in many cases also reduce the burning velocity of premixed flames. Rosser, Wise and Miller (1959) have shown that the burning velocity of methane/air flames is reduced by methyl bromide and by HBr, and that in both cases there is a simultaneous reduction in the strength of the OH radiation at 3064 Å, although there is no significant change in flame temperature. The inhibition is again attributed to removal of OH

$$OH + HBr = |H_2O + Br$$

The inhibition of flames of carbon monoxide (containing a trace of hydrogen or moisture) burning with air, O_2 or N_2O by halogens, especially CCl_4 and Cl_2 has been studied by Palmer and Seery (1960). Detailed rate constants have been listed and it has been shown that the main inhibitory reaction, in this case

$$OH + HCl = H_2O + Cl,$$

has a collision efficiency of at least 0.1.

Spectra of flames supported by perchloric acid ($HClO_4$) have been studied by Cummings and Hall (1965) and Hall and Pearson (1969). With methane a two-stage flame was obtained, the bright first stage showing strong cool-flame emission (CH_2O) and some OH and CH. The second-stage flame was enhanced by adding air or oxygen and then showed the usual C_2, CH and OH emission (CN at 3883 Å is also visible on the published spectra), but no band systems involving chlorine were noted. For perchloric acid/ethane flames bands of HCO also occurred. For methyl alcohol the cool flame bands were the main feature. Burning velocities were high, up to 4.5 m/s for CH_4, and there appeared to be direct reaction leading initially to formaldehyde production. These flames are of interest because of the use of ammonium perchlorate in propellants. Hall and Pearson conclude that the perchloric acid has a dual role, producing highly reactive chlorine-oxygen species and also oxygen, which account for the two-flame structure, the hydrochloric acid retarding the second stage. In ammonium perchlorate propellants the formation of nitric oxide is attributed to reaction of chlorine-oxygen species with the ammonia, and a three-stage structure may be possible, reaction with nitric oxide providing the third stage.

Flames Supported by Halogens

The flames of hydrogen burning with fluorine, chlorine or bromine show the strong continua due to recombination of normal and excited ($^2P_{\frac{1}{2}}$) halogen atoms. The chemiluminescent excitation of the vibration-rotation bands of HF, HCl and HBr in the infra-red has already been discussed (page 241). In addition Kitagawa (1937, 1938) found that flames of H_2 burning in chlorine and in bromine gave emission bands of Cl_2 and Br_2 at the red end of the visible spectrum; for details see Appendix.

Spectra of organic fuels burning in F_2 and ClF_3 have been discussed at the end of Chapter VIII, but fluorine and ClF_3 will support flames with many inorganic substances as well. Perhaps the most surprising is the pale blue flame given by impinging a jet of fluorine on to the surface of cold water; this shows the Schu-mann—Runge bands of O_2 in emission, but their appearance is rather unusual because of their low effective rotational temperature (see Plate 6e). The flame also shows OH bands. In view of the low rotational temperature and the conditions of this flame, the emission must be due to chemiluminescence, as thermal emission of O_2 always requires a very hot flame. Recombination of ground-state O atoms usually gives the Herzberg bands of O_2 (Broida and Gaydon 1953b) and it seems probable that for the F_2/H_2O flame $O(^1D)$ atoms may be involved; the $B\ ^3\Sigma$ upper state of the Schumann—Runge system is known to correlate with $O(^1D) + O(^3P)$. The flame of peroxide of hydrogen with F_2 gives a similar spectrum, but the H_2O/ClF_3 flame apparently gives only continuous emission.

For full details of these flames supported by F_2 and ClF_3 the original papers (Durie, 1951, 1952b; Skirrow and Wolfhard, 1955) must be consulted. Ammonia, with both F_2 and ClF_3, shows the α-band of NH_2 and NH; the absence of OH enables new weaker bands of NH to be seen. Carbon monoxide gives a continuous spectrum, but a mixture of CO and H_2 burning with ClF_3 shows OH and ClO_2 (this assignment to ClO_2 needs confirmation; the bands could be ClO or ClO_2 self-absorption). Fluorine and sulphur give weak S_2 on a very strong continuum. CS_2 with ClF_3 gives only S_2 and a continuum, but, on addition of hydrogen, bands of CS and C_2 appear. Fluorine reacting with iodine or bromine gives new band systems of IF and BrF. Flames of F_2 with HBr and HI also show BrF and IF bands, and in the presence of atmospheric oxygen, bands of OH and either BrO or IO appear. Fluorine and phosphorus trichloride give only a strong continuum, but fluorine and phos-phorus tri-iodide give IF and the 3425 Å band of I_2 as well as a continuum. All these flames supported by F_2 and ClF_3 readily excite spectra of impurities, especially bands of S_2, SH, CS and Cu halides.

Although there is some doubt whether hydrocarbons etc. give genuine flames with atomic hydrogen (see page 212) many organic halides do give quite bright flames (Gaydon and Wolfhard, 1952). Thus carbon tetrachloride and atomic hydrogen give a continuum, which is strongest in the violet, with bands of C_2, CH, CCl and C_3; both the red and violet systems of CN occur as impurity. For the flames, with atomic hydrogen, of some other halides the spectra in the visible region are rather complex and not completely identified.

Atomic hydrogen also reacts with halogens to give strong infra-red chemiluminescent excitation of the hydrogen halide; thus $H + Br_2$ gives HBr excited to $v = 6$, $H + Cl_2$ gives HCl to $v = 6$ and the reaction $H + Cl + M = HCl + M$ also leads to some HCl with $v = 7$ (Cashion and Polanyi, 1960). These reactions have given valuable information about detailed energy distribution in the products and have also been considered as the basis of possible chemical lasers (see also page 242).

Flames of Sulphur, Hydrogen Sulphide and Carbon Disulphide

The flames of sulphur and of hydrogen sulphide were studied by Fowler and Vaidya (1931) and for both flames the main feature was found to be the extensive system of S_2 (see Appendix) in the blue and near ultra-violet. Emission bands of SO (see Appendix) also appeared weakly. Absorption bands of SO_2 were usually observed against the continuous background from the flames, but emission bands of SO_2 have not been reported. The flame of H_2S also showed OH emission at 3064 Å, but the similar band of SH at 3226 Å has not been reported in this flame. Lines of atomic sulphur have not been observed, but as the resonance lines lie outside the accessible quartz region and the triplet at 4695 Å requires high excitation and lies among the S_2, the absence of S lines cannot be taken as evidence for the absence of atomic sulphur. However, a chain mechanism involving S and O seems less likely than one using SO radicals, such as

$$H_2S + SO = H_2O + S_2$$
$$S_2 + O_2 = S_2O_2 = 2\,SO + 26 \text{ kcal/mole}$$

The bands of S_2 are a very persistent impurity in hydrogen flames, including the hydrogen/fluorine and hydrogen/chlorine flames, and in some early observations Brinsley and Stephens (1946) detected the bands with one part of sulphur in 4×10^6 of hydrogen. Mr H. G. Crone suggested to the author that the reaction might be

$$H + H + S_2 = H_2 + S_2^*$$

and pointed out that this reaction gave sufficient energy but without much excess. The bands excited in this way occur best in low-temperature flames with excess hydrogen, or with a diffusion flame of air burning in an atmosphere of hydrogen. An alternative mechanism for S_2 excitation would be

$$S + S + M = S_2^* + M$$

and this is known to occur in atomic flames (see later) leading to excitation of S_2 to $v' = 9$. However the very strong excitation of trace quantities of sulphur in flames seems to be restricted to rich hydrogen ones, and the spectrum always shows the $v' = 0$ progression as the dominant one.

This excitation of S_2 in hydrogen flames has now acquired importance for detecting sulphur compounds as pollutants. Brody and Chaney (1966) used an interference filter (at about 3940 Å) and photomultiplier to detect sulphur emission from a small rich hydrogen/air flame to identify sulphur compounds separated by gas chromatography; they were able to record down to 0.6 parts per million; later Adlard and Matthews (1971) used a similar method to "finger-print" oil pollutants from their sulphur compounds. Stevens et al (1969) modified the method slightly to give continuous recording of the sulphur content of air and were able to detect 5 parts in 10^9 for SO_2 in air. Barynin and Wilson (1972) used a flame of air burning in a hydrogen atmosphere and were able to detect SO_2 pollution down to $5 \mu g/m^3$ or about 1.7 parts in 10^9. Their instrument had a short response time (80% response in 1 sec) and could be used to follow fluctuations in SO_2 concentration due to atmospheric turbulence*. The light emission is roughly proportional to the square of the sulphur concentration and although the detection limit was around $5 \mu g/m^3$, it was possible to detect a change of only $1 \mu g/m^3$ at a concentration of $60 \mu g/m^3$ of SO_2.

The spectrum of the normal hot flame of carbon disulphide was also studied by Fowler and Vaidya (1931), and found to show strong S_2 bands on a continuous background. Weak emission bands of SO were also observed, with absorption bands of SO_2 and S_2 superposed on the emission continuum. The band of CS at 2575 Å has not been found in this flame (Eméléus, 1926; Skirrow and Wolfhard, 1955). The presence of S_2 and absence of CS suggests that the main step in the oxidation is reaction of O_2 and CS_2 to form S_2 and an oxide of

---

*Dr Wilson has pointed out to the author that the rapid response of this detector is difficult to reconcile with calculations on the rate of formation of S_2 based on collision theory at these very low concentrations; it could be an interesting problem to determine the various rate constants.

carbon, followed by formation of SO by oxidation of the sulphur. The spectrum of the flame with N_2O is similar to that with air.

The cool "phosphorescent" flame of carbon disulphide was also studied by Eméleus (1926) and Fowler and Vaidya (1931). It was slightly different from the hot flame; bands of S_2 were the main feature, but the band system of CS, with strong (0,0) head at 2575 Å, was also present and the SO bands were relatively stronger. The occurrence of CS in the ultra-violet, despite the much lower temperature, suggests that the chemical reactions in the cool flame differ from those in the normal flame.

Bright atomic flames also occur with sulphur compounds. Atomic hydrogen and H_2S gives strong S_2 bands with emission from the vibrational levels $v' = 6, 7, 8$ and 9 being particularly strong. This is attributed (Fair and Thrush, 1969) to the reactions

$$H + H_2S = H_2 + HS \tag{1}$$
$$H + HS\ = H_2 + S \tag{2}$$
$$S(^3P) + S(^3P) + M = S_2 \,(\text{B}\,^3\Sigma_u^- \, v \leqslant 9) + M \tag{3},$$

and the rate-constant k_3 for the last step is found to be $(1.0 \pm 0.2) \times 10^{15}$ cm^6 mole^{-2} sec^{-1}, with M = argon. There is also some excitation to $v' = 10$ which has a different pressure dependence and this is thought to be due to the bimolecular inverse pre-dissociation

$$S + S = S_2 \,(\text{B}\,^3\Sigma_u^- \,, v = 10).$$

The H + CS_2 flame (Gaydon and Wolfhard, 1952) also shows the S_2 bands with excitation to $v' = 9$ and 10. Atomic oxygen gives flames with CS_2, H_2S and COS (Sharma, Padur and Warneck, 1965), all of which show continuous emission between about 2600 and 5000 Å which is similar to that of the afterglow of SO_2 and is attributed to the association $SO + O = SO_2 + h\nu$. The O + CS_2 flame also gives strong infra-red chemiluminescence (Hancock, Ridley and Smith, 1972) in which CO is strongly excited to $v'' = 14$, and slightly to even higher levels, leading to strong infra-red emission around 1.8 and 2.6 μm; there is also OCS emission around 4.8 μm. The CO excitation leads to population inversion and may have laser applications. It is attributed to

$$O + CS_2\ = SO + CS$$
followed by
$$O + CS\ = CO + S.$$

Flames with Added SO_2, SO_3 or H_2S

The spectra of various flames to which small quantities of SO_2, SO_3 or H_2S had been added were studied by Gaydon and Whittingham (1947).

The outer cones of premixed flames with added SO_2 showed a strong continuous emission which gave a blue or violet colour to the flame. This is probably due to an association reaction between SO_2 and atomic oxygen to form SO_3, but the $SO + O$ reaction may also contribute. With some organic flames, especially ethylene, addition of SO_2 led to some S_2 emission from the outer cone. The addition of small quantities of SO_3 did not have any marked effect on the spectrum of the outer cone.

The premixed hydrogen/air flame developed a sort of pale blue inner cone when SO_2 or SO_3 were added, this showing the S_2 bands, as noted in the previous section. Neither SO nor SH bands were observed in any part of a hydrogen flame.

When SO_2 or SO_3 were introduced into a fully aerated Bunsen flame (manufactured gas) or similar flame of methane, ethylene or methyl alcohol, the inner cone showed strong SO bands; these were probably produced by reaction of SO_2 with C_2 or CH; such reactions would not be sufficiently exothermic to produce the SO radicals in the electronically excited state, but their emission from the inner-cone region of fully aerated flames, rather than from the outer cone or rich flames, was probably due to the high excitation in the inner cone and the oxidizing conditions. In these fully aerated flames both CS and SH were weak or absent.

With under-aerated Bunsen or hydrocarbon flames, addition of SO_2 or SO_3 had a marked effect on the formation of luminous carbon particles. With SO_3 in premixed flames (but not diffusion flames) the carbon formation was increased to a remarkable extent. SO_2, on the other hand, tended slightly to reduce carbon formation, and a flame which was initially just luminous could be rendered non-luminous by addition of from 2 to 5 per cent of SO_2. The non-luminous flame produced in this way showed strong CS bands and also weak S_2 and SH; this was the first record of the SH band in emission (for details see Appendix), although it had previously been obtained in absorption under special conditions by Lewis and White (1939). The occurrence of SH seemed to be associated with that of CS, and among the several possible reactions by which it could be formed, the reaction

$$CS + OH = CO + SH^*$$

seemed most likely. SH was not observed in flames of H_2S or of sulphur/hydrogen mixtures unless organic compounds were also present.

Effects of adding H_2S were generally similar to those of adding SO_2. There was again a slight tendency to suppress carbon formation, and bands of CS, SH, SO and S_2 were observed in the inner cone. In the fully aerated Bunsen flame, CS and SH were more noticeable than with SO_2 added to a similar flame.

Phosphorus in Flames

Early studies of the spectra of a number of flames involving phosphorus were made by Eméleus and Downey (1924), Eméleus (1925) and Eméleus and Purcell (1927). Phosphorus burning in air showed a strong continuum in the visible and bands in the ultra-violet which, from the measurements, were obviously the β and γ systems of PO. The β system is very complex, with heads degraded in both directions; the main sequence is strongest near 3270 Å and has a diffuse head near 3240 Å. The γ system consists of double-double headed bands, shaded to shorter wavelengths, with strongest sequence heads as 2636.3 (0,2), 2555.0 (0,1), 2477.9 (0,0) and 2396.3 Å (1,0). With the flame at reduced pressure, so that the temperature was only recorded as 125°C, the continuous emission was found to be less intense. Eméleus also studied the glow of the slow oxidation of phosphorus, one of the first recorded cases of chemiluminescence (hence the word phosphorescence!). This again showed the PO bands but the visible continuum was less strong than in the normal flame. The glow of phosphorus trioxide was similar.

When small quantities of phosphorus compounds are introduced into a premixed flame there is strong emission in the visible region of the spectrum. Herrmann and Alkemade (1963) make several references to continuous emission in the visible and apparently attribute it to oxide particles, although P_2O_5 sublimes at about 300°C and other phosphorus oxides are also volatile or unstable. Mavrodineanu and Boiteux (1965) show a spectrogram with an apparent continuum. However tracings of the spectrum by Brody and Chaney (1966) and Dagnall, Thompson and West (1968) show that the visible emission has a more banded structure with a strong maximum around 5280 Å and weaker maxima at 5100 and 5600 Å. The 5280 Å emission has been used by these authors for flame-photometric estimation of phosphorus, and can detect down to 1×10^{-12} gm/sec of phosphorus entering the flame; response is linear with P concentration. Studies by Lam Thanh and Peyron (1963, 1964) of the glow of moist atomic hydrogen reacting with phosphorus, and of the isotope shift using deuterium, show that the bands are due to the triatomic emitter HPO (see Appendix, under PHO). Both vibrational and rotational analyses have been made. It seems almost certain that the apparently continuous emission from phosphorus compounds in hydrogen or organic flames must be due to HPO, but the higher temperature makes the structure more complex and apparently diffuse.

Eméléus also studied the normal flame of phosphine, both at 1 atm and at reduced pressure, and noted that this also showed the bands due to PO, with the band at 3270 Å (PO β system) relatively stronger than in the flame of phosphorus. The OH bands at 3064 and 2811 Å were also observed, but there was no mention of the PH band near 3400 Å; this PH band could perhaps be masked by the rather wide complex PO structure extending from 3270 Å and might account for the enhancement of this. Semenoff (1938) has proposed an oxidation mechanism for phosphine involving PH and atomic oxygen. Complex bands of PH_2 (see Dixon, Duxbury and Ramsay, 1967) between 3600 and 5500 Å are known in absorption following flash-photolysis of phosphine, but do not appear to have been reported from flames.

The flame of atomic oxygen with phosphorus vapour (P_4) and with phosphine have been studied by Guénebaut, Couet and Houlon (1964), Davies and Thrush (1968) and Verma and Broida (1970). The main features of both these flames are due to the β and γ systems of PO, but a new weaker system of PO, bands of P_2, probably PO_2 and also unknown bands have been reported.

Flames Containing Boron

Boron, despite its toxicity, is of interest as a high energy fuel. Boranes, besides having high flame temperatures and very high burning velocities, have extraordinarily wide limits of flammability, and quite small additions of boranes considerably widen the flammability region of organic fuels. For a summary of the combustion properties of boron and boranes see Gaydon and Wolfhard (1970).

The greenish flame of burning boron, and also an ordinary flame coloured green by addition of a boron compound, show regularly spaced waves of diffuse bands. These waves have maxima around 6390, 6200, 6030, 5800, 5450, 5180, 4930, 4710 and 4520 Å. They are usually referred to as the boron "fluctuation bands", and were initially attributed to B_2O_3 but have now been assigned to BO_2 (Kaskan and Millikan, 1960). The strength of the fluctuation band at 5470 Å (or 5450) was measured in absorption and the OH and an infra-red emission band of HBO_2 at 2040 cm^{-1} were studied under various conditions and it was concluded that the reaction

$$OH + HBO_2 \rightleftharpoons H_2O + BO_2$$

was always quickly equilibrated. In these flames bands of the α system of BO (see Appendix) in the visible region are also emitted.

When finely divided boron is heated in air in a low-temperature furnace, there is a violet glow (Porter and Dows, 1956) the spectrum of which shows emission between 4250 and 4500 Å, with red-degraded heads near 4403 and 4344 Å, and also a second region of diffuse emission between 4750 and 5100 Å. The emitter is believed to be B_2O_2.

The flame of diborane, B_2O_6, was studied by Parker and Wolfhard (1956); the premixed flame with air has a bright inner cone which emits the α bands of BO, the BH band, with a strong narrow red-degraded Q head at 4331.6 Å, and the atomic boron lines at 2496.8 and 2497.7 Å. The upper part of the flame was green and showed only the fluctuation bands of BO_2.

For pentaborane, B_5H_9, it was found (Berl et al, 1956) that a diffusion flame of the borane diluted with excess nitrogen and burning in air showed mainly the α system of BO, with the fluctuation bands (BO_2) weakly present; this flame had a temperature of only about 500 K and the emission must be chemiluminescent. In pentaborane/air explosions the emission was solely that of the fluctuation bands; emission from BO, BH, OH or BN was not found. The BO_2 emission might be thermal.

The oxidation of diborane was studied in a furnace and when initiated by flash photolysis by Carabine and Norrish (1967). The time-history of absorption bands of BO, BO_2, BH and OH was investigated, and it was concluded that a degenerate chain-branching mechanism, similar to that for hydrocarbons, was involved. The OH radical took part in the principal reactions.

Explosion flames of diborane with nitric oxide (Roth, 1958) show continuous emission in the visible and near ultra-violet and this is believed to be due to the formation of excited NO_2 molecules. The fluctuation bands of BO_2 are superposed on the continuum. The reaction is very rapid and it seems that as many as four reaction zones could occur.

Metals in Flames

When metals or their compounds are introduced into flames, lines of the free metal atoms are usually obtained and in many cases bands of oxides, hydrides or hydroxides are also observed. Usually the metallic line spectrum is restricted to a few lines with low excitation energy, especially resonance lines like the extremely persistent yellow lines of sodium or the infra-red resonance lines of potassium. In the inner cones of organic flames, a more complete development of the atomic spectrum may occur because of the high effective electronic excitation temperature in this region (see page 183).

Flames containing alkali metals show strong resonance lines and continuous emission. Some very old papers by Hartley (1894, 1907) give a clear account of the spectra of these flames. In an oxy-hydrogen flame Li gives continuum from 4600–3200, Na from 6020 to 3600 and weakly to 3320, and K from 5700 to 3400 Å. These continua were previously attributed by the author to ion-electron recombination processes of the type

$$Na^+ + e^- = Na + h\nu$$

but it is now known (see page 95) that the main reaction is of the type

$$Na + OH = NaOH + h\nu$$

Hartley also speaks of superposed complex band structure. This, from the description, appears to be due to the hydrides, LiH, NaH and KH. No oxide spectra of these metals are known.

The brightly coloured flames obtained by adding salts of alkaline earth metals have been known for a very long time, and the bright red flames with strontium and green ones with barium are the basis of many firework compositions and coloured flares. These colorations were at first believed to be due to bands of the oxides, but James and Sugden (1955) suggested, from studies of variation with flame composition, that the monohydroxides CaOH, SrOH and BaOH were responsible. Studies of isotope shifts using deuterium (Gaydon, 1955; Charton and Gaydon, 1956) showed that the main bands were indeed due to the monohydroxides, although in some cases weaker bands of the diatomic oxide and possibly of dimers such as Sr_2O_2 also occurred. These conclusions received further support from the work of van der Hurk, Hollander and Alkemade (1973) who studied the excitation in acetylene and carbon monoxide flames. For calcium, the main emission bands are centred near 5540 and 6230 Å and are due to CaOH; they are very persistent impurities in the spectra of flames and explosions. For strontium the main bands around 6060, 6470, 6690 and 6820 Å are due to SrOH, but some weaker bands may be due to SrO or Sr_2O_2. For barium strong complex bands of BaOH occur around 4870 and 5120 Å, but weaker red-degraded bands of the diatomic oxide BaO also occur at 4784.1, 4965.4, 5012.4, 5086.7 and 5214.7 Å.

When magnesium salts are introduced into an oxy-hydrogen flame a moderately strong sequence of simple red-degraded bands of MgO appears in the green at 5007.3 Å. Two strong groups of diffuse bands around 3700 and 3830 Å, due to MgOH (Pesic and Gaydon, 1959) give this flame a violet colour; bands of MgO also occur in this region, but in the flame the MgOH emission is dominant. The flame of burning magnesium powder or magnesium ribbon also shows the oxide bands superposed on a very strong continuum emitted by hot oxide particles (Edse et al, 1963). The burning of metals is dealt with more fully by Gaydon and Wolfhard (1970).

Similarly, aluminium shows a simple red-degraded system of AlO in the green, with (0,0) band at 4842 Å, and continuous emission from particles of Al_2O_3. Aluminium itself has some virtues as a high-energy fuel, and aluminium foil burning in oxygen is used in flash lamps. The very bright continuum from the oxide particles has a brightness temperature and colour temperature around the boiling point of the oxide, about 3900 K. These spectra are discussed by Edse et al (1963) and Brzustowski and Glassman (1964).

The green colour obtained when copper is present in flames must be known to everyone. The main emission is a diffuse band at 5350 Å extending to 5550; this, and a weaker band 6150–6250, are due to CuOH (Bulewicz and Sugden, 1956). Bands of CuO in the orange and CuH at 4280 Å may also be obtained; the use of these bands to study concentrations of OH and H in hydrogen flames has been mentioned (page 118). Bands of CuCl in the violet are a persistent impurity in flames of carbon monoxide, and those of CuF may occur when fluorine is present.

Other metals which show bands of hydroxides are manganese, with diffuse bands of MnOH between 3500 and 4000 Å (Padley and Sugden, 1959a), iron, with bands of FeOH from 3530–3580 and 3630–3675 Å and tin, with a band of SnOH at 4850 Å. Many of these transition metals are strong catalysts for the recombination of free radicals in flames, probably by cycles involving metal oxide and hydroxide intermediates (Bulewicz and Padley, 1971a), and thus act as flame inhibitors.

Both the normal and cool flames of arsenic show only a strong continuous spectrum from 4300 to 4900 Å (Eméleus, 1927, 1929).

The flame of selenium burning in oxygen shows a large number of bands superposed on a continuous background from the green to the violet. These are attributed to Se_2 and SeO_2 (Eméleus and Riley, 1933). A similar spectrum has been obtained from flames of a number of substances (NH_3, CS_2, C_2H_4, acetone, benzene, alcohols) burning in the vapour of selenium dioxide. These flames do not show OH or other emission bands in the ultra-violet and are quite different from the flames of these fuels with oxygen.

Certain organo-metallic compounds are of special interest because of their influence on knock in engines. The effect of lead ethyl on engine spectra has already been discussed in Chapter XI. Egerton and Rudrakanchana (1954) gave a detailed description of the diffusion flames of some organo-metal compounds and of diffusion flames of methane, hydrogen, etc. containing these compounds. They used a flat-flame burner and found that many of the flames had a complex structure; the spectra of the various zones were examined to gain information about the decomposition and reaction processes.

Diffusion flames of lead tetramethyl gave lines of Pb and bands of PbO (see Appendix), CH, C_2 and OH.

Zinc dimethyl, burning as a diffusion flame, showed a continuum due to hot particles, lines of Zn and bands of ZnO (at 3435 Å), ZnH (4301, 4240 Å etc.), CH and OH. This compound was spontaneously inflammable, ignition occurring via an unstable peroxide. In the diffusion flame, however, once ignition had started, there appeared to be thermal decomposition to Zn and methyl radicals.

Cadmium dimethyl gave only continuous emission, without CdO or CdH bands, but in the flame of hydrogen containing cadmium dimethyl some CdH bands were found.

Flames containing iron carbonyl give very strong bands of FeO in the yellow, orange and red (see Pearse and Gaydon, 1965 and the Appendix), and these colour the flame yellow. This yellow colour can easily be mistaken for a luminous soot zone, but is readily distinguished by its characteristic spectrum. Carbon monoxide which has been stored in steel cylinders frequently contains iron carbonyl as impurity and the flame then emits FeO bands. The diffusion flame of iron carbonyl shows these FeO bands and also continuous emission from hot oxide particles. With iron carbonyl in a low pressure oxy-acetylene flame (Gaydon and Wolfhard, 1948) there is a well marked fore zone of greenish colour, the spectrum showing a continuum in the yellow to blue-green. The main reaction zone, which is well separated from this fore zone, is bright blue-green and shows strong Fe lines as well as the usual C_2, CH etc. This main zone is enclosed below by a yellow sheath showing strong FeO bands, this sheath also extending upwards from the edge of the flame. The continuum in the fore zone presumably indicates some process in the decomposition of the carbonyl, but the formation of FeO below the main hydrocarbon reaction zone is unexpected; perhaps this is another case of FeO and FeOH taking part in catalytic removal of free radicals.

When nickel carbonyl is introduced into a Bunsen flame, bands of NiH and NiO are obtained (Pearse and Gaydon, 1965); the NiH bands in the yellow to red are degraded to the red and have very open rotational structure; they occur best just above the tip of the inner cone. The NiO bands extend over most of the visible region but are strongest in the green and are emitted mainly by the outer cone of the flame. A number of Ni lines occur, especially in the inner cone. The diffusion flame of pure nickel carbonyl shows only continuous emission (Egerton and Rudrakanchana, 1954), while the diffusion flame of the carbonyl mixed with methane shows weak Ni lines but no NiO or NiH.

Chapter XIII

Flame Spectrophotometry

One of the first applications of the study of flame spectra was for qualitative analysis, and indeed some elements were first discovered from their characteristic flame spectra, e.g. rubidium by Bunsen and Kirchhoff in 1861. Quantitative spectrochemical analysis now has frequent and important applications in both industry and research, especially for trace elements. The main sources for excitation of spectra are arcs, sparks and flames. The spectrum of an element, as developed in a flame, is relatively simple, consisting normally of only a few resonance lines; identification of the lines is thus much easier than with the more complicated arc or spark spectra, and confusion with lines of other elements is less likely. With such simple spectra it is often sufficient to use spectrographs or monochromators of low resolving power, or even in some cases just interference colour filters. Also flames are much steadier sources than arcs or sparks and are capable of higher precision for quantitative measurements. Since the previous edition of this book was written there have been major improvements in the techniques of flame photometry, especially in the use of atomic absorption and atomic fluorescence by C. T. J. Alkemande, A. Walsh, T. S. West and J. B. Willis and their colleagues, and these techniques are commented on here.

There is an extensive literature on flame spectrophotometry and a number of books on the subject; the author is familiar with those by Mavrodineanu and Boiteux (1965), Gilbert's translation of Herrmann and Alkemade (1963) and the composite work edited by Dean and Rains (1969). These books between them give an excellent account of the experimental techniques and discuss the advantages and limitations. The author has no practical experience in spectrochemical analysis, but the understanding of flame processes which comes from many years research on flame spectra serves as a valuable background for some comments on problems of spectrophotometry.

After a summary of some of the basic principles, comments will be made on the effects of the combustion processes and departures from equilibrium in flame gases on the reliability of the methods and the best choice of flame type and conditions. A full review of the literature will not be attempted.

The Choice of Flame Type

The two main requirements of the flame are that it shall have the right temperature and that the spectrum of the flame itself shall not interfere with observation of the lines being measured.

Most early work, dating from Bunsen in 1859, was done with simple Bunsen-type flames of manufactured gas or natural gas with air. These flames readily give the spectra of alkali metals, which have a very low excitation energy, but are not hot enough to produce the spectra of other elements. The inner cone region may give rather stronger excitation but is unsuitable because of the various band systems of C_2, CH, OH and HCO which are emitted from this part of the flame. The outer cone of the Bunsen flame does show some OH and CO-flame emission but for flames of either methane or of manufactured gas containing a fair amount of hydrogen this emission is not very strong and is not a serious limitation.

The acetylene/air flame is rather hotter (up to about $2300°C$) and was used in the early days with success by Lundegårdh and is still the basis of some commerical instruments. The higher temperature enables the spectra of the alkaline earth metals and a number of other elements (Cu, Cr, Mn, In, Tl etc.) to be obtained; Mavrodineanu and Boiteux list 34 elements which Lundegårdh observed in the acetylene/air flame. Background flame radiation can be significantly reduced by using a separated flame, either of the Smithells type (Kirkbright, Semb and West, 1967) or with a nitrogen shield (Hobbs et al, 1968). However, for some of these elements sensitivity is still low and there are other elements which require still hotter flames. Flames with oxygen are much hotter and oxy-coal gas, oxy-propane, oxy-butane, oxy-acetylene and oxy-hydrogen have been used by various workers. Even rare-earth metals can be analysed with these flames, and Gilbert (1964) lists 76 metals and non-metals which can be determined by some form of flame photometry.

Although these flames with oxygen have the advantage, for some purposes, of being very hot, they also have very high burning velocities, which create practical difficulties in flame stability and flame control. Thus in recent years the acetylene/nitrous oxide flame has been much advocated and used, because although very hot its burning velocity is not very different from that of the much cooler acetylene/air flame. This is illustrated by the following values of maximum burning velocity, S_u, and adiabatic flame temperature, taken partly from a review by Walsh (1966)

Flame	S_u cm/sec	$T, °C$
C_2H_2/air	160	2300
C_2H_2/O_2	1130	3050
C_2H_2/N_2O	180	2955
H_2/O_2	1180	2810

This acetylene/nitrous oxide flame is probably the best general purpose flame and is hot enough to atomize (in the true sense of the word) most elements, so that it can be used for atomic absorption and atomic fluorescence as well. Some studies have also been made of the hydrogen/nitrous oxide flame, but although easy to maintain it seems less suitable because for some reason it is slow to atomize elements with stable oxides (Willis, Fassel and Fiorino, 1969).

The oxy-cyanogen flame is extremely hot because of the high dissociation energies of the products, CO and N_2; the mixture $C_2N_2 + O_2$ has an adiabatic flame temperature of about 4835 K (Thomas, Gaydon and Brewer, 1952). It has, however, a fairly low burning velocity, and therefore appears suitable for use with those elements which have stable or refractory oxides and cannot be atomized in low temperature flames. This flame has been used by several authors (Baker and Valee, 1955; Fuwa et al, 1959; Robinson, 1961) and indeed has been found to give the expected strong excitation. The high price and toxicity of cyanogen are major disadvantages; also this flame is rapidly cooled if too much solution is sprayed into it. The flame of carbon subnitride (C_4N_2) with oxygen is even hotter, up to about 5260 K. Other exotic flames which have been suggested are hydrogen/fluorine (4300 K) and hydrogen/perchloryl fluoride (ClO_3F) which may have a temperature of around 3550 K. Willis (1968) has reviewed the use of high temperature flames for atomic absorption photometry and favours the acetylene/nitrous oxide flame for most purposes.

Generally flames of organic fuels such as hydrocarbons all suffer to some extent from emission of the CO-flame spectrum which reduces the signal-to-noise ratio, but, away from the immediate vicinity of the inner cone the gases are usually in good chemical and thermal equilibrium and so results tend to be reproducible. Careful control of mixture strength may be important as, for those metals which readily form oxides, the degree of dissociation of oxide to free atoms may vary rapidly in changing even slightly from oxidizing to reducing conditions in the flame; reducing conditions, i.e. a fuel-rich mixture, will be best for studying these metals. Hydrogen flames, apart from practical difficulties due to the high burning velocity, have the advantage of being without the CO-flame or other background emission, but, as we have seen in Chapter V, are much more subject to anomalous effects due to departures from equilibrium. These departures easily upset the reliability of quantitative measurements, although they may in some cases increase the sensitivity either by giving strong chemiluminescent excitation or by shifting the chemical equilibrium (e.g. the excitation of Li at the base of hydrogen flames, page 118). The spectrograms published by Mavrodineanu and Boiteux clearly show that several elements (Pt, Ru, Sn, Ph, V) are much more strongly excited in the inner cone of the acetylene/air flame than in the outer cone, and this is also shown quantitatively by the measurements of excitation temperature in low-pressure flames (page 184). Gilbert (1963) has particularly emphasized the increased sensitivity obtained by chemiluminescent excitation; he favours the reaction zone of rich hydrogen/air flames for exciting some species and notes that SnO is dissociated to Sn by hydrogen atoms; excitation of metals in hydrogen flames has also been discussed in Chapter V. For atomic absorption or fluorescence, however, although high atom concentrations may exist in the reaction zone schlieren-type disturbances due to refractive-index gradients are likely to be troublesome (p. 28).

Although it is usually advantageous to use a hot flame, because this increases the excitation, favours dissociation to free atoms and reduces chemical interference from formation of refractory compounds, there are some disadvantages. When small quantities of easily ionized metals, like potassium, are introduced into a flame, a high proportion of the atoms is ionized and not available as neutral atoms to emit or absorb radiation. This ionization varies rapidly with temperature; above 3000 K small amounts of potassium are almost completely ionized (>95%). Also at high temperature emission spectra tend to be more fully developed and complex so that there is a greater risk of error due to chance overlapping of lines, and the continuous background from the flame emission may also worsen the signal-to-noise ratio.

Line spectra are mostly used for spectrophotometry of metals, but for many non-metals the line spectrum is not readily excited, but it is then sometimes possible to use molecular band spectra. The use of the HPO band at 5280 Å to estimate phosphorous (page 314) and of S_2 in hydrogen flames for sulphur (page 310) have already been mentioned. Halogens may be estimated from the emission of a diatomic halide band of a deliberately added metal e.g. chlorine by CuCl bands when copper is added and fluorine from SrF (Burrows and Horwood, 1963). In other cases (B, Si) oxide bands may be used. For the excitation of these band spectra flames at not too high temperature are required as the molecules tend to dissociate in very hot flames. The intensities of the band systems are especially likely to be sensitive to departures from equilibrium in the flame gases and indeed in some cases, e.g. S_2, depend on it. Care is therefore needed in making any quantitative measurements and the possibility of other constituents of the sample affecting the rate of attainment of equilibrium in the flame and thus affecting the excitation processes should be considered; it is, of course, nearly always possible to calibrate empirically using standard specimens of similar general composition to the unknown sample. Some metals, especially the rare-earth metals, have very stable oxides and are best analysed from the band spectra of these diatomic oxides, but for these it is still desirable to use a fairly high temperature flame.

The Burner and Spray System

In early work, e.g. by Bunsen and Kirchhoff, the sample was introduced into the flame on a platinum loop, a method useful for spot tests but obviously not satisfactory for quantitative determinations. Ramage was able to make roughly quantitative analyses by putting weighed samples in a filter-paper spill and feeding it into a flame at a controlled rate, and some other methods of introducing solid samples have been used. However, the present practice is to make up solutions of known concentrations and spray these into the flame. The text books already referred to discuss in detail the burners and spray systems (sometimes referred to as atomizers and sometimes, to prevent confusion with the process of dissociation to free atoms, as nebulizers). The methods of making up the solutions do not appear to be dealt with at all well; for biological samples the preparation of the solution may be fairly straightforward, but for some mineral specimens, such as those containing silicates, there are obvious problems. For reliable results a consistent sample, a reproducible spray system and a steady flame are necessary.

If the spray device is separate from the burner, there will be some loss of solution in the connection between the spray and burner, and this is usually collected and fed back to the flame; provided the comparison standard samples are of similar consistency this will not cause any error but it could happen that a viscous biological sample might behave differently from an inorganic comparison sample. In some devices the solution is injected directly into the flame gases; thus Warren (1952, 1959) has described a water-cooled burner for propane/oxygen or hydrogen/oxygen in which the fuel burns on a ring of small holes and the solution is injected with oxygen under pressure into a central hole; an outer sheath of nitrogen may be used to reduce the CO-flame or NO + O continuum radiation and to prevent atmospheric dust from entering the flame. Samples as small as 0.2 ml may be used to give a scan, either in emission or absorption, across a spectrum line. For efficient atomization of refractory substances it seems that the spray must produce very small droplets (Willis, 1967) as larger droplets are slow to evaporate; in some cases the molecular dissociation of oxides may also be slow (Willis, 1970).

The usual practice in spectrophotometry is to make measurements relative to those of standard samples, but Shelton and Walsh (1958) have considered the possibility of absolute determinations based on predetermined transition probabilities (f values) flame dimensions and concentrations of solution in the flame. They have considered the use of such a method with atomic absorption photometry, using a light chopper to determine the absorption coefficient in the centre of the line. They assume a Doppler contour for the spectrum lines, and list f values for twelve elements. The measurement of the optical path length through the flame and the rate of use of solution and gas flow rates (to get the concentration in the flame) may be limitations; the use of an internal standard of a known line of an added element might serve as a check.

Technical details of burner design, spray devices and flow-metering of gases are given most fully in the book by Mavrodineanu and Boiteux. Usually the steadiest flames are of the laminar premixed type, but, for flames using oxygen, turbulent diffusion ones have a safety advantage because they cannot strike back. For a laminar flame it is necessary to use small burner holes as the flow becomes unsteady and turbulent above a critical burner diameter, and the burner length must also be sufficient to give an established smooth flow pattern (see *Flames*, p. 14). The flame must be screened from draughts. Gas flows to the burner and spray system should be controlled with good valves. For gases which are stored as liquids under pressure (e.g. propane, butane) uneven boiling of the liquid in the gas bottle may cause flow irregularities; a ballast volume between two valves will steady the flow. Flowmeters and spray devices are both quite sensitive to temperature changes, which affect the viscosity, and temperature control is not easy because, after the flame is started, parts near the burner tend to warm up, while pressure reducing valves on gas bottles tend to cool.

Recording Methods

The intensities of the spectrum lines may be measured either photographically or photo-electrically. The photographic method has the advantage that the whole spectrum, or at least a large part of it, is recorded simultaneously, so that many elements may be measured together and unexpected elements may be detected. Spectroscopic interference of one element with another, by superposition of a band spectrum for example, can also be seen. The photographic method thus has some big advantages for preliminary study of an unknown sample. Both the accuracy and range of intensity covered by a photographic plate are, however, quite limited, and time and care are required to put comparison intensity standards on the plate and to photometer it. Some of the difficulties have been discussed in Chapter II. Thus for routine analyses of specimens of known type, photoelectric methods are much to be preferred and are in more common use.

For photoelectric measurements, it is possible to use either a simple photo-cell, when the light intensity is high, or a more elaborate photomultiplier system, often with chopped light beam and tuned a.c. amplification. The highest sensitivity and accuracy is obtained with a photomultiplier and monochromator of high resolving power, but simpler instruments are often quite adequate. For high sensitivity the ratio of signal strength to instrumental noise and flame background is important. An electronic system which stores the signal for an appreciable time helps to average out noise and flame fluctuations and so gives greater accuracy. For a complete analysis it is usual to scan slowly through the spectrum, but for routine analyses it is possible to use several photomultipliers to study several elements simultaneoulsy, or to serve as a control on an internal standard line. Photomultipliers should be selected having high quantum efficiency of the cathode and a high ratio of sensitivity to dark current. For spectrum lines in the red a photomultiplier which responds to red light must of course be used, but these usually have a worse ratio of sensitivity to dark current than those responding only to short wavelengths.

For photographic work fairly fast spectrographs of moderate dispersion covering a wide spectrum range are suitable, medium quartz prism and quartz-Littrow instruments being frequently used. For photoelectric recording smaller grating or prism monochromators are adequate, although if simultaneous measurements are to be made on a number of lines the mechanical problems of attaching several slits and photomultipliers make the use of a large-dispersion grating advisable. For ultimate sensitivity high dispersion and high light-gathering power are desirable to reduce flame background and improve signal-to-noise ratio.

For straightforward work on easily excited elements, such as the alkali metals, it is possible to use interference colour filters instead of monochromators. For atomic absorption spectrophotometry with a background source giving a simple line spectrum of the element being studied, interference filters or even simple colour filters may suffice; the background light must of course be chopped and a tuned amplifier used. For atomic fluorescence, when the emitted light is very weak but consists normally of only a single line or atomic multiplet a colour filter is sufficient and gives less loss of light; for ultra-violet lines it may even suffice to use a "solar blind" photomultiplier which is insensitive to most of the flame emission (Larkins, 1971).

The Relation between Line Intensity and Concentration

The strength of a spectrum line is a complex function of the concentration of the element in the original solution. It depends on the proportion of the total number of atoms of the element which are present in the flame as free neutral atoms and on the effective excitation temperature and is modified by self-absorption of the radiation. Thus in practice it is necessary to plot a "curve of growth" or "working curve" using standard solutions in which the element is present in known concentration.

At very low concentrations, self-absorption will, of course, be negligible, so that initially the line intensity will be proportional to the concentration of free neutral atoms. At higher concentrations self absorption will limit the strength of the line, especially in the centre of the line contour. With very high resolving power, so that the line contour is fully resolved, we should find that the brightness in the centre of the line approaches a limit set by the black-body emission at the flame temperature, but that the line continues to increase in width as the concentration rises (or the path length through the flame becomes greater). Thus with small dispersion we should still find some increase in total line intensity even at quite high concentration. The actual variation of total intensity with concentration will depend on the contour of the spectrum line, which is normally determined partly by its Doppler width and partly by collision or pressure broadening; in special cases hyperfine structure of the spectrum line, due to nuclear spin and isotope effects, may add a further complication. The mathematics of the subject, considering Doppler and pressure-broadening have been treated by a number of investigators. In simple cases, such as the alkali-metal resonance lines (James and Sugden, 1955) there is an initial stage at very low concentration in which the curve of growth is linear, and then a region at rather higher concentration in which the intensity varies roughly as the square root of the concentration. This is well shown for the sodium D line at 5890 Å in Figure XIII.1 (taken from Beckman Instruments, after Herrmann and Alkemade (1963)) in which the log of the intensity is plotted against the log of concentration. It can be seen that up to a concentration of two parts per million the slope is 1.0, corresponding to the linear relationship, and above 20 parts per million is 0.5, in agreement with the square-root law. The actual curve, and position of the break from

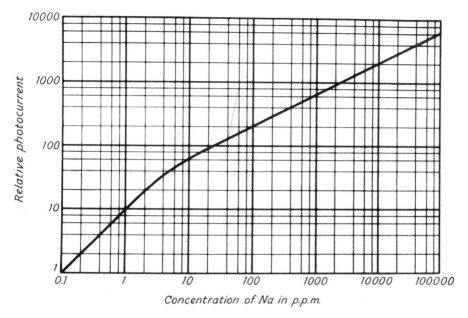

Fig. XIII.1. Curve of growth for the 5890 Å line of sodium, in logarithmic coordinates.

linear to square root, depends on flame dimensions and many other parameters. In flames of non-uniform temperature, reversal in the cooler outer layers may modify the law, and in very hot flames ionization processes may alter the initial slope (see later).

In atomic absorption photometry the curve of growth is different from that in emission. Measurements are usually made with a background source which emits a fine atomic spectrum line, so that the absorption is determined for the centre of the contour of the flame line. At high concentrations the absorption in the line centre is almost complete so that the accuracy of the determination by this method falls. It is usual to plot the absorbance or optical density (i.e. $\log I_0/I$) against concentration; at low concentration this curve again should be linear, but bends over at high concentration (see for example Mavrodineanu and Boiteux, 1965, page 197, and Sullivan and Walsh, 1966).

For atomic fluorescence we have to consider weakening of the exciting light beam by absorption and also of the emitted fluorescence by self-absorption. The curve of growth is initially linear, but at higher concentration the emission reaches a maximum and then actually falls again. This difficulty may, of course, be avoided in practice by working with a more dilute solution of the sample or reducing the flow rate of solution to the spray.

The number of free neutral atoms in a flame may differ from the total number of atoms sprayed into it because of the formation of molecular compounds and because of ionization. The formation of simple molecular compounds, like oxides, hydrides and hydroxides, affects the total strength of emission but does not cause the curve of growth to depart appreciably from the linear form. Thus if p_A denotes the partial pressure of free atoms of the element A, p_{AB} the partial pressure of the molecule AB, and p_B the partial pressure of B (which may be oxygen atoms, hydrogen atoms, hydroxyl radicals, etc.), then

$$\frac{p_A \times p_B}{p_{AB}} = K_p \qquad\qquad \text{(XIII, 1)}$$

where K_p is an equilibrium constant which depends on temperature and dissociation energy and can be evaluated. The partial pressures of the element being analyzed p_A and of the molecule p_{AB} are usually very small, of the order of parts per million, while the concentrations of free atoms in the flame (i.e. O, H or OH) are usually much higher, at least of the order parts per thousand. Thus for flames in equilibrium p_B is not appreciably affected by addition of the element A, and so the ratio p_A/p_{AB} is constant. Perhaps in some non-equilibrium flames, where H atoms are in excess, addition of the sample may catalyse the H-atom recombination and produce some disturbance. Also the formation of diatomic molecules from the element, such as Na_2 from Na, could affect the curve of growth at high concentrations, but in practice is not important for the concentration range used in flame photometry. The formation of compounds does, however, affect the variation of line strength with flame temperature, as higher temperatures favour dissociation to free atoms.

The formation of ionized atoms has a more important effect on the curve of growth. A rather similar equilibrium exists

$$\frac{p_A^+ \times p_e}{p_A} = K_i \qquad\qquad \text{(XIII, 2)}$$

where p_A^+, p_e and p_A are the partial pressures of ionized atoms, free electrons and neutral atoms. In the equilibrium region away from the reaction zone the pure flame gases are not ionized and if A is the only easily ionized metal present, then $p_e = p_A^+$. Thus

$$\frac{(p_A^+)^2}{p_A} = K_i.$$

In this case the proportion of atoms which is ionized obviously varies with the partial pressure of these atoms. Thus when small quantities of easily ionized metals (K, Cs, Rb, Na, Ba) are added to hot flames the tendency is for the number of neutral metal atoms to increase more rapidly than proportionally to the concentration of metal sprayed into the flame. It is well known that the curve of growth for potassium may at low concentrations show a marked upward curvature; addition of an excess of another easily ionized metal, such as sodium, will suppress this curvature because it maintains p_e constant independent of the potassium concentration (Herrmann and Alkemade, 1963, page 55). In flames of organic fuels there is abnormally high ionization in the reaction zone, and this persists to some extent well above the inner cone; the high concentration of free electrons and the possibility of charge transfer from positive ions might produce anomalies in the curve of growth for potassium etc. in this region of the flame.

Interference by one Element with the Estimation of Another

It is well known that the presence of one element may affect the strength of the spectrum lines of another element, thus complicating the spectrochemical analysis. Thus addition of sodium may strengthen the potassium lines; addition of Al weakens the Ca line. Effects may be even more complex; thus addition of Sr reduces the weakening effect of Al on Ca. Anionic substances may also be troublesome, e.g. the effect of phosphate or sulphate formation in reducing Ca and other alkaline-earth metals; these are particularly troublesome as the anionic substances rarely show by their own spectrum. Most of these interferences involve alterations to the concentrations of free atoms and so are important for atomic absorption and atomic fluorescence photometry as well as for emission.

Most interferences can be explained as due to one of four types, overlapping of spectra, alteration of the chemical equilibria, alteration of the ionization equilibria, and physical effects connected with spraying and evaporation.

Spectrum lines may be overlapped either by bands in the flame (e.g. the most sensitive lines of Sn, Si, Mg, Pt and Bi all lie among the strong OH bands) or by lines or bands of other elements from the sample. It is this spectroscopic interference which limits the use of colour filters; thus emission of the orange band of CaOH will interfere with Na estimation if a simple colour filter is used. Most of this type of interference can be eliminated by use of large dispersion monochromators, although a few chance coincidences may still occur; for example, the bismuth line at 3067.7 Å falls exactly on the R_2 (9) line of OH which can cause trouble even in atomic fluorescence. Overlapping band spectra of oxides etc. and continuous emission can be reduced in importance by using large dispersion but may not be entirely eliminated. Spectroscopic interference tends to be less with relatively low temperature flames, as very hot flames excite many more spectrum lines of each element and increase the risk of a chance overlap.

Chemical interference may be caused by the formation of new molecules in the gas phase, by formation of refractory compounds or by interaction with the free atoms and radicals of the flame. New compounds are most likely to be formed by anionic substances. Thus halogens tend to remove alkali metal atoms by formation of relatively stable alkali-halides. The well-known interference with alkaline-earth metals by phosphorus, and to a less extent by sulphur, is attributed to formation of refractory phosphates and sulphates respectively. Interferences of this type can be reduced or entirely suppressed by adding an excess of another element (a "releasing agent"); thus addition of excess Ca suppresses phosphate interference with Sr, or alternatively addition of excess Sr suppresses phosphate interference with Ca. Aluminium shows strong interference with calcium estimation, probably due to formation of a refractory calcium-aluminium-oxygen complex. In a few cases the interference may take the form of an enhancement instead of a reduction e.g. perchloric acid strengthens calcium lines. The extent of the interference often varies with height in the flame, especially when the formation of refractory compounds is involved, e.g. Al on Ca. Chemical interferences involving flame radicals are probably less important but are most likely to occur when hydrogen flames are used for the excitation; thus Bulewicz, James and Sugden (1956) attribute the influence of Cl on Li to changes in H-atom concentration modifying the formation of LiOH rather than to LiCl formation. All these chemical interferences tend to be reduced by using very hot flames. They are discussed at length in the various books, and Herrmann and Alkemade give a seven-page table documenting about 350 specific chemical interferences.

The alteration to the ionization equilibrium is most important with the alkali metals, and to a lesser extent Ba and Sr, and these may seriously interfere with each other. The partial pressure of free electrons, p_e in equation XIII, 2 will be increased by addition of a second alkali metal, and so the equilibrium will be shifted in the direction of an increased partial pressure of free neutral atoms, p_A. Thus addition of Na increases the strength of the K lines and also alters the shape of the curve of growth. These ionization interferences are greatest with very hot flames in which a high proportion of atoms are ionized and are best avoided by using relatively low-temperature flames, which is the opposite of the way of avoiding chemical interferences.

Physical interferences are mainly due to processes of droplet evaporation and to the rate of escape of volatile materials from any residual refractory particles. This is sometimes known as the "matrix" effect, and is discussed by Alkemade (see Dean and Rains, 1969, page 115). He quotes the different effects of aluminium nitrate and aluminium chloride on calcium, and the change of interfering effect with flame height. It seems that in addition to possible compound formation the evaporation of calcium from the involatile alumina matrix is involved. Some additives, such as halogens, which form volatile compounds, may assist evaporation from the matrix and so have an enhancing interference effect. These physical interferences tend to be reduced by good nebulization (i.e. a very fine spray) and by using very hot flames.

All these interferences of one element with another are a great nuisance in flame spectrophotometry. In general they prevent quantitative analysis of a sample by a single straightforward recording of its spectrum, either in emission or absorption. At best it is necessary to analyse elements separately and at worst can reduce the subject from a science to an art. For routine analyses of similar samples spectrophotometry is of great value, but these interferences make initial quantitative analysis of a completely unknown sample difficult and laborious.

Atomic Absorption

In recent years developments in atomic absorption have led to its increased use, and for accurate work it is tending to replace emission measurements. In emission small fluctuations in the flame temperature may cause error; in general a one per cent temperature fluctuation causes a change of between 10 and 20 per cent in light emission. For absorption work, using hot flames giving nearly complete atomization, temperature fluctuations have quite a minor effect. Also errors, in emission, due to variation in chemiluminescent excitation processes are avoided. The main advantage, however, is for elements whose sensitive spectrum lines lie at short wavelengths, roughly below about 4000 Å. Flame emission is then very weak because of the Boltzmann factor, $\exp(-E/kT)$, but absorption is still strong, and sensitivity still high. A minor limitation, discussed on page 329, is the flattening of the curve of growth when measuring high concentrations.

The usual practice is to measure the absorption against a background source which emits the atomic line being used. The light from this source is either chopped mechanically or the power supply to the source is modulated so that a.c. amplification or a phase-sensitive detector can be used to give the absorption irrespective of flame emission; fluctuations in flame emission from flame turbulence may however give some signal contributing to noise and limiting the sensitivity. If the background source emits broad spectrum lines then the flame absorption is restricted to the central region of the broad line, and sensitivity is reduced. The major advance has been in the development of special discharge lamps and hollow cathode lamps which give a simple spectrum of very fine lines (Sullivan and Walsh, 1966; Lowe, 1970). The method is highly selective and spectral interferences are rare so that low-dispersion fast monochromators or even filters may be used. The use of atomic absorption has greatly increased the number of elements which can be estimated at low concentrations in flames. Possible minor troubles to look out for are generalized absorption in the flame by particles or by hot oxygen molecules and when working near the reaction zone weakening of the light beam by schlieren-type deflections. If very long path lengths are used troubles due to generalized absorption, especially from molecules from the matrix, become increasingly important (Koirtyohann and Pickett, 1965).

Atomic Fluorescence

When a flame, containing the sprayed sample, is illuminated by a beam of light from a discharge lamp, resonance fluorescence may occur if the same element is present in the flame as in the discharge lamp. It is again usual to chop the illuminating light beam or to modulate electronically the discharge lamp (Browner, Dagnall and West, 1969). The method is thus rather similar to atomic absorption, but is even more selective and free from spectral interference. The strength of fluorescence is limited by molecular quenching in the flame gases and may vary a little with gas composition. The main problem is to obtain sufficient light and a reasonable signal-to-noise ratio. Flame radiation contributes to the noise and may be reduced by using a separated flame (Larkins, 1971) since the inter-conal gases have very low luminosity.

It is important to have a bright illuminating source. In early work the hollow-cathode lamps giving fine lines, as used for atomic absorption, were employed, but for fluorescence there is no need to have such fine lines and brighter discharge lamps are now used. With ordinary spectrum discharge lamps only one element at a time can be studied, but some work has been done to develop multi-element lamps (Cresser and West, 1970a, Marshall and West, 1970). Some measurements have also been made using a continuum source (a xenon high-pressure lamp) and this will excite all suitable elements (Cresser and West, 1970b); this excitation would seem to be less selective and need the use of a monochromator and might suffer error from light scattered by droplets or matrix particles in the flame.

The curve of growth, as already noted, is initially linear but reaches an intensity maximum and then falls. Zeegers and Winefordner (1971) have made a theoretical and experimental study of the curve of growth for magnesium. The fluorescence intensity and the concentration at which the maximum occur varies rapidly with position in the flame; this suggests the need to mount the burner and optics carefully and firmly to maintain reproducible alignment on the same part of the flame. Interference by other elements seems to be small with atomic fluorescence (Goodfellow, 1966); this is because the method usually employs hot flames which give good atomization and because, to keep on the linear part of the curve of growth, dilute solutions are often used and chemical interferences are then less important.

The atomic fluorescence method seems particularly good for estimating elements which are only present in trace amounts, and is most suitable for those whose sensitive lines lie at short wavelengths, below 2500 Å. Larkins (1971), using a "solar-blind" detector found that atomic fluorescence was superior to atomic fluorescence for Au, Cd, Co, Hg, Mg, Ni, Sb and Zn. Dagnall, Thompson and West (1967) have also used this method successfully for the non-metals selenium (the 2040 Å line) and tellurium (2143 Å), and it is also good for beryllium.

Effects of Flame Disequilibria

It has been pointed out, especially in Chapter X, that flame gases are not in complete chemical or thermal equilibrium, and several references to these effects have been made in this chapter. Abnormally high electronic excitation and other departures from equilibrium occur in the reaction zones of most organic flames. In carbon monoxide flames there may be a persistent excess concentration of free oxygen atoms, and in hydrogen flames atomic hydrogen, atomic oxygen and hydroxyl may all be in excess. We have also commented on the abnormal ionization in organic flames.

In emission the strength of spectrum lines will certainly be affected by abnormal excitation effects and in absorption and fluorescence the populations of free atoms will depend on chemical equilibrium being established. For spectrochemical analysis in which the test sample is compared with standard samples of varying composition, errors should not be large provided care is taken to use the same type of flame and exactly the same part of it. A few small "second order" errors might occur. It is known that the chemical disequilibrium in hydrogen flames can be reduced by quite small quantities of some additives (Arthur and Townend, 1954) and it is probable that the abnormal ionization and excitation in organic flames is also sensitive to some impurities; thus any difference between the composition of sample and standard solution may cause trouble; some transition metals seem to cause recombination of excess radicals through the intermediate formation of mono-hydroxides. It has been noted that in flames containing both Na and Li the Na emission varies little with height, but that of Li is sensitive to the hydroxide formation higher in the flame; if a flame is adjusted for height using sodium light this might not be sufficiently sensitive for use on lithium. The excess-radical anomaly, which affects lithium, is most noticeable with hydrogen flames at low temperature, but does also occur with acetylene/air flames (Hollander, 1964).

For emission spectrophotometry ultimate sensitivity may be increased by using the chemiluminescent excitation in the inner cone of organic flames, but for reliability this region should be firmly avoided. For atomic absorption hot flames are best, and low-temperature hydrogen flames, which show chemical disequilibria, should be avoided. For atomic fluorescence the advantage of the hydrogen flame having low emission itself may to some extent outweigh the uncertainties due to abnormal radical concentrations, but low-temperature flames of hydrogen with air should not be used.

Appendix

I. BAND SPECTRA EMITTED BY FLAMES

In the following Tables the wavelengths (λ) in international angstroms in air (1 Å = 10^{-10} metre), wave numbers in vacuum, in cm^{-1}, relative intensities (I) and assignment of vibrational quantum numbers (v', v'') are given for those band systems which frequently occur in emission in flames. Intensities, in the previous edition, were given on a visual scale of 0 to 10, but while this system had some advantages it tended to give a false impression that such intensities were quantitative; for this revised edition intensities are indicated as

vs	very strong
s	strong
m	moderately strong
w	rather weak
vw	very weak

Where bands form well marked groups or sequences the letter S before the vibrational assignment indicates the sequence head. Measurements refer to band heads unless otherwise stated, sometimes the type of band forming the head is noted. The direction of degradation is usually stated in the preamble to each system; where it is necessary to state this for individual bands the letters R, V, M and O are used respectively to indicate that the band is degraded to longer wavelengths (red) to shorter wavelengths (violet), has a maximum of intensity but no head, or that the measurement is of the band origin. The tables are usually set out so that the sequence heads precede the other bands of the sequence, i.e. in increasing order of wavelength for red-degraded systems and in decreasing wavelengths for violet-degraded systems.

The band systems listed here are those that are readily emitted by organic flames and also those that are of frequent occurrence when added metals or non-metals are present. A few more molecules are included than in the previous edition, but there are, of course, very many other band systems which can be excited in flames, such as metal oxides and halides, which are not listed here. Much fuller data for identification of molecular spectra and references to the literature are given by Pearse and Gaydon (1965), and the book edited by Rosen (1970) gives both wavelength tables and molecular constants for diatomic molecules.

Band systems are often known by a convenient name, such as that of the discoverer, or by an indication of its character, and the electronic transition is indicated following this name. The ground state is generally referred to as x and excited states in order as A, B, c etc; where a molecule shows two multiplicities the convention is to use capital letters for those states having the same multiplicity as the ground state and lower case letters a, b, c etc. for the other multiplicity. (C_2, for historical reasons, is an exception). In this edition values of the oscillator strength (f value) for the electronic transition are given where readily available, and references to data for Franck—Condon factors (vibrational intensity distribution) are quoted for the main systems. Some derived molecular constants are listed for a few molecules in Appendix III.

AlO

Green System A $^2\Sigma^+$–x $^2\Sigma^+$

Emitted from burning aluminium powder or foil (especially photo-flash bulb), from flames containing Al salts, and recently (Gole and Zare, 1972) observed in chemiluminescence of Al reacting with ozone. Jensen and Jones (1972) have studied flame equilibria involving AlO and Al (OH)$_2$.

Single-headed bands degraded to the red, forming marked sequences. Full data, including Franck—Condon factors, in Tyte and Nicholls (1964). Outstanding bands only:

λ	ν	I	v', v''	λ	ν	I	v', v''
4470.5	22362	w	S 2, 0	4842.3	20646	vs	S 0, 0
4494.0	22245	m	3, 1	4866.4	20543	s	1, 1
4516.4	22135	m	4, 2	4888.8	20449	m	2, 2
4537.6	22032	m	5, 3	5079.4	19682	s	S 0, 1
4557.6	21935	w	6, 4	5102.1	19594	s	1, 2
4648.2	21508	s	S 1, 0	5123.3	19513	m	2, 3
4672.0	21398	s	2, 1	5337.1	18732	w	S 0, 2
4694.6	21295	m	3, 2	5357.7	18659	m	1, 3
4715.5	21201	w	4, 3	5376.8	18593	m	2, 4

BO

α System, $A\,^2\Pi - X\,^2\Sigma^+$

Double-double headed bands, degraded to the red. These occur in flames of boron, boranes and hot flames to which boron compounds have been added; the BO_2 bands often partly mask them. The following table gives all four heads of the three strongest bands and the R_2 heads only of some other bands*.

λ	ν	I	v', v''	λ	ν	I	v', v''
3374.7	29624	w	5, 0	4339.4	23038	vs	1, $1^S R_{21}$
				4341.9	23025		R_2
3511.2	28472	m	4, 0	4363.4	22911		R_1
				4365.9	22898		Q_1
3662.2	27290	s	3, 0				
				4585.7	21801	vs	0, $1^S R_{21}$
3829.9	26103	s	2, 0	4588.8	21786		R_2
				4612.7	21673		R_1
4015.0	24900	s	1, $0^S R_{21}$	4615.4	21661		Q_1
4017.1	24887		R_2				
4035.5	24773		R_1				
4037.4	24761		Q_1	4718.7	21187	s	1, 2
4124.1	24241	m	2, 1	5011.7	19948	s	0, 2
4227.5	23648	m	0, 0	5513.0	18134	s	0, 3

BO_2

Boron oxide fluctuation bands, $A\,^2\Pi_u - X\,^2\Pi_g$

These occur in flames of boron, boranes and those containing boron compounds. They consist of waves of narrow red-degraded bands and

*These measurements all refer to the most abundant isotopes $B^{11}O^{16}$

usually appear diffuse. These waves show maxima in the regions 4520, 4710, 4930, 5180, 5450, 5800, 6030, 6200 and 6390 Å. See page 315).

Johns (1961) has observed them in absorption during flash photolysis of $BCl_3 + O_2$ and has made a rotational analysis. The following are the R_1 and R_2 heads of the strongest bands of $B^{11}O_2$.

λ	ν	$A\,^2\Pi_u$ $v_2\ v_2{}^l\ v_3$	$X\,^2\Pi_g$ $v_1\ v_2{}^l\ v_3$	λ	ν	$A\,^2\Pi_u$ $v_1\ v_2{}^l\ v_3$	$X\,^2\Pi_g$ $v_1\ v_2{}^l\ v_3$
4929.3	20281	$2\ 0^0 0$	$0\ 0^0 0\ R_1$	5456.8	18321	$0\ 0^0 0$	$0\ 0^0 0\ R_1$
4941.3	20232		R_2	5470.9	18274		R_2
4965.4	20134	$0\ 4^0 0$	$0\ 0^0 0\ R_1$	5790.7	17264	$0\ 0^0 0$	$1\ 0^0 0\ R_1$
4973.6	20101		R_2	5813.2	17198		R_2
5168.8	19341	$1\ 0^0 0$	$0\ 0^0 0\ R_1$	6171.6	16199	$0\ 0^0 0$	$2\ 0^0 0\ R_1$
5180.7	19297		R_2	6202.2	16119		R_2
5196.1	19240	$0\ 2^0 0$	$0\ 0^0 0\ R_1$	6376.6	15678	$0\ 0^0 0$	$0\ 0^0 2\ R_1$
5207.2	19199		R_2	6396.0	15630		R_2

$B_2 O_2$?

Porter and Dows (1956) observed a violet glow of boron oxidizing in air, giving emission between 4250 and 4500 Å and between 4750 and 5100 Å. The first group showed red-degraded heads at 4403 and 4344 Å.

BaO
$A\,^1\Sigma^+ - X\,^1\Sigma^+$

This system of single-headed red-degraded bands occurs with barium salts in flames, but they are much weaker than the BaOH. A selection of a few outstanding bands only is given.

λ	ν	I	v', v''	λ	ν	I	v', v''
5349.7	18688	s	4, 0	5976.3	16728	w	0, 0
5492.7	18201	vs	3, 0	6039.6	16553	vs	1, 1
5644.1	17713	vs	2, 0	6224.7	16061	m	0, 1
5701.0	17536	s	3, 1	6291.0	15891	s	1, 2
5805.1	17221	m	1, 0	6493.1	15397	vs	0, 2
5864.5	17047	vs	2, 1	6782.8	14739	s	0, 3

BaOH

The strong green colouration of flames containing barium salts is due to BaOH. Diffuse bands with maxima at 4870 and 5120 Å are superposed on an apparent continuum from 4700 to 5200 Å. There is also emission in the near infra-red, with strong maxima at 8280 and 8670 and a weaker maximum at 7400 Å.

Br_2

Orange-Red System, $^3\Pi - x^1\Sigma$

These bands occur in the outer cones of flames containing bromine, the flames being coloured orange-red. This is the main system and consists of a very large number of bands, which are degraded to the red, although this may not be obvious on small dispersion. The following is a list of the strongest bands, with averaged wavelengths from references quoted by Pearse and Gaydon (1965); λλ 5588, 5603, 5725, 5752, 5826, 5864, 5942, 5957, 6071, 6120, 6168, 6189, 6220, 6239, 6263, 6291, 6312, 6342, 6415, 6472, 6546.

2900 Å Band

In the inner cone of organic flames containing bromine Coleman and Gaydon found a strong band at 2900 Å. With small dispersion it appears diffuse and shaded to shorter wavelengths, commencing at 2930 Å and being strongest at 2910. There are similar but much weaker bands at 3330, 3120 and 2750 Å.

BrO

Ethyl Bromide Flame Bands, $A^2\Pi - x^2\Pi$

This system was discovered by Vaidya; measurements given here are by Coleman and Gaydon (1947), with revised vibrational assignments from Durie and Ramsay (1958). The bands occur in an oxygen/methyl bromide or ethyl bromide flame and in an oxy-hydrogen flame containing Br_2. The bands are degraded to the red and many of the heads are slightly diffuse.

λ	ν	I	v', v''	λ	ν	I	v', v''
3958	25258	w	1, 4	4270	23413	vs	0, 6
3999	24999	w	2, 5	4349	22987	w	2, 8
4029	24813	w	0, 4	4398	22731	vs	0, 7
4069	24569	m	1, 5	4533	22055	vs	0, 8
4109	24330	w	2, 6	4673	21394	s	0, 9
4147	24107	m	0, 5	4817	20754	w	0, 10
4186	23882	w	1, 6	4856	20587	vw	1, 11
4225	23662	vw	2, 7	4965	20135	vw	0, 11

C_2

At least eleven systems of C_2 are known, of which five or six occur in flames. The ground state is usually denoted $x^1 \Sigma$; the lowest triplet state is only 610 cm^{-1} above this and is denoted $x\,^3\Pi$.

The Swan System, $A\,^3\Pi_g - x\,^3\Pi_u$

This is by far the strongest system and occurs in the inner cones of all hydrocarbon flames and most other organic flames. It is responsible for the greenish colour of the inner cone. The main bands appear to be single headed and degraded to the violet, forming marked sequences. A few very weak "tail bands" are degraded to the red. These data are from Phillips and Davis (1968). Tyte, Innanen and Nicholls (1967) have also given full data and good photographs of this system, and Franck–Condon factors.

λ	ν	I	v', v''	λ	ν	I	v', v''
6762.4	14784	vw	1, 4	5447.7	18351	vw	5, 6
6675.9	14975	vw	2, 5	5165.2	19355	vs	S 0, 0
6599.1	15149	vw	3, 6	5129.4	19490	m	1, 1
6533.6	15301	vw	4, 7	5097.7	19611	w	2, 2
6481.8	15424	vw	5, 8	5070.9	19715	vw	3, 3
6444.7†	15512	vw	6, 9	4996.7*	20008	vw	13, 12
6191.3	16147	w	S0, 2	4911.0*	20357	vw	12, 11
6122.2	16330	w	1, 3	4836.1*	20672	vw	11, 10
6059.7	16498	w	2, 4	4770.2*	20958	vw	10, 9
6004.9	16648	w	3, 5	4737.1	21104	vs	S 1, 0
5959.0	16777	vw	4, 6	4715.3	21202	s	2, 1
5923.9	16876	vw	5, 7	4697.6	21282	s	3, 2
5901.0†	16942	vw	6, 8	4684.9	21339	w	4, 3
5635.5	17740	s	S0, 1	4680.2†	21361	vw	6, 5
5585.5	17899	s	1, 2	4678.6	21368	vw	5, 4
5540.7	18043	m	2, 3	4382.2	22813	vw	S 2, 0
5501.9	18170	w	3, 4	4371.4	22869	w	3, 1
5470.3	18276	vw	4, 5	4364.9	22903	w	4, 2

*Tail band, degraded to red, only observed in very hot flames.
†Bands with $v' = 6$ are part of the "High Pressure" system and are specially strengthened in some sources, such as high-pressure discharges and some atomic flames.

Many measurements of the oscillator strength have shown a lack of agreement, but a direct determination of the radiative lifetime by W. H. Smith (1969) found $170 \pm 20 \times 10^{-9}$ sec. for the (0, 0) Swan band, from which $f = 0.0178 \pm 0.0018$ is derived.

For rotational energy level data see Phillips and Davis (1968).

Fox–Herzberg System, $B^3\Pi_g - X^3\Pi_u$

This is a weak system emitted only by hot flames such as oxy-acetylene and usually only detected with large dispersion. The bands are strongly degraded to the red and the heads are not prominent. No intensities are available, but the (0, 3) and (0, 4) bands are strongest and the (0, 5) and (0, 6) may be fairly strong.

λ	ν	v', v''	λ	ν	v', v''
2378.2	42036	4, 1	2731.6	36598	0, 2
2429.9	41141	3, 1	2772.1	36063	1, 3
2486.3	40208	2, 1	2855	35015	0, 3
2527.9	39547	3, 2	2896.4	34516	1, 4
2589.0	38613	2, 2	2987	33468	0, 4
2656.3	37635	1, 2	3129	31949	0, 5
2698.8	37043	2, 3	3283	30451	0, 6

Rotational data are given by Phillips (1949) and Franck–Condon factors by Halmann and Laulicht (1966).

Mulliken's System, $d^1\Sigma_u^+ - x^1\Sigma_g^+$

Although involving the ground state of C_2, the high excitation energy limits its occurrence to relatively hot flames, in which it is easily observed. The band is hardly degraded at all so that the normal appearance is of P and R branches spreading out from the origins, giving diffuse maxima on each side of this origin region. The (0, 0) band is strongest and other bands of this sequence are superposed. The (0, 1) band is very weak. The following are band *origins*,

λ_0	ν_0	v', v''
2312.7	43227	0, 0
2314.0	43201	1, 1
2315.4	43176	2, 2
2316.8	43151	3, 3
2414.8	41399	0, 1

Rotational data are given by Landsverk (1939) and Franck–Condon factors by Halmann and Laulicht (1966). W. H. Smith (1969) found a radiative lifetime of 14.6×10^{-9} sec., giving $f = 0.055 \pm 0.006$ for the (0, 0) band.

Phillips Near Infra-red System, $b^1 \Pi_u - x^1 \Sigma_g^+$

These bands are the most prominent feature of hydrocarbon flame spectra in the photographic infra-red. They are degraded to longer wavelengths and the R branches form good heads; the strong Q branches do not form true heads but the bands may appear close double-headed under small dispersion. The following are the R heads (except for the (1, 0) band).

λ	ν	v', v''	λ	ν	v', v''
7714.6	12959	3, 0	8980.5	11132	3, 1
7907.7	12642	4, 1	10147*	9852*	1, 0
8108.2	12330	5, 2	12070.2	8283	0, 0
8750.8	11424	2, 0	15484.1	6456	0, 1

*Band origin.

Rotational data are given by Phillips (1948) and Franck-Condon factors by Halmann and Laulicht (1966).

Deslandres d'Azambuja System, $c^1 \Pi_g - b^1 \Pi_u$

This is a weak system due to a transition between two excited electronic states. It is often masked by CN bands but does occur in hot flames of hydrocarbons. Single-headed bands, degraded to shorter wavelengths, except where indicated otherwise.

λ	ν	I	v', v''	λ	ν	I	v', v''
4102.3	24370	vs	S 0, 1	3607.3	27714	s	S 1, 0
4068.1	24574	m	1, 2	3599.3	27775	w	4, 3
4041.9	24735	w	2, 3	3592.9	27825	s	2, 1
4026.9	24826	vw	3, 4	3587.6	27866	m	3, 2
3852.2	25952	vs	S 0, 0	3431.9‡	29129	vw	5, 3
3825.6	26132	m	1, 1	3405.1*	29359	vw	4, 2
3689.0‡	27100	vw	6, 5	3399.8	29406	w	S 2, 0
3617.9‡	27632	vw	5, 4	3398.1	29419	w	3, 1

*Origin of headless band.
‡Degraded to the red.

Rotational data are given by Phillips (1948) and Franck–Condon factors by Halmann and Laulicht (1966).

Ballik and Ramsay Infra-red System, $A'^3\Sigma_g^- - X^3\Pi_u$

These bands (see Ballik and Ramsay, 1963) although not apparently recorded in flame spectra are likely to occur as a strong feature of the infra-red spectrum because of their low excitation energy; they have the same final state as the Swan system. They are degraded to longer wavelengths.

λ	ν	v', v''
11724	8527	2, 0
14075	7103	1, 0
17675	5656	0, 0
24745	4040	0, 1

C_3

4050 Å Comet-Head Group, $^1\Pi_u - X\,^1\Sigma_g^+$

This band, lying between the two main CH bands, has been observed just above the tip of the inner cone of a very rich C_2H_2/O_2 flame (see Plate 5*e*), in flames supported by fluorine and in flames with moist atomic hydrogen. Under large dispersion it shows sharp heads and rotational fine structure, but with small dispersion may show a number of diffuse intensity maxima; the red-degraded head at 4050 Å is the most prominent feature, and another red-degraded head lies at about 4072 Å. The band was first observed in the spectra of comets, and later in interrupted discharges through CH_4, and was initially attributed to CH_2 but is now definitely assigned to C_3 (Clusius and Douglas, 1954). Kiess and Broida (1956) have measured the band in flames and Gausset et al (1965) have made a rotational analysis. The band extends from about 3600 to 4200 Å. The following are among the more conspicuous features (mostly red-degraded heads).

λ	ν	I	λ	ν	I
3966.9	25201	w	4038.0	24758	m
3994.6	25027	w	4041.7	24735	w
4007.6	24946	m	4049.8	24686	vs
4008.7	24939	m	4065.0	24594	w
4018.4	24880	m	4072.4	24550	s
4022.7	24852	m	4093.8	24421	w
4029.1	24812	m			

An f value of 0.13 has been found by Brewer and Engelke (1962); see also Jessen and Gaydon (1969).

CCl

$A^2\Delta - x^2\Pi$ *System*

Observed by Vaidya in flames containing CH_3Cl, $CHCl_3$ or CCl_4. Each band has about four heads and is degraded to shorter wavelengths. Rotational data by Gordon and King (1961) and Verma and Mulliken (1961). Promiment heads,

λ	ν	I	v', v''	λ	ν	I	v', v''
2927.7	34147	vw	P_2 0, 2	2789.5	35838	s	P_2 1, 1
2916.1	34284	vw	Q_1 0, 2	2788.3	35854	s	P_2 0, 0
2856.8	35000	vw	P_2 1, 2	2778.8	35976	m	Q_1 1, 1
2856.7	35002	w	P_2 0, 1	2777.6	35992	s	Q_1 0, 0
2846.2	35131	vw	Q_1 1, 2	2723.9	36700	vw	P_2 1, 0
2845.9	35134	w	Q_1 0, 1	2713.9	36886	vw	Q_1 1, 0

CF

$A^2\Sigma^+ - x^2\Pi$ *System*

Observed in inner cones of flames containing fluorine compounds. Double-double headed bands, resembling the NO γ-system, degraded to shorter wavelengths. Data from Andrews and Barrow (1951), Mann, Broida and Squires (1954) and Porter, Mann and Acquista (1965). All heads of strong bands and first heads of others,

λ	ν	I	v', v''	λ	ν	I	v', v''
2558.2	39078	s	$^0P_{12}$ 0, 3	2404.1	41584	vs	$^0P_{12}$ 0, 1
2557.3	39092		P_2	2403.1	41600		P_2
2553.3	39153		P_1	2399.6	41661		P_1
2552.5	39165		Q_1	2398.8	41674		Q_1
2479.4	40321	vs	$^0P_{12}$ 0, 2	2332.0	42868	m	$^0P_{12}$ 0, 0
2478.4	40336		P_2	2308.7	43302	w	$^0P_{12}$ 1, 1
2474.7	40397		P_1	2242.2	44586	w	$^0P_{12}$ 1, 0
2473.9	40409		Q_1				

CH

The 4315 Å and 3900 Å systems are among the strongest features of the spectrum of the inner cone of a Bunsen flame and all organic flames. The 3143 Å occurs readily in hot flames. For all systems full rotational data are given by Moore and Broida (1959) and for detailed identification of lines the atlas by Bass and Broida (1961) is valuable. Franck–Condon factors for all systems are listed by Hallmann and Laulicht (1966). Detailed descriptions of all three systems are given on page 144 and large-dispersion spectra are reproduced in Plate 3.

4315 Å System, $A^2\Delta-X^2\Pi$
Degraded to the violet. The strong Q and weaker P heads are listed,

λ	ν	I	v', v''	λ	ν	I	v', v''
4941.2	20232	vw	P 0, 1	4385	22799	m	P 0, 0
4913.5	20346	vw	P 1, 2	4323	23126	m	Q 2, 2
4890.3	20443	w	Q 0, 1	4314.2	23173	vs	Q 0, 0
4858.8	20576	vw	Q 1, 2				

The oscillator strength has been determined by Bennett and Dalby (1960) and Fink and Welge (1967); $f = 0.005 \pm 0.0005$.

3900 Å System, $B^2\Sigma^--X^2\Pi$
Degraded to the red. R and Q heads,

λ	ν	I	v', v''	λ	ν	I	v', v''
3627.2	27562	vw	R 1, 0	4025.3	24836	w	R 1, 1
3633.3	27515	vw	Q 1, 0	4033.8	24783	vw	Q 1, 1
3871.4	25823	vs	R 0, 0	4495.5	22238	vw	R 1, 2
3886.4	25723	s	Q 0, 0				

The oscillator strength for the (0, 0) band (Fink and Welge, 1967) is $f = 0.0028 \pm 0.0005$.

3143 Å System, $C^2\Sigma^+-X^2\Pi$
Not clearly degraded. Q maxima,

λ	ν	I	v', v''
3144	31797	vs	0, 0
3157	31667	m	1, 1

The oscillator strength of the (0, 0) band is $f = 0.006 \pm 0.001$ (Linevsky, 1967).

CHO see HCO

CH_2O, Formaldehyde
Emeléus's Cool Flame Bands, $^1A'' - ^1A_1$

See page 154. These occur in cool flames of ether, acetaldehyde and most higher hydrocarbons, and also in the normal flame of methanol. Under small dispersion they appear as a number of narrow, approximately equally spaced bands, either headless or degraded to the red. For analysis see Brand (1956). See Plate 2e

λ	ν	I	λ	ν	I
3405	29359	vw	4347	22997	s
3544	28207	w	4434	22547	s
3679—98	27173—036	s	4551—69	21971—881	s
3763—77	26569—469	m	4673—95	21393—294	m
3846.5	25991	s	4821	20735	w
3952	25297	vs	4947	20207	vw
4044	24721	m	5097	19613	vw
4121	24258	s	5227	19125	vw
4220—40	23687—579	vs			

CN
Violet System, $B\,^2\Sigma^+ - X\,^2\Sigma^+$

These bands occur strongly in the inner cones of flames of fuels containing combined nitrogen or burning with N_2O, and weakly in

λ	ν	I	v', v''	λ	ν	I	v', v''
4606.2	21704	vw	S 0, 2	3883.4	25743	vs	S 0, 0
4216.0	23712	m	S 0, 1	3871.4	25823	vs	1, 1
4197.2	23819	m	1, 2	3861.9	25887	s	2, 2*
4181.0	23911	m	2, 3	3854.7	25935	s	3, 3
4167.8	23987	w	3, 4	3851.0	25960	m	4, 4
4158.1	24043	w	4, 5	3590.4	27844	s	S 1, 0
4152.4	24076	w	5, 6	3585.9	27879	s	2, 1
				3583.9	27894	m	3, 2

*When the (0, 0) and (1, 1) bands are masked by CH the (2, 2) may be outstanding as the head of a group of three bands.

some other organic flames with air, such as C_2H_2/air. Strong sequences of single-headed bands degraded to shorter wavelengths. Some weak "tail" bands are degraded to the red, but these seldom appear in flames. Data from Brocklehurst et al (1972). See Plate 4e.

For rotational data see Douglas and Routly (1955), and for Franck–Condon factors see Brocklehurst et al (1972). Bennett and Dalby (1962) give the oscillator strength $f = 0.027 \pm 0.0003$.

Red System, $A\,^2\Pi - X\,^2\Sigma^+$

The strongest bands of this system lie well down in the photographic infra-red, and are emitted readily by flames under the same conditions as the Violet system, but weaker bands extending into the green are only obtained from the flame of cyanogen and from hot flames of fuels containing combined nitrogen. In active nitrogen the bands are excited very strongly, with a different intensity distribution. The bands are degraded to longer wavelengths and under small dispersion appear triple headed; only the first, R_2, heads are listed below. Data from Brocklehurst et al (1971).

λ	ν	I	v', v''	λ	ν	I	v', v''
5239.3	19081	vw	7, 1	6925.8	14435	vw	3, 0
5354.1	18672	vw	8, 2	7088.7	14103	vw	4, 1
5473.3	18265	vw	9, 3	7259.0	13772	vw	5, 2
5597.9	17859	vw	10, 4	7437.3	13442	vw	6, 3
5729.9	17447	vw	6, 1	7872.7	12699	m	S 2, 0
5858.2	17065	vw	7, 2	8067.1	12393	w	3, 1
5992.6	16682	vw	8, 3	8270.8	12087	w	4, 2
6192.1	16145	vw	4, 0	9140.6	10937	vs	S 1, 0
6332.2	15788	w	5, 1	9381.1	10657	s	2, 1
6478.5	15432	w	6, 2	10925.5	9150	vs	S 0, 0
6631.3	15076	w	7, 3	14074	7108	s	S 0, 1

Rotational data are given fully by Davis and Phillips (1963), Franck–Condon factors by Brocklehurst et al (1971) and an f value of 0.0034 ± 0.0003 by Jeunehomme (1965).

CO

For a discussion of spectra of this molecule see page 155. Of the many known systems of CO only four occur weakly in flames. The Fourth Positive system is the most important, and a few bands of the

Cameron system are listed here; for other systems see Pearse and Gaydon (1965) or references given on page 157, and Krupenie (1966).

Fourth Positive System, $A\,^1\Pi - X\,^1\Sigma^+$
This occurs as the result of chemiluminescent excitation in the inner cone of hydrocarbon flames and also in flames of carbon suboxide. The strongest bands are in the vacuum ultra-violet and only weak bands are accessible with quartz optics or through air. Single-headed bands degraded to the red. The following is a selection of the large number of bands; wavelengths less than 2000 Å are in vacuum.

λ_{vac}	ν	I	v', v''	λ_{air}	ν	I	v', v''
1493.6	66952	w	3, 1	2011.8	49691	vw	5, 11
1510.4	66206	w	4, 2	2025.8	49348	vw	2, 9
1525.8	65542	m	2, 1	2046.3	48853	vw	3, 10
1544.2	64758	m	0, 0	2067.7	48347	vw	4, 11
1560.1	64097	s	1, 1	2089.9	47834	vw	5, 12
1576.7	63425	s	2, 2	2113.1	47310	w	6, 13
1597.1	62612	s	0, 1	2137.0	46780	vw	7, 14
1630.4	61335	s	2, 3	2150.2	46493	vw	4, 12
1653.0	60495	vs	0, 2	2173.0	46005	vw	5, 13
1669.7	59892	vs	1, 3	2196.8	45506	vw	6, 14
1712.2	58405	vs	0, 3	2221.5	45001	vw	7, 15
1729.3	57829	vs	1, 4	2238.3	44664	vw	4, 13
1747.2	57234	s	2, 5	2261.7	44200	vw	5, 14
1774.9	56341	s	0, 4	2286.1	43729	w	6, 15
1792.4	55792	vs	1, 5	2311.5	43249	vw	7, 16
1810.8	55224	s	2, 6	2338.0	42759	vw	8, 17
1841.5	54305	m	0, 5	2381.6	41975	vw	6, 16
1859.4	53780	m	1, 6	2407.6	41552	vw	7, 17
1878.3	53239	m	2, 7	2433.9	41058	vw	8, 18
1897.8	52692	m	3, 8	2463.2	40585	vw	9, 19
1930.7	51795	w	1, 7				
1950.1	51280	w	2, 8				
1990.9	50229	w	4, 10				

Pilling, Bass and Braun (1971) have measured f values of a number of bands, leading to $f_{el} \sim 0.15$ and a radiative lifetime of about 0.8 nsec.

Cameron System, $a^3\Pi - x\,^1\Sigma^+$

Degraded to the red, each band showing about five close heads. The following are the first, R_3, heads of the bands observed in rich oxy-acetylene flames

λ	ν	v', v''
2257.7	44279	0, 2
2369.0	42199	0, 3
2491	40132	0, 4

Vibration-Rotation Bands (Ground state, $x\,^1\Sigma^+$)

These are strongest in hot flames such as hydrocarbon/oxygen and carbon monoxide/oxygen, and occur in the interconal gases and outer cone. Bands show P and R branches and form R heads, but the (1, 0) sequence is both self-absorbed and overlapped by CO_2 so that the R branch and heads are rarely visible. Measurements of heads and origins,

λ_{head} (μm)	ν_{head} (cm^{-1})	λ_{origin} (μm)	ν_{origin} (cm^{-1})	band v', v''
2.2929	4360.0	2.3467	4260.1	$2 \rightarrow 0$
2.3221	4305.2	2.3762	4207.2	$3 \rightarrow 1$
2.3519	4250.8	2.4064	4154.4	$4 \rightarrow 2$
2.3823	4196.5	2.4373	4101.7	$5 \rightarrow 3$
*		4.6644	2143.3	$1 \rightarrow 0$
*		4.7228	2116.8	$2 \rightarrow 1$
*		4.7825	2090.4	$3 \rightarrow 2$

*Masked by CO_2.

CO_2

Carbon Monoxide Flame Bands, $^1B_2 - x\,^1\Sigma_g^+$

For details of appearance see page 129 and Plates 1e, 2d and 7. They

λ	ν	λ	ν	λ	ν	λ	ν
3912	25555	4335	23060	4646	21520	4980	20075
4035	24775	4344	23015	4674	21390	5129	19490
4045	24715	4411	22665	4768	20965	5169	19340
4093	24425	4527	22085	4798	20835	5276	18950
4104	24360	4553	21955	4893	20430	5318	18800
4156	24055	4567	21890	4933	20265	5430	18410
4260	23465						

occur in CO flames and in outer cone of Bunsen-type flames. Very complex structure from 3000 to beyond 5000 Å is superposed on a continuum. Under small dispersion the listed maxima of narrow bands may be observed; under higher dispersion the structure is complex and many bands are double.

Vibration-Rotation Bands (Ground state x $^1\Sigma_g^+$)
For a description of the Infra-red spectrum see Chapter IX. Bands are emitted by the inter-conal gases. Bands are slightly degraded to longer wavelengths, but structure is complex and only some form good heads. Self absorption modifies the band at 4.26 μm. Origins and heads of strongest bands. See also page 226.

$\lambda_0(\mu m)$	$\nu_0(cm^{-1})$	$\lambda_H(\mu m)$	$\nu_H(cm^{-1})$	$v_1'v_2'^{l'}v_3'$ $v_1''v_2''^{l''}v_3''$
2.6912	3714.7			$1\,0^0\,1 - 0\,0^0\,0$
2.7672	3612.8			$0\,2^0\,1 - 0\,0^0\,0$
4.2557	2349.2	4.1713	2396.7	$0\,0^0\,1 - 0\,0^0\,0$
4.2790	2336.7	4.1912	2385.3	$0\,1^1\,1 - 0\,1^1\,0$
4.2955	2327.4	4.2110	2374.1	$0\,2^0\,1 - 0\,2^0\,0$
4.2970	2326.6	4.2057	2377.1	$1\,0^0\,1 - 1\,0^0\,0$
4.3014	2324.2	4.2140	2372.4	$0\,0^0\,2 - 0\,0^0\,1$
13.8699	720.79			$1\,0^0\,0 - 0\,1^1\,0$
14.9794	667.40			$0\,1^1\,0 - 0\,0^0\,0$

CS
A $^1\Pi$−x $^1\Sigma^+$ *System*
Occurs in the inner cones of hydrocarbon flames containing SO_2 or H_2S and in cool flame of CS_2. Degraded to the red, marked sequences of close double-headed bands, the R and Q heads being separated by about 2 Å. Strongest R heads,

λ	ν	I	v', v''	λ	ν	I	v', v''
2507.3	39871	m	S 1, 0	2605.9	38363	m	2, 2
2523.2	39620	m	2, 1	2662.6	37546	s	S 0, 1
2538.7	39378	m	3, 2	2677.0	37344	s	1, 2
2575.6	38814	vs	S 0, 0	2693.2	37120	m	2, 3
2589.6	38604	s	1, 1	2708.9	36904	w	3, 4

For rotational data see Lagerqvist et al (1958).

CaOH

These bands are a persistent impurity in flame spectra, especially of explosion flames. For details see Gaydon (1955) and Van der Hurk et al (1973).

Green Band, 5550 Å.
This is a fairly narrow band with complex rotational structure. There are no sharp heads but strong lines or maxima in the structure occur at λλ5554.3, 5547.4, 5546.4, 5542.9, 5539.8 and 5538.6.

Orange-red Band, 6230 Å
This is quite diffuse and wider. The strong central region shows maxima at λλ6207.5, 6220.0, 6235.2, 6248.1, 6259.2, 6261.4 and 6264.0. There are less strong bands around 6033 and 6453 Å.

Cl_2

Bands of the resonance system of Cl_2 were observed by Kitagawa (1937) in emission from H_2 burning in Cl_2. Narrow bands, degraded to the red. Prominent heads λλ5716, 5761, 5821, 5842, 5882, 5898, 5942, 6003, 6073, 6134 and 6201.

The inner cones of organic flames containing chlorine also show a wide diffuse band of Cl_2 at about 2580 Å; see page 306.

ClO

$A\,^2\Pi - X\,^2\Pi$ *System*
Observed in flames containing chlorine compounds, especially with Cl_2 in H_2/O_2. Degraded to the red, with rather complex heads; some bands are diffuse due to predissociation.

λ	ν	I	v', v''	λ	ν	I	v', v''
3652.5	27371	m	1, 5	4114	24300	vs	1, 9
3729.5	26805	s	0, 5	4154	24066	w	2, 10
3761	26581	vs	1, 6	4241	23573	m	1, 10
3841	26027	vs	0, 6	4283	23341	s	2, 11
3874	25806	vs	1, 7	4373	22861	w	1, 11
3957	25265	s	0, 7	4417	22633	w	2, 12
3991	25049	s	1, 8	4459	22420	w	3, 13
4078	24515	m	0, 8				

CuCl

Bands of CuCl frequently appear as impurity, especially in flames of CO, in furnaces and in coal fires. In the visible region there are five overlapping systems, denoted A to E, due to the transitions $A^1\Pi-X^1\Sigma^+$, $B^1\Pi-X^1\Sigma^+$, $C^1\Sigma^+-X^1\Sigma^+$, $D^1\Pi-X^1\Sigma^+$ and $E^1\Sigma^+-X^1\Sigma^+$. All systems are degraded to the red and form marked sequences. The groups of pairs of bands formed by systems D and E are strong and characteristic.

λ	ν	I	System	v', v''	λ	ν	I	System	v', v''
4187.9	23871	w	E	2, 0	4515.9	22137	m	D	0, 2
4210.9	23741	w	D	2, 0	4755.7	21021	vw	C	1, 0
4259.0	23473	s	E	1, 0	4788.5	20877	vw	B	1, 0
4280.9	23353	vs	D	1, 0	4846.9	20625	w	C	0, 0
4333.2	23071	vs	E	0, 0	4881.5	20479	w	B	0, 0
4353.9	22961	vs	D	0, 0	4946.1	20212	vw	C	0, 1
4412.4	22657	s	E	0, 1	4982.2	20065	vw	B	0, 1
4433.8	22547	vs	D	0, 1	5152.4	19403	m	A	1, 0
4493.8	22246	m	E	0, 2	5262.4	18997	w	A	0, 0

CuH

Single-headed red-degraded bands of the $A^1\Sigma^+ - X^1\Sigma^+$ system occur in flames, especially hydrogen flames, when copper is present.

λ	ν	I	v', v''
4279.6	23360	vs	0, 0
4327.7	23100	m	1, 1

CuOH

The green colour of flames containing copper is due to emission by CuOH. In the green there are four broad diffuse bands with maxima near 5220, 5400, 5660 and 5760 Å. There is another broad band in the orange around 6200 Å.

FeO

Bands in the orange and infra-red are emitted by flames containing iron compounds, especially iron carbonyl. They may occur as impurity. The orange bands are complex, but are mostly degraded to the red. Some have been analysed into three systems, all involving the ground state (x). Strongest heads, degraded to the red except where indicated.

λ	ν	I	System	v', v''	λ	ν	I	System	v', v''
5531.4	18073	w			5903.0	16935	m	C — X	0, 1
5582.8	17907	m	D — X	0, 0	5911*	16912	w		
5614.0	17807	m	C — X	0, 0	5919*	16890	w		
5621.3	17784	w			5974.6	16732	m	C — X	1, 2
5624.1	17775	w			6097.3*	16396	vs		
5646.6	17704	m	D — X	1, 1	6109.9	16362	vs	B — X	1, 2
5789.8	17266	vs	B — X	0, 0	6180.5	16175	vs	D — X	0, 2
5807.4	17214	vw	B — X	1, 1	6218.9	16075	vs	C — X	0, 2
5819.2	17179	m	B — X	2, 2	6524.1	15323	vw	D — X	0, 3
5868.1	17036	vs	D — X	0, 1	6566.7	15224	vw	C — X	0, 3

*maximum of narrow headless band.

The strongest bands in the infra-red have heads at 8112, 8137, 8230, 8302, 8578, 9120, 9286, 9676, 9893, 10102, 10553, 10803 and 11916 Å.

FeOH
Two diffuse groups of bands 3530–3580 Å and 3630–3675 Å occur when iron carbonyl is introduced into a hydrogen flame. Jensen and Jones (1973) have studied flame equilibria involving FeOH and FeO.

HCO
Vaidya's Hydrocarbon Flame Bands
Complex bands, degraded to the red in the inner cone of hydrocarbon flames, especially ethylene, and in chilled flames and engine

λ	ν	I	Prog.	λ	ν	I	Prog.	λ	ν	I	Prog.
2516.8	39720	vw	A_2	2857.8	34982	m	A_1	3501.0	28551	s	A_0
2531.8	39486	vw	A_0	2947.4	33918	s	A_0	3538.7	28251	vw	A_3
2584.1	38687	w	A_1	3013.7	33172	s	A_1	3587.5	27867	s	A_1
2639.8	37870	vw	A_2	3114.8	32095	vs	A_0	3674.6	27206	vw	A_2
2658.1	37610	w	A_0	3186.0	31378	vs	A_1	3729.3	26807	m	A_0
2714.7	36825	m	A_1	3298.2	30311	vs	A_0	3823.3	26148	w	A_1
2773.8	36041	vw	A_2	3376.3	29610	vs	A_1	3986.0	25081	vw	A_0
2796.1	35754	m	A_0	3457.2	28917	vw	A_2	4088.2	24454	w	A_1

exhaust glows. For description and a discussion of the analysis see page 150. The following bands are assigned to the progressions A_0, A_1, A_2 and A_3. See Plates 4b and 8.

Weak unassigned bands were measured by Vaidya at 2436.5, 2550.3, 2751.5, 3215.5, 3417.4 and 3635.6 Å. In hot flames he observed an additional "B" progression with heads at 2704.5, 2780.4, 3001.5, 3359.0, 3472.5, 3569.2, 3697.7 and 3802.7 Å. Dixon attributes a progression of bands with heads at 2422.2, 2487.4, 2537.0, 2661.3 and 2796.8 Å to another excited electronic state, $\tilde{C}$.

HCl
Vibration-Rotation Bands (Ground State $x\,^1\Sigma^+$)

See page 241. Bands with single P and R branches, weakly degraded to longer wavelengths, occur in chlorine/hydrogen flames. R heads,

$\lambda(\mu m)$	$\nu(cm^{-1})$	band
1.7138	5833	$2 \to 0$
1.7787	5620	$3 \to 1$
3.1428	3181.0	$1 \to 0$

HF
Vibration-Rotation Bands (Ground State $x\,^1\Sigma^+$)

See page 242. Bands with single P and R branches, degraded to longer wavelengths. R heads,

$\lambda(Å)$	ν	band	$\lambda(Å)$	ν	band	$\lambda(\mu m)$	ν	band
5485.3	18225	$5 \to 0$	7696.0	12990	$7 \to 3$	1.2505	7994.6	$2 \to 0$
5742.1	17402	$6 \to 1$	8088.2	12360	$8 \to 4$	1.308	7643.2	$3 \to 1$
6024.2	16595	$7 \to 2$	8517.3	11738	$9 \to 5$	1.3695	7299.7	$4 \to 2$
6326.1	15883	$8 \to 3$	8666.4	11536	$3 \to 0$	1.4355	6964.4	$5 \to 3$
6655.1	15022	$9 \to 4$	9068.0	11025	$4 \to 1$	1.6065	6636.1	$6 \to 4$
6686.3	14952	$4 \to 0$	9497.5	10526	$5 \to 2$	2.2696	4404.9	$1 \to 0$
6999.2	14284	$5 \to 1$	9959.1	10038	$6 \to 3$	2.3742	4210.8	$2 \to 1$
7334.5	13630	$6 \to 2$				2.4876	4020.3	$3 \to 2$

H_2O
Vibration-Rotation Bands

These very complex bands are responsible for much of the infra-red emission from flames. See page 224. The infra-red bands do not form heads. The fundamentals at 1.87, 2.66 and 6.26 μm are very strong. Origins,

$\lambda_0 (\mu m)$	ν_0	$v_1' \, v_2' \, v_3' - v_1'' \, v_2'' \, v_3''$
1.1351	8807.0	1 1 1 → 0 0 0
1.3782	7253.6	1 0 1 → 0 0 0
1.8752	5331.2	0 1 1 → 0 0 0
2.6618	3755.8	0 0 1 → 0 0 0
2.7337	3657.1	1 0 0 → 0 0 0
3.1722	3151.5	0 2 0 → 0 0 0
6.2698	1594.5	0 1 0 → 0 0 0

In a flame of oxygen burning in hydrogen these bands extend into the visible (see page 107 and Plate 1c). Structure is complex. The following are strong features from Kitagawa (1936) and Gaydon (1942b); a few more recent assignments of band heads are indicated.

$\lambda(\text{Å})$	ν	I	$v_1' \, v_2' \, v_3'$	$v_1'' \, v_2'' \, v_3''$	$\lambda(\text{Å})$	ν	I	$v_1' \, v_2' \, v_3'$	$v_1'' \, v_2'' \, v_3''$
5806.9	17216	w			6468.0	15456	s		
5861.6	17055	w			6490.4	15403	s		
5880.2	17002	m			6516.8	15341	vs		
5900.2	16944	m			6574.5	15206	s		
5948.8	16805	m			6628.6	15082	m		
5988.8	16693	m			6919.0	14449	m	1 0 3 →	0 0 0
6165.7	16214	s	1 1 3 →	0 0 0	7164.5	13954	m	3 0 1 →	0 0 0
6202.2	16118	s			7229	13697	m		
6220.0	16073	m			8097	12347	s	2 1 1 →	0 0 0
6255.1	15983	s			8916	11213	m	0 0 3 →	0 0 0
6321.6	15814	s			9277	10776	vs	2 0 1 →	0 0 0
6377.1	15677	s			9333	10712	m		
6457.5	15482	s	3 1 1 →	0 0 0	9669	10339	s		

I_2

Thermal emission of the main system, responsible for the violet colour of iodine vapour, is believed to lead to bands in the red and near infra-red, but details of the appearance of these numerous red-degraded bands in flames do not seem to be available. In the inner cones of organic flames containing iodine, there is a strong diffuse band at about 3425 Å.

IO

Methyl Iodide Flame Bands, $A^2\Pi - X^2\Pi$

First observed by Vaidya to occur just above the inner cone of a flame of methyl iodide, these bands occur well in oxy-hydrogen flames with added iodine. Long progressions of red-degraded bands. Those with $v' = 1$, 4 and 5 are diffuse, due to predissociation.

λ	ν	I	v', v''	λ	ν	I	v', v''
4189.0	23865	m	5, 0	5131.2	19483	vs	0, 3
4268.2	23423	s	4, 0	5208.5	19194	m	2, 5
4355.6	22953	vs	3, 0	5307.3	18837	vs	0, 4
4396.7	22738	m	4, 1	5493.2	18199	vs	0, 5
4448.9	22471	s	2, 0	5533	18069	s	1, 6
4487.2	22279	vs	3, 1	5689.7	17571	s	0, 6
4586.2	21798	vs	2, 1	5730	17447	s	1, 7
4693.5	21300	vs	1, 1	5900	16945	m	0, 7
4730.3	21134	s	2, 2	5939	16832	m	1, 8
4844.5	20636	vs	1, 2	5973.4	16736	s	2, 9
4964.2	20139	m	0, 2	6192.2	16145	w	2, 10
5002.3	19985	m	1, 3	6231.5	16043	w	3, 11

MgO

Green System, $B^1\Sigma^+ - X^1\Sigma^+$

In oxy-hydrogen flame containing Mg salts and in flame of burning magnesium ribbon or powder. Degraded to the violet, with single heads and marked sequences.

λ	ν	I	v', v''	λ	ν	I	v', v''
5206.0	19203	w	S 0, 1	4974.5	20097	s	3, 3
5192.0	19255	w	1, 2	4962.1	20147	m	4, 4
5007.3	19965	vs	S 0, 0	4949.5	20198	m	5, 5
4996.7	20008	vs	1, 1	4935.3	20257	w	6, 6
4985.9	20051	s	2, 2				

MgOH

In flames containing Mg salts there are two strong groups of complex band structure around 3600 and 3830 Å. The following are strong features, either degraded to the violet (V) or maxima (M).

λ	ν	I	deg	λ	ν	I	deg	λ	ν	I	deg
3696	27050	s	V	3801	26300	m	V	3834	26070	vs	M
3703	27000	vs	V	3807	26260	vs	M	3848	25980	w	V
3707	26970	s	M	3810	26240	vs	V	3855	25930	w	V
3719	26880	s	V	3815	26200	s	M	3859	25910	w	V
3731	26790	m	V	3822	26160	s	M	3882	25750	w	V

MnO

Emitted from flames containing manganese salts and also from furnace flames, including steel furnaces. Bands are degraded to red and form marked sequences. The following are the strong heads only,

λ	ν	I	v', v''	λ	ν	I	v', v''
5158.9	19379	m	S 2, 0	5582.2	17909	vs	S 0, 0
5192.0	19255	m	3, 1	5608.4	17825	s	1, 1
5227.7	19124	m	4, 2	5853.1	17080	vs	S 0, 1
5359.6	18653	s	S 1, 0	5879.5	17004	vs	1, 2
5390.0	18548	s	2, 1	5910.1	16916	m	2, 3
5423.2	18434	s	3, 2	6148.3	16260	w	S 0, 2
				6174.8	16191	w	1, 3

MnOH

In flames containing manganese salts a number of diffuse bands are attributed to MnOH (Padley and Sugden, 1959; Gilbert, 1961). In the violet region there are bands with maxima at 3650, 3750, 3820, 3880, 3960, 4050 and 4140 Å, and in the near infra-red at 8010, 8110, 8365, 8580 and 8710 Å. Bands 2550–2650 may also be due to MnOH

NCN

A complex group of bands around 3290 Å are emitted weakly by cyanogen flames and more strongly by atomic nitrogen reacting with organic compounds; also in absorption in flash photolysis. The bands, due to a transition $A\,^3\Pi_u - X\,^3\Sigma_g^-$ of linear NCN, have been analysed by Herzberg and Travis (1964). The strongest heads in the complex (0, 0, 0–0, 0, 0) sequence are at 3271.3, 3280.4, 3283.8 and 3290.1 Å or 30561, 30475, 30443 and 30385 cm^{-1}.

NH

3360 Å System, $A^3\Pi - X^3\Sigma^-$

See page 158 and Plate 3e. The bands are barely degraded at all so that the Q branches form strong narrow maxima, with the triple P and R branches spreading out, forming in some cases very weak heads. The bands occur strongly in flames involving combined nitrogen and weakly in hot flames of hydrocarbons with air. They are usually partly masked by OH, but may be seen clear of OH in flames supported by fluorine.

λ	ν	I	deg	branch	v', v''	λ	ν	I	deg	branch	v', v''
3023.0	33070	vw	R	R	1, 0	3336.0	29967	vw	R	R	2, 2
3047.3	32806	vw	R	Q_2	1, 0	3360	29753	vs	M	Q	0, 0
3050.3	32775	vw	R	Q_1	1, 0	3370	29665	s	M	Q	1, 1
3119.5	32047	vw	R		2, 1?	3383	29551	w	M	Q	2, 2
3137.0	31868	vw	R		2, 1?	3635.0	27502	vw	V	Q	1, 2?
3281.0	30470	w	R	R	0, 0	3637.7	27482	w	V	Q	0, 1?
3317.0	30139	w	R	R	1, 1?	3676.3	27194	vw	V	P	0, 1?

For a rotational analysis see Dixon (1959). Smith and Liszt (1971) have given Franck-Condon factors and also an oscillator strength for the (0, 0) band, $f = 0.00745 \pm 0.0015$; the radiative lifetime is 455 ± 90 nsec; see also Lents (1973).

NH$_2$

Ammonia α Band, $^2A_1 - ^2B_1$

This band occurs strongly in the oxy-ammonia flame and in hydrogen/nitrous oxide flames and less strongly in some other flames containing hydrogen and combined nitrogen. It is a many-line system extending throughout the visible and near infra-red but is strongest in the yellow and green. Under small dispersion Fowler and Badami (1931) noted apparent heads at 6652*, 6470, 6302*, 6042*, 5870, 5713*, 5575, 5436*, 5384 and 5265 Å; the stronger bands are marked with an asterisk. There is also a head at 7350 Å. Dressler and Ramsay (1959) have made a full vibrational and rotational analysis from the absorption spectrum obtained in flash photolysis; their very large dispersion plates indicate violet-degraded heads at 6302.0, 6033.6, 6020.3, 5977.5, 5962.5, 5708.5 and 5167.7 Å. See page 123 and Plate 1d.

NO

γ System, $A^2\Sigma^+ - X^2\Pi$

These bands, well down in the ultra-violet, occur in the premixed flames H_2/N_2O, NH_3/O_2 and organic flames supported by N_2O or of fuels containing combined nitrogen (e.g. nitrobenzene, analine, pyridine). The system consists of double-double headed bands, degraded to shorter wavelengths. Main heads,

λ	ν	I		v', v''	λ	ν	I		v', v''
2859.5	34961	m	$^0P_{12}$	0, 5	2478.7	40331	vs	$^0P_{12}$	0, 2
2858.9	34980		P_2		2477.4	40353		P_2	
2849.6	35083		P_1		2471.2	40454		P_1	
2848.2	35099		Q_1		2470.1	40473		Q_1	
2722.2	36724	m	$^0P_{12}$	0, 4	2370.2	42178	vs	$^0P_{12}$	0, 1
2720.8	36743		P_2		2368.9	42201		P_2	
2713.2	36845		P_1		2363.2	42302		P_1	
2712.0	36862		Q_1		2362.2	42321		Q_1	
2595.7	38514	s	$^0P_{12}$	0, 3	2269.3	44052	m	$^0P_{12}$	0, 0
2594.3	38534		P_2		2268.0	44078		P_2	
2587.5	38636		P_1		2262.9	44177		P_1	
2586.3	38654		Q_1		2261.8	44198		Q_1	

For details of rotational structure see Deezsi (1958). Farmer, Hasson and Nicholls (1972) have found an oscillator strength $f_{(0, 0)} = 0.0004 \pm 0.00002$.

β System, $B^2\Pi - X^2\Pi$

This is a very weak system, rarely prominent in flames. It consists of double headed bands, degraded to the red. The first heads of the strongest bands are at 2885.2 (0, 6), 3034.9 (0, 7), 3198.0 (0, 8), 3376.4 (0, 9), 3572.4 (0, 10) and 3788.5 Å (0, 11). The oscillator strength (Hasson and Nicholls, 1971) is only $f_{(0, 0)} = 2.46 \times 10^{-8}$; $f_{(4, 0)} = 1.16 \times 10^{-5}$.

O_2

Schumann–Runge System, $B^3\Sigma_u^- - X^3\Sigma_g^-$

This is the main resonance system of oxygen, but the system origin lies well down in the ultra-violet (2025 Å). It is emitted by hot flames such as oxy-hydrogen and carbon monoxide/oxygen and weakly by organic flames with oxygen. In diffusion flames it occurs on the oxygen side (see Plate 1a & b). The bands show open rotational structure and are strongly degraded to the red so that heads and origins are close together and heads are not usually prominent. In hot flames the (0, 12), (0, 13), (0, 14) and (0, 15) heads are most readily observed. See Plate 1b. The bands are also emitted during the reaction between fluorine and water (page 308 and Plate 6e); under these conditions heads are more prominent. In the following table approximate measurements of heads (λ_H in Å) and intensities (I_F) for the fluorine flame are given, with wavenumbers for the band origins (ν_0 in cm^{-1}). Data are from Durie (1952b) and Hébert, Innanen and Nicholls (1967).

λ_H	ν_0	I_F	v', v''	λ_H	ν_0	I_F	v', v''	λ_H	ν_0	I_F	v', v''
2440	40948	vw	3, 7	2923	34197	s	1, 11	3671	27216	vs	0, 16
2480	40306	vw	2, 7	2984	33509	s	0, 11	3743	26715	m	1, 17
2528	39553	w	3, 8	3039	32892	m	1, 12	3841	26027	s	0, 17
2570	38911	m	2, 8	3104	32204	s	0, 12	3914	25550	s	1, 18
2613	38246	m	1, 8	3232	30923	vs	0, 13	4021	24862	s	0, 18
2663	37536	m	2, 9	3370	29665	vs	0, 14	4096	24408	m	1, 19
2710	36872	m	1, 9	3434	29117	vw	1, 15	4179	23956	m	2, 20
2762	36182	m	0, 9	3517	28429	vs	0, 15	4294	23291	vw	1, 20
2813	35522	m	1, 10	3582	27902	w	1, 16	4375	22863	w	2, 21
2870	34826	s	0, 10								

For summarized data, including rotational data, Franck–Condon factors and oscillator strengths see Hébert et al (1967).

The Atmospheric Bands, $b^1\Sigma_g^+ - x^3\Sigma_g^-$
See page 128. This strongly forbidden system occurs in carbon monoxide/oxygen flames and weakly in the outer cone of oxyhydrocarbon flames. Degraded to longer wavelengths.

λ	ν	I	v', v''	λ	ν	I	v', v''
6867.2	14558	vw	1, 0	7879.2	12688	vw	3, 3
7593.7	13165	vs	0, 0	8597.8	11628	m	0, 1
7683.8	13011	s	1, 1	8697.8	11494	vw	1, 2
7779.0	12851	w	2, 2	8803	11357	vw	2, 3

The Herzberg Bands, $A^3\Sigma_u^+ - X^3\Sigma_u^+$

Another forbidden system which has not been reported in flames but occurs in the afterglow of oxygen and thus probably in some atomic flames; also the night-sky glow. Single headed bands degraded to the red. Strongest heads at 3211 (3, 4), 3370 (3, 5), 3453 (2, 5), 3542 (3, 6), 3633 (2, 6), 3734 (1, 6), 3829 (2, 7) and 4309 Å (0, 8).

OH

3064 Å System, $A^2\Sigma^+ - X^2\Pi$

One of the strongest features of most flame spectra. See page 99 and Plates 3a and 6. Degraded to the red, with open rotational structure. Bands show four heads of comparable strength and the (0, 0) also shows a very weak S-form head.

λ	ν	I		v', v''	λ	ν	I		v', v''
2444	40904	vw	R_1	3, 0	3021.3	33089	vs	$^0R_{21}$	0, 0
2608.5	38325	w	R_1	2, 0	3063.6	32632		R_1	
2613.4	38253		R_2		3067.7	32589		R_2	
2613.4	38253		Q_1		3078.4	32475		Q_1	
2622.1	38127		Q_2		3090.4	32349		Q_2	
2677.3	37340	w	R_1	3, 1	3121.6	32025	m	R_1	1, 1
2681.8	37277		Q_1		3126.3	31977		R_2	
2683.1	37259		R_2		3134.6	31893		Q_1	
2691.1	37148		Q_2		3146.6	31772		Q_2	
2811.3	35560	s	R_1	1, 0	3184.7	31391	w	R_1	2, 2
2816.0	35501		R_2		3190.1	31338		R_2	
2819.1	35461		Q_1		3195.9	31282		Q_1	
2829.0	35338		Q_2		3208.5	31159		Q_2	
2875.3	34769	m	R_1	2, 1	3253.9	30724	vw	R_1	3, 3
2880.6	34705		R_2		3260.2	30664		R_2	
2882.3	34684		Q_1		3263.4	30634		Q_1	
2892.7	34560		Q_2		3276.7	30510		Q_2	
2945.0	33946	m	R_1	3, 2	3428.0	29163	w	R_1	0, 1
2951.2	33875		R_2		3432.0	29129		R_2	
2951.3	33873		Q_1		3458.5	28906		Q_1	
2962.4	33747		Q_2		3472.2	28792		Q_2	
					3483.7	28697	vw	R_1	1, 2
					3488.4	28658		R_2	
					3509.0	28490		Q_1	
					3523.4	28374		Q_2	

Full rotational data are given by Dieke and Crosswhite (1962). For Franck—Condon factors see page 101 and Anketell and Learner (1967). For oscillator strength see Bennett and Dalby (1964) and Anketell and Pery—Thorne (1967); $f_{(0,0)} = 0.0008$. Rouse and Engleman (*J.Q.S.R.T.*, 13, 1503 (1973)) give 0.00096.

Vibration-Rotation Bands (Ground state, $x^2\Pi$)
Thermal emission from the outer cones and burnt gases of most flames. Usually partly masked by H_2O and CO_2 ; the strongest band, the (1, 0) at 2.80 μm, is particularly masked by H_2O. The bands have open structure with P, Q and R branches. They are slightly degraded to longer wavelengths and the shorter wavelength bands form R heads. The Q branch forms an intensity maximum or diffuse head near the origin. Heads and origins,

$\lambda_H(\text{Å})$	ν_H	$\lambda_0(\text{Å})$	ν_0	Band
7461.4	13399	7521.5	13291	$4 \to 0$
7849.3	12736	7911.0	12637	$5 \to 1$
8278.3	12076	8341.7	11985	$6 \to 2$
9653	10356	9788.0	10214	$3 \to 0$
10143	9856	10273	9722	$4 \to 1$
13909	7188	14336	6974	$2 \to 0$
14606	6844	15047	6644	$3 \to 1$
15369	6505	15824	6318	$4 \to 2$
		28007	3569.6	$1 \to 0$
		29369	3404.0	$2 \to 1$

In the reaction between atomic hydrogen and ozone, Meinel's night-sky bands, involving vibration-rotation bands from levels up to $v = 9$, occur. For references see Pearse and Gaydon (1965). Under these lower temperature conditions the Q heads are more prominent; the following are among the strongest: 6258 (9, 3), 7284.2 (8, 3), 7755.8 (9, 4), 7918.5 (5, 1), 8344 (6, 2), 8829.4 (7, 3), 9400 (8, 4) and 10,000 Å (9, 5).

PHO
See page 314. In normal flames the structure appears diffuse with a strong maximum at 5280 Å and weaker maxima at 5100 and 5600 Å. The following measurements of band maxima are for the

cooler flame of atomic hydrogen and phosphorus, by Lam Than and Peyron (1963, 1964) who have also made a rotational analysis.

λ	ν	I	$v'_1\ v'_2\ v'_3$	$v''_1\ v''_2\ v''_3$	λ	ν	I	$v'_1\ v'_2\ v'_3$	$v''_1\ v''_2\ v''_3$
4879	20488	w	0 1 3	0 1 0	5597	17860	s	0 0 0	0 0 0
5020	19915	w	0 1 1	0 0 0	5698	17544	w	0 0 1	0 1 0
5097	19614	m	0 1 2	0 0 0	5849	17091	w	0 1 0	0 0 2
5249	19047	vs	0 1 0	0 0 0	5923	16879	m	0 0 0	0 0 1
5425	18427	m	0 0 2	0 1 0?	5991	16687	m	0 0 0	0 1 0
5535	18062	m	0 1 0	0 0 1	6101	16386	w	0 0 1	0 2 0

PO

β System, $A^2\Sigma^+ - X^2\Pi$

These occur in flames of phosphorus burning in air and in organic flames containing phosphorus compounds such as $POCl_3$. This is a very complex system with heads degraded to the red (R) and to shorter wavelengths (V). The strongest bands are in the (0, 0) sequence which under small dispersion appears to start at about 3240 Å and gets weaker towards the red. Main heads

λ	ν	I	deg	head	v', v''	λ	ν	I	deg	head	v', v''
3218.1	31065	w	R	R_1	7, 6	3346.2	29876	s	R	$^QR_{12}$	6, 6
3246.2	30796	vs	V	$^QP_{21}$	0, 0	3362.1	29735	w	R	R_1	8, 8
3255.3	30710	s	V	$^QP_{21}$	1, 1	3365.9	29701	s	R	$^QR_{12}$	7, 7
3270.5	30568	vs	V	P_2	0, 0	3379.8	29579	m	V	$^QP_{21}$	0, 1
3302.9	30268	m	R	R_1	5, 5	3387.6	29511	w	R	$^QR_{12}$	8, 8
3311.8	30186	m	R	$^QR_{12}$	4, 4	3387.9	29508	m	V	$^QP_{21}$	1, 2
3321.0	30103	s	R	R_1	6, 6	3397.8	29422	w	V	$^QP_{21}$	2, 3
3328.2	30037	s	R	$^QR_{12}$	5, 5	3405.7	29354	m	V	P_2	0, 1
3340.6	29926	s	R	R_1	7, 7	3414.1	29282	m	V	P_2	1, 2

For rotational data see Singh (1959) and Meinel and Krauss (1966).

γ System, $B^2\Sigma^+ - X^2\Pi$

Occurrence as for β system. Degraded to shorter wavelengths, double-double headed bands forming good sequences. All four heads of the strong (0, 0) band and the first ($^0P_{12}$) heads of the other bands are listed.

λ	ν	I	v', v''	λ	ν	I		v', v''
2721.5	36734	vw	0, 3	2477.9	40345	vs	$^{O}P_{12}$	0, 0
2706.8	36933	w	4, 1	2476.4	40369		P_2, Q_2	
				2464.2	40568		P_1	
2692.4	37131	w	2, 5	2462.9	40590		Q_1	
2636.3	37921	s	0, 2	2468.3	40501	m		1, 1
2623.4	38107	s	1, 3	2396.3	41718	s		1, 0
2610.7	38293	vw	2, 4	2387.9	41864	s		2, 1
2555.1	39126	vs	0, 1	2320.6	43079	vw		2, 0
2543.9	39297	s	1, 2	2313.7	43208	w		3, 1
2533.0	39467	w	2, 3	2306.9	43334	w		4, 2

Rotational data for some bands are given by Rao (1958) and Coquart et al (1967).

PbO

Three systems of red-degraded bands occur in flames containing lead ethyl or other lead compounds, The transitions are; system A, A $0^+ - $ x $^1\Sigma^+$; system B, B $1-$x $^1\Sigma^+$; system D, D $1-$x $^1\Sigma^+$. The stronger heads only are listed.

λ	ν	I	System	v', v''	λ	ν	I	System	v', v''
3209.2	31151	m	D	2, 0	4983.8	20051	m	B	0, 3
3264.4	30625	m	D	1, 0	5138.2	19457	m	A	1, 1
3341.8	29915	m	D	1, 1	5162.3	19366	m	B	0, 4
3401.9	29387	vs	D	0, 1	5331.1	18753	m	A	1, 2
3485.7	28681	vs	D	0, 2	5459.4	18312	s	A	0, 2
4229.0	23640	vs	B	3, 0	5677.8	17608	vs	A	0, 3
4317.1	23157	vs	B	2, 0	5910.7	16914	s	A	0, 4
4410.4	22667	s	B	1, 0	6160.5	16228	s	A	0, 5
4553.7	21954	vs	B	1, 1	6250.8	15993	vs	A	1, 6
4658.0	21463	vs	B	0, 1	6427.7	15553	s	A	0, 6
4816.9	20755	s	B	0, 2					

S_2

Main System, B $^3\Sigma_u^- - $x $^3\Sigma_g^-$

A very extensive system of red-degraded bands, mostly of comparable intensity. In ordinary flames bands of the $v' = 0$ progression are strongest; in flames supported by atomic hydrogen bands up to $v' = 10$ are stronger. The following are the main heads of a limited selection from the strong bands; some bands have weaker heads to shorter wavelengths.

λ	ν	v', v''	λ	ν	v', v''	λ	ν	v', v''
*2799.0	35717	10, 0	3416.8	29259	1, 4	4193.8	23838	1, 12
*2829.3	35334	9, 0	3500.5	28559	1, 5	4311.0	23190	1, 13
*2860.2	34952	8, 0	3587.4	27867	1, 6	4433.6	22549	1, 14
*2888.1	34615	9, 1	3645.2	27426	0, 6	4610.0	21686	2, 16
*2920.4	34232	8, 1	3740.0	26730	0, 7	4747.6	21057	2, 17
*2989.7	33438	6, 1	3837.3	26053	0, 8	4790.8	20868	3, 18
3024.8	33050	5, 1	3939.1	25379	0, 9	4842.2	20646	4, 19
3091.7	32335	5, 2	4045.8	24710	0, 10	4990.1	20034	4, 20
3369.4	29670	2, 4	4157.2	24048	0, 11	5036.8	19848	5, 21

*Rather weak in ordinary flames but strong in atomic flames.

SH

$A^2 \Sigma^+ - X^2 \Pi$

In the inner cones of some organic flames containing sulphur compounds. Not in flame of $H_2 S$. A single band resembling the OH band and degraded to the red. This band is predissociated. Rotational analysis by Ramsay (1952). Heads,

λ	ν		v', v''
3236.6	30888	R_1	0, 0
3240.7	30849	Q_1	
3279.3	30486	Q_2	

SO

$B^3 \Sigma^- - X^3 \Sigma^-$

These bands are a prominent feature of flames containing sulphur or oxides of sulphur. Single headed and degraded to the red, the strong bands form a long $v' = 0$ progression.

λ	ν	I	v', v''	λ	ν	I	v', v''
2622.2	38125	w	2, 2	3164.7	31589	vs	0, 9
2664.8	37515	w	1, 4	3271.0	30563	vs	0, 10
2744.0	36432	m	1, 5	3383.2	29549	vs	0, 11
2791.3	35815	w	0, 5	3502.1	28546	s	0, 12
2827.4	35358	m	1, 6	3628.2	27554	m	0, 13
2877.7	34740	s	0, 6	3675.6	27199	m	1, 14
2968.4	33678	vs	0, 7	3761.6	26577	w	0, 14
3064.1	32627	vs	0, 8	3810.8	26234	m	1, 15

SrOH

The strong red colour of flames containing strontium salts is mainly due to SrOH, although some SrO bands also occur. Bands are wide and complex but the strongest at 6060 Å shows structure on the long-wave side with close double-headed bands degraded to the red. For recent observations see van der Hurk et al (1973). The following are positions of maxima,

λ	ν	I	λ	ν	I	λ	ν	I
6060	16497	vs	6460	15475	m	6820	14658	vs
6200	16124	vw	6590	15170	w	7070	14140	m
6260	15970	vw	6710	14899	vs	7220	13846	w

II. ABSORPTION SPECTRA

In the following tables are listed briefly some of the absorption spectra of frequent interest in combustion work. These are mainly extracted from Pearse and Gaydon (1965) where fuller data and references to the literature will be found. Wavelengths only are listed since the emphasis is usually on identification rather than analysis. Intensities are again listed as vs = very strong, etc. For details of analysis of polyatomic spectra see Herzberg (1966).

In flames only OH absorption is readily observed, but many of the systems listed in Appendix I may also be detected in absorption in the appropriate conditions if the specially refined techniques discussed on page 24 are employed. In absorption the intensity of these systems may be different, since strong absorption usually occurs only for bands of low final vibrational quantum number, v'', while for emission bands with low upper-state quantum number, v', occur most readily. Bands of polyatomic molecules tend to become diffuse at high temperature and may merge into regions of continuous absorption. Many more systems may be observed in flash photolysis. Some bands, e.g. of benzene and SO_2, which often occur as impurities are included here.

CH_3

In absorption in the reaction zone of low-pressure flames of methane, methyl ether, acetone and acetaldehyde etc. Also in flash photolysis. Diffuse double bands, shaded slightly to the red. The strongest band has maxima at 2163.6 and 2157.6 Å, and there is a weaker band at 2202 and 2188 Å. The strongest bands are far down in the ultra-violet 1502.9, 1498.9, 1496.7, 1385.6 and 1382.6 Å. See Gaydon, Spokes and van Suchtelen (1960), Harvey and Jessen (1973) and Herzberg and Shoosmith (1956).

C_2H_2, Acetylene

There is a relatively weak system of absorption bands around 2300 Å, superposed on weak continuous absorption below 2350 Å. Bands are degraded to the red, and in some cases there are two or more close heads; only the first heads of each group are listed,

λ	I	λ	I	λ	I	λ	I	λ	I
2377.0	vw	2343.4	vw	2314.2	w	2288.3	vs	2255.7	w
2371.4	vw	2340.5	w	2309.3	vw	2285.3	vs	2249.5	w
2355.9	vw	2326.6	vw	2300.5	vw	2266.9	m	2247.8	w
2352.0	vw	2320.1	m	2291.9	vw	2260.7	w	2245.8	w

In the near infra-red there is a band with intensity maxima at 7874 and 7901 Å and weaker bands at 7956 and 8622 Å.

C_2H_5

A diffuse absorption band with maxima at 2242 and 2228 Å is provisionally assigned to the ethyl radical (Gaydon, Spokes and van Suchtelen, 1960). It occurs in the reaction zone of low-pressure flames of ether and ethane.

C_6H_6, Benzene

The absorption by benzene is very strong and frequently appears as an impurity. Manufactured gas shows benzene absorption slightly.

Spokes (1959) has also reported benzene absorption in rich-limit flames of ether and hexane. The bands form well-marked sequences which are degraded to the red. The following are the sequence heads, with intensities in parentheses, 2363.5 (w), 2415.9 (m), 2471.0 (vs), 2528.6 (vs), 2589.0 (s) and 2667.1 Å (w).

CH_2O, Formaldehyde

These bands occur in absorption during the slow combustion of hydrocarbons, ethers and other fuels, and have also been observed (Spokes, 1959) in rich-limit flames of hydrocarbons, methanol and other fuels, especially between the first and second-stage cool flames. This is part of the same $^1A'' - {}^1A_1$ system as the emission bands (see page 349) which, however lie at longer wavelengths and show little correspondence. Under small dispersion the bands appear degraded to the red, often occurring in pairs and forming sequences. The following wavelengths are from old measurements by Henri and Schou; sequence heads are denoted "S". See Plate 4d.

λ	I	λ	I	λ	I	λ	I
2706	w	2874	m	3033	vs S	3164	s
2747	m	2931	vs S	3051	s	3250	s S
2787	m	2948	m	3085	m	3288	m
2839	vs S	2979	m	3135	s S	3387	w S

CH_3CHO, Acetaldehyde

These have been observed in slow combustion, preceding formation of formaldehyde. The bands are less sharp than those of formaldehyde. Most bands are slightly shaded to the red, others apparently quite diffuse. Beyond 3200 Å bands merge into continuous absorption which is strongest at 2900 Å. The following are the limits of the strong bands, from Schou (1929a)

$\lambda\lambda$	I	$\lambda\lambda$	I	$\lambda\lambda$	I	$\lambda\lambda$	I
3399–3381	w	3305–3300	w	3241–3234	m	3207–3202	m
3377–3363	vw	3296–3290	m	3231–3229	m	3199–3196	s
3359–3344	w	3281–3274	m	3222–3217	m	3191–3180	vs
3342–3329	w	3268–3258	s	3216–3213	m	3178–3172	s
3320–3315	m	3254–3247	m				

C_2H_5CHO, Propionaldehyde

Observed by Newitt and Baxt (1939) during the early stages of slow oxidation of propane. Diffuse bands 3400—3250 Å merging into a continuum at shorter wavelengths. Measurements of the extent of main bands from Schou (1929*b*).

$\lambda\lambda$	I	$\lambda\lambda$	I	$\lambda\lambda$	I
3371—3363	w	3322—3316	s	3277—3272	s
3344—3340	m	3298—3294	vs	3269—3262	s
3336—3332	m	3289—3285	s	3259—3249	s
3331—3325	m				

C_3H_6O, Acetone

Ketones usually show absorption around 2800 Å. Although acetone itself has not been detected in combustion processes, Bowen and Thompson (1934) found that the normal continuous absorption 3200—2400 Å at 1 atm. broke up at reduced pressure into four groups, each of about 25 diffuse bands; these groups showed maxima at 3150, 2900, 2710 and 2570 Å.

HCO

Vaidya's hydrocarbon flame bands have not been found in absorption, but Herzberg and Ramsay (1955) have observed another system in flash photolysis. Degraded to the red, strongest heads 5195.6, 5624.0, 5860, 6138.0, 6436 and 6766.3 Å.

HNO

In flash photolysis complex red-degraded bands have been observed by Dalby (1958). The strongest band is in the region 7341—7718 Å with prominent heads at 7379.3, 7416.4, 7452.7, 7487.3, 7520 and 7542 Å.

HNO_2

Observed by Newitt and Outridge (1938) in absorption by explosion flames of moist CO with N_2O and NO. Some of their bands were due to NO_2. Measurements of narrow diffuse bands by Porter (1951)

λ	I	λ	I	λ	I
3844	w	3417	s	3277	w
3690	vs	3387.5	w	3203	vw
3542.5	vs	3306	w	3177	vw
3509	vw				

H_2O

In the ultra-violet, water vapour shows absorption between 1860 and 1450 Å with a maximum at 1655 Å; often it causes a fairly sharp cut-off at 1800 Å. There is more absorption at still shorter wavelengths, and in the infra-red the vibration-rotational bands (see page 357) give very strong absorption.

NO

Bands of the γ system (for details see page 362) also occur readily in absorption in hot flame gases, but with a different intensity distribution, bands with $v'' = 0$ being strongest. The following are the first heads ($^0P_{12}$) in absorption,

λ	I	v', v''	λ	I	v', v''
2595.7	vw	0, 3	2269.4	s	0, 0
2478.7	vw	0, 2	2154.9	vs	1, 0
2370.2	w	0, 1	2052.4	s	2, 0

These γ bands require an equivalent path length of about ½ mm at 1 atm. for satisfactory observation. Much stronger bands of the δ system, (0, 0) at 1914.2 Å, and the ϵ system, (0, 1) at 1949.7 Å, lie at shorter wavelengths.

NO_2, Nitrogen Dioxide

Visible System, $A\,^2B_1 - x\,^2A_1$

Bands have sharp rotational structure but heads are not obvious. At shorter wavelengths the bands are diffuse due to predissociation and merge into a continuum in the near ultra-violet. With long path lengths absorption extends to the red as far as 9000 Å. The following are outstanding maxima (M) or red-degraded heads (R); abridged from Pearse and Gaydon.

λ	I	λ	I	λ	I	λ	I	λ	I
4081	R w	4304	R m	4480	M vs	4630	R m	4945	M w
4102	M w	4350	M m	4545	R w	4740	R m	5027	M w
4133	M m	4390	R s	4580	M w	4795	M w	5048	M w
4270	R m	4448	R s	4605	M w	4880	R w	5095	M w

Ultra-Violet System, $B^2 B_2 - X^2 A_1$

Degraded to the red. The longer wavelength bands are sharp with open rotational structure, but shorter-wavelength bands are diffuse.

λ	I	λ	I	λ	I
2351	vs	2390	s	2446.7	m
2363	w	2419	vs	2459.3	m
2372	s	2430	w	2491.4	m

O_2

Absorption by the Schumann–Runge bands of O_2, $B^3 \Sigma_u^- - X^3 \Sigma_g^-$, (see page 20) sets the limit to spectroscopic work in air in the ultra-violet. The following are the wavelengths, in vacuo, of the red-degraded heads as observed at room temperature,

λ_{vac}	I	v', v''	λ_{vac}	I	v', v''	λ_{vac}	I	v', v''	λ_{vac}	I	v', v''
1767.7	w	15, 0	1815.6	vs	10, 0	1900.1	vw	8, 1	1946.7	w	3, 0
1774.3	m	14, 0	1830.1	vs	9, 0	1901.9	m	5, 0	1959.5	vw	5, 1
1782.3	s	13, 0	1845.8	vs	8, 0	1918.3	vw	7, 1	1971.4	vw	2, 0
1792.0	vs	12, 0	1863.0	vs	7, 0	1923.5	w	4, 0	1997.5	vw	1, 0
1803.1	vs	11, 0	1881.7	s	6, 0	1938.0	vs	6, 1			

With heated oxygen the absorption extends to much longer wavelengths, as far as 2600 Å in hot flames with excess oxygen. In this region the heads are less pronounced and the appearance is of a large number of lines of the rotational fine structure (plate 4c); it is seen best with large dispersion. Some heads can be picked out by following branches to the heads, and the following are wavelengths (in air) of some which may be detected,

λ	v', v''	λ	v', v''	λ	v', v''
2017.3	8, 3	2101.1	7, 4	2250	4, 5
2020.7	5, 2	2125.0	6, 4	2282	3, 5
2037.6	7, 3	2150.8	5, 4	2359	3, 6
2060.0	6, 3	2167.9	7, 5	2396	2, 6
2079.7	8, 4	2193.4	6, 5	2441	3, 7
2084.2	5, 3	2221.0	5, 5		

For further details on this system see Hebért, Innanen and Nicholls (1967)

O_3

Ozone absorption is not observed in flames, but it occurs as a pollutant following combustion processes. The usual appearance is a fairly sharp cut off of the ultra-violet, between 3000 and 3600 Å, according to the path length and concentration, with a few narrow bands showing clearly in the neighbourhood of the cut off. They are often clearest 3200–3400 and may appear weakly degraded to the red. Wavelengths of maxima,

λ	I	λ	I	λ	I	λ	I
3089.5	s	3177.0	s	3227.2	vs	3304.1	w
3114.3	s	3194.8	m	3249.7	s	3311.5	m
3137.4	vs	3201.0	s	3255.5	m	3338.5	w
3156.1	s	3221.5	vs	3279.8	s	3374.1	w

SO_2

Flames containing sulphur or its compounds often show strong SO_2 absorption, and it is sometimes observed superposed on the continuous emission from a flame. It is also important as an atmospheric pollutant. Bands are degraded to the red but have complex structure. The following measurements of maxima are from Clements (1935); the heads usually lie about 1 Å to the violet of these maxima,

λ	I	λ	I	λ	I	λ	I
2646.6	w	2789.4	s	2923.1	s	3087.7	w
2685.0	w	2797.0	s	2924.8	s	3108.4	w
2727.5	w	2815.5	s	2937.7	s	3129.5	w
2734.6	m	2828.1	s	2943.8	vs	3131.3	w
2738.1	w	2832.3	s	2961.2	vs	3151.8	vw
2751.2	w	2852.0	vs	2980.0	vs	3159.0	vw
2754.6	w	2868.9	s	3001.0	vs	3167.0	vw
2765.2	w	2887.7	vs	3022.1	vs	3173.0	vw
2772.0	w	2900.9	m	3043.3	s	3181.1	vw
2780.0	m	2906.5	s	3065.9	m	3190.9	vw

Organic Compounds

All the lower *paraffins* and saturated *alcohols* are transparent throughout the visible and quartz ultra-violet.

Most organic *acids* and *peroxides* show continuous absorption at the ultra-violet end of the spectrum, the strength of the absorption increasing to shorter wavelengths.

Most *aldehydes* show diffuse bands around 3200–3000 Å.

Ketones show a region of continuous absorption around 2800 Å.

Glyoxal shows banded absorption 4600–3600 Å and diffuse bands 3200–2300 Å

Most simple *benzene derivatives* show banded absorption in the region 2600–2400 Å.

Brief information about a large number of organic compounds which might be formed as combustion products have been given by Egerton and Pidgeon (1933) and Ubbelohde (1935b). For more recent information, references and theoretical interpretations on both organic and inorganic molecules see Herzberg (1966). It should be remembered that the appearance of the absorption spectrum of a polyatomic molecule often changes rapidly with temperature, becoming more complex and usually merging finally into a continuum at higher temperature.

III. SOME ATOMIC AND MOLECULAR ENERGY LEVELS AND CONSTANTS

In this table brief data for some of the low energy states of carbon, hydrogen, nitrogen, oxygen and simple molecules involving these elements are listed. All values are expressed in wave numbers (cm^{-1}) except the dissociation energies which are in electron volts.

The first column, headed "State", gives the electronic type of the energy level for the atom or molecule.

The second column gives the energy of the level measured above the ground state; for molecules this is T_0, the energy above the lowest actual vibrational and rotational level of the ground state. (Some books quote the derived constant T_e corresponding to the minima of the potential energy curves.) For some multiplets of diatomic molecules which are intermediate between Hund's coupling cases a and b, the coupling constant A is also given in this column.

In the third column, for diatomic molecules the vibrational constant ω_e is tabulated; this is the derived constant for infinitesimal amplitude (see page 72). For polyatomic molecules the values of v_1, v_2 etc are the observed first vibrational energy intervals. The fourth column, for diatomics only, gives the derived second vibrational constant $x_e \omega_e$.

The fifth column gives rotational constants. For diatomics the derived constant B_e for infinitesimal amplitude is again listed, except for NH where values of B_0, the observed values for the lowest vibrational level, are given. For polyatomics the observed values A_0, B_0, and C_0 for the lowest vibrational level are given, where known.

The last column gives the dissociation energy, D_0, in electron volts. This is the energy required to produce free atoms with zero kinetic energy (i.e. at $0°$ K) from molecules initially in the lowest electronic, vibrational and rotational level. For polyatomic molecules this is the energy completely to dissociate the molecule to free atoms, e.g. for H_2O to $H + H + O$.

Data have been taken from standard books such as Herzberg (1950, 1966), Moore-Sitterley (1949), Rosen (1970) and Gaydon (1968). Some additional data for C_2 are from Marenin and Johnson (1970) and Messerle and Krauss (1967); for NH, Lents (1973); CH, Herzberg and Johns (1969); CO, Krupenie (1966); N_2, Lofthus (1962); O_2, Hébert, Innanen and Nicholls (1967).

The values of derived molecular constants, such as ω_e and $x_e \omega_e$ are very sensitive to the accuracy of the experimental data and to the number of terms used in the series expansion. The variation in values between the early edition of this book (or *Spectroscopy and Combustion Theory*) and the present edition is surprisingly great. Many figures have been rounded because of this obvious lack of accuracy. In the table, doubtful values are given in parentheses. Where high precision is important the reader is advised to consult original papers rather than to rely on compilations of this type.

	State	Energy T_0, cm^{-1}	ω_e or ν cm^{-1}	$x_e\omega_e$	Rotl. const. B_e or A, B, C	Dissocn. Energy, e.V.
C	$^3P_0$	0				
	$^3P_1$	16				
	$^3P_2$	43				
	$^1D_2$	10,194				
	$^1S_0$	21,648				
	^5S	33,735				
C_2	$x^1\Sigma_g^+$	0	1854.8	13.4	1.8198	6.11 ± 0.04
	$x^3\Pi_u$	610	1641.3	11.7	1.6325	
	$A'^3\Sigma_g^-$	6443	1470.4	11.2	1.4985	
	$b^1\Pi_u$	8145	1608.2	12.1	1.6161	
	$A^3\Pi_g^-$	19,988	1788.2	16.4	1.7527	
	$c^1\Pi_g$	34,240	1809.1	15.8	1.7834	
	$B^3\Pi_g$	40,416	1106.6	39.3	1.1922	
	$d^1\Sigma_u^+$	43,228	1829.6	14.0	1.8334	
C_3	$X^1\Sigma_g^+$	0	ν_1 (1230)			13.9
			ν_2 63.1		B_0 0.4305	
			ν_3 2040			
	$A^1\Pi_u$	24,675	ν_1 1086			
			ν_2 308		B_0 0.4124	
			ν_3 ?			
CH	$X^2\Pi$	0($A=28$)	2858.5	63.0	14.457	3.47
	$A^2\Delta$	23,217	2930.7	96.6	14.934	
	$B^2\Sigma^-$	25,698	(2250)		13.39	
	$C^2\Sigma^+$	31,778	2840.2	125.9	14.603	
CH_2O	X^1A_1	0	ν_1 2766		A_0 9.4053	15.6
			ν_2 1746		B_0 1.2954	
			ν_3 1500		C_0 1.1342	
			ν_4 1167			
			ν_5 2843			
			ν_6 1251			
	a^3A''	25,194				
	A^1A''	28,188				
CN	$X^2\Sigma^+$	0	2071.1	13.8	1.8993	7.89
	$A^2\Pi$	9,117	1812.5	12.6	1.7151	
	$B^2\Sigma^+$	25,799	2144.8	12.2	1.9696	
CO	$X^1\Sigma^+$	0	2169.8	13.3	1.9313	11.09
	$a^3\Pi$	48,474	1743	14	1.6911	
	$a'^3\Sigma^+$	55,354	1230.6	11.0	1.3453	
	$d^3\Delta$	60,647	1152.6	7.3	1.3099	

Bardwell, J., & Hinshelwood, Sir Cyril, 1951, *P.R.S.*, **205**, 375.
Barnard, J. A., & Watts, A., 1969, *12th Comb. Symp.*, p. 365.
Barusch, M. R., Crandall, H. W., Payne, J. Q., & Thomas, J. R., 1951a, *Industr. Engng Chem.*, **43**, 2764.
―――― Neu, J. T., Payne, J. Q., & Thomas, J. R., 1951b, *Industr. Engng Chem.*, **43**, 2766.
Barynin, J. A. M., & Wilson, M. J. G., 1972, *Atmosph. Environ.*, **6**, 197.
Basco, N., & Norrish, R. G. W., 1961, *P.R.S.*, **260**, 293.
Bass, A. M. & Broida, H. P., 1953, *Nat. Bur. Stand. Circular*, No. 541.
―――― & ―――― 1959, *Nat. Bur. Stand. Monograph*, No. 24.
Bawn, C. E. H., & Garner, W. E., 1932, *J. Chem. Soc.*, p. 129.
Bayes, K. D., 1967, *11th Comb. Symp.*, p. 1189.
―――― & Jansson, R. E. W., 1964, *P.R.S.*, **282**, 275.
Becker, K. H., & Bayes, K. D., 1968, *J.C.P.*, **48**, 653.
―――― Kley, D., & Norstrom, R. J., 1969, *12th Comb. Symp.*, p. 405.
Bell, J. C., & Bradley, D., 1970, *C. & F.*, **14**, 225.
Benedict, W. S., Bass, A. M., & Plyler, E. K., 1954, *J. Res. Nat. Bur. Stand.*, **52**, 161.
―――― Bullock, B. W., Silverman, S., & Grosse, A. V., 1953, *J. Opt. Soc. Amer.*, **43**, 1106.
―――― Herman, R. C., & Silverman, S., 1951, *J.C.P.*, **19**, 1325.
―――― & Plyler, E. K., 1954, *Energy Transfer in Hot Gases*, Nat. Bur. Stand. Circular **523**, 57.
Bennett, R. G., & Dalby, F. W., 1960, *J.C.P.*, **32**, 1716.
―――― & ―――― 1962, *J.C.P.*, **36**, 399.
―――― & ―――― 1964, *J.C.P.*, **40**, 1414.
Berl, W. G., Gayhart, E. L., Olsen, H. L., Broida, H. P., & Shuler, K. E., 1956, *J.C.P.*, **25**, 797.
Blades, A. T., 1967, *11th Comb. Symp.*, p. 1189.
Bleekrode, R., 1966, *J.C.P.*, **45**, 3153.
―――― & Nieuwpoort, W. C., 1965, *J.C.P.*, **43**, 3680.
Boers, A. L., Alkemade, C. T. J. & Smit, J. A., 1956, *Physica*, **22**, 358.
Bone, W. A., Fraser, R. P., & Winter, D. A., 1927, *P.R.S.*, **114**, 402.
―――― Fraser, R. P., & Witt, F., 1927, *P.R.S.*, **114**, 442.
―――― & Newitt, D. M., 1927, *P.R.S.*, **115**, 41.
―――― & Outridge, L. E., 1936, *P.R.S.*, **157**, 234.
Bonhoeffer, K. F., & Eggert, J., 1939, *Z. Angew, Phot. Wiss. u. Tech.*, **1**, 43.
―――― & Haber, F., 1928, *Z. phys. Chem. A*, **137**, 263.
―――― & Harteck, P., 1928, *Z. phys. Chem. A*, **139**, 64.
Bonne, U., Homann, K. H., & Wagner, H. G., 1965, *10th Comb. Symp.*, p. 503
Boothman, D., Lawton, J., Melinek, S. J., & Weinberg, F. J., 1969, *12th Comb Symp.*, p. 969.
Botha, J. P., & Spalding, D. B., 1954, *P.R.S.*, **225**, 71.
Bowen, E. J., & Thompson, H. W., 1934, *Nature*, **133**, 571.
Bradley, D., & Matthews, K. J., 1967, *11th Comb. Symp.*, p. 359.
Brand, J. C. D., 1956, *J. Chem. Soc.*, p. 858.
Brewer, L., & Engelke, J. L., 1962, *J.C.P.*, **36**, 992.
Brinsley, F., & Stephens, S., 1946, *Nature*, **157**, 622.

	State	Energy T_0, cm^{-1}	ω_e or ν cm^{-1}	$x_e\omega_e$	Rotl. const. B_e or A, B, C	Dissocn. Energy, e.V.
	$e^3\Sigma^-$	63,709	1113.7	9.6	1.2848	
	$A^1\Pi$	64,746	1515.6	17.3	1.6116	
	$b^3\Sigma^+$	83,832	(2188)		1.986	
	$B^1\Sigma^+$	86,918	(2082)		1.961	
CO_2	$X^1\Sigma_g^+$	0	ν_1 1388 ν_2 667 ν_3 2349		B_0 0.3902	16.5_5
	A^1B_2	(46,700)				
H	$^2S_{1/2}$	0				
	^2P	82,259				
H_2	$^1\Sigma_g^+$.	0	4405.3	125.3	60.87	4.47733
HCO	X^2A'	0	See p. 150		A_0 22.36 B_0 1.494 C_0 1.401	12.4
	$A^2A''(\Pi)$	9294	ν_1 3316 ν_2 802 ν_3 1813		linear B_0 1.338	
	C	(39,500)	See p. 152			
H_2O	X^1A_1	0	ν_1 3657 ν_2 1595 ν_3 3756		A_0 27.877 B_0 14.512 C_0 9.285	9.50
N	$^4S_{1\frac{1}{2}}$	0				
	$^2D_{2\frac{1}{2}}$	19,223				
	$^2D_{1\frac{1}{2}}$	19,231				
	$^2P_{\frac{1}{2},\,1\frac{1}{2}}$	28,840				
N_2	$X^1\Sigma_g^+$	0	2358.1	14.2	1.9987	9.76
	$A^3\Sigma_u^+$	49,756	1460.6	13.9	1.4545	
	$B^3\Pi_g$	59,310	1735.4	15.2	1.6375	
	$a'^1\Sigma_u^-$	67,738	1530.0	12.0	1.480	
	$a^1\Pi_g$	68,951	1694.1	13.9	1.6155	
NH	$X^3\Sigma^-$	0	3266	78.5	B_0 16.345	3.2
	$a^1\Delta$	a	3314	63	B_0 16.45	
	$b^1\Sigma^+$	$a+8502$	3355	74.4	B_0 16.430	
	$A^3\Pi$	29,072	3188	87.5	B_0 16.322	
	$c^1\Pi$	$a+31,289$	2503	194	B_0 14.16	
NH_2	X^2B_1	0	ν_2 1497.2		A_0 23.73 B_0 12.94 C_0 8.17	
	$A^2A_1(\Pi)$	10,249	ν_1 3325 ν_2 633		linear B_0 8.78	

	State	Energy T_0, cm^{-1}	ω_e or ν cm^{-1}	$x_e\omega_e$	Rotl. const. B_e or A, B, C	Dissocn. Energy, e.V.
NO	$X^2\Pi_{1/2}$	0	1903.0	14.0	1.7046	6.50
	$X^2\Pi_{1\frac{1}{2}}$	121	1903.7	14.0	1.7046	
	$A^2\Sigma^+$	44,200	2371.3	14.5	1.9977	
	$B^2\Pi_{1/2}$	45,488	1036.9	7.5	1.076	
	$B^2\Pi_{1\frac{1}{2}}$	45,516	1038.3	7.5	1.178	
NO$_2$	X^2A_1	0	ν_1 (1320)		A_0 8.0012	9.6$_2$
			ν_2 750		B_0 0.4336	
			ν_3 1618		C_0 0.4104	
	A^2B_1	<15,000	ν_2 (896)		B 0.370	
	B^2B_2	40,126	ν_1 1184		A_0 4.132	
					B_0 0.402	
					C_0 0.366	
N$_2$O	$X^1\Sigma^+$	0	ν_1 2224		linear	
			ν_2 589		B_0 0.419	
			ν_3 1285			
O	$^3P_2$	0				
	$^3P_1$	158				
	$^3P_0$	226				
	$^1D_2$	15,868				
	$^1S_0$	33,792				
O$_2$	$X^3\Sigma_g^-$	0	1580.2	12.0	1.4456	5.115
	$a^1\Delta_g$	7,882	1509	12.9	1.4264	
	$b^1\Sigma_g^+$	13,121	1432.7	13.9	1.4004	
	$A^3\Sigma_u^+$	35,006	(805)	(13)	0.917	
	$B^3\Sigma_u^-$	49,349	700.4	8.0	0.8184	
O$_3$	X^1A_1	0	ν_1 1110		A_0 3.5535	6.15
			ν_2 705		B_0 0.4452	
			ν_3 1042		C_0 0.3948	
	A	10,000	ν_2 567			
	B	16,625				
	C	28,447	ν_1 (636)			
			ν_2 (352)			
	D	33,000	ν_2 (300)			
OH	$X^2\Pi$	0($A=-140$)	3725.2	82.8	18.871	4.40
	$A^2\Sigma^+$	32,682	3184.3	97.8	17.355	

References

Abbreviations of titles of journals are given in the usual forms except for the following shortened abbreviations for a few which occur frequently:

C. & F.	*Combustion and Flame.*
Comb. Symp.	*Symposium (International) on Combustion.* The 3rd, 4th and 8th were published by Williams, Wilkins and Co., Baltimore; the 5th and 6th by Reinhold, New York; the 7th by Butterworths, London; the 9th by Academic Press, New York; and the 10th and subsequent ones by The Combustion Institute, Pittsburgh.
J.C.P.	*Journal of Chemical Physics.*
J.Q.S.R.T.	*Journal of Quantitative Spectroscopy and Radiative Transfer.*
P.R.S.	*Proceedings of the Royal Society* (London), Series A.
T.F.S.	*Transactions of the Faraday Society.*

Adlard, E. R., & Matthews, P. H. D., 1971, *Nature, Phys. Sci.*, **233**, 83.
Affleck, W. S., & Fish, A., 1967, *11th Comb. Symp.*, p. 1003.
Agnew, W. G., & Agnew, J. T., 1956, *Industr. Engng Chem.*, 48, 2224.
———————— & Wark, K., 1955, *5th Comb. Symp.*, p. 766.
Andrews, E. B., & Barrow, R. F., 1951, *Proc. Phys. Soc.*, **64**, 481.
Anketell, J., & Learner, R. C. M., 1967, *P.R.S.*, **301**, 355.
———— & Pery-Thorne, A., 1967, *P.R.S.*, **301**, 343.
Anlauf, K. G., Maylotte, D. H., Pacey, P. D., & Polanyi, J. C., 1967, *Phys. Letters*, **24A**, 208.
Arthur, J. R., & Townend, D. T. A., 1954, *Energy Transfer in Hot Gases*, Nat. Bur. Stand. Circular 523, 99.
Ausloos, P., & Van Tiggelen, A., 1953a, *4th Comb. Symp.*, p. 252.
———— & ———— 1953b, *Bull. Soc. Chim. Belg.*, **62**, 223.

Bailey, C. R., & Lih, K. H., 1929, *T.F.S.*, **25**, 29.
Bailey, H. C., & Norrish, R. G. W., 1952, *P.R.S.*, **212**, 311.
Baker, M. R. & Vallee, B. L., 1955, *J. Opt. Soc. Amer.*, **45**, 773.
Ball, G. A., 1955, *5th Comb. Symp.*, p. 366.
Ballik, E. A., & Ramsay, D. A., 1959, *J.C.P.*, **31**, 1128.
———— & ———— 1963, *Astrophys. J.*, **137**, 61.
Barat, P., Cullis, C. F., & Pollard, R. T., 1972, *P.R.S.*, **329**, 433.

Brocklehurst, B., Hébert, G. R., Innanen, S. H., Seel, R. M., & Nicholls, R. W., 1971, *Identification Atlas of Molecular Spectra*, **8**, York University.
——— ——— ——— & ——— 1972, *Identification Atlas of Molecular Spectra*, **9**, York University.
Brody, S. S., & Chaney, J. F., 1966, *J. Gas Chromatogr.*, **4**, 42.
Broida, H. P., & Carrington, T., 1955, *J.C.P.*, **23**, 2202.
——— Everett, A. J., & Minkoff, G. J., 1954, *Fuel*, **33**, 251.
——— & Gaydon, A. G., 1953a, *T.F.S.*, **49**, 1190.
——— & ——— 1953b, *P.R.S.*, **218**, 60.
——— & ——— 1954, *P.R.S.*, **222**, 181.
——— & Heath, D. F., 1957, *J.C.P.*, **26**, 223.
——— & Kostkowski, H. J., 1955, *J.C.P.*, **23**, 754.
——— & ——— 1956, *J.C.P.*, **25**, 676.
——— Levedahl, W. J., & Howard, F. L., 1951, *J.C.P.*, **19**, 797.
——— & Shuler, K. E., 1952, *J.C.P.*, **20**, 168.
——— & ——— 1957, *J.C.P.*, **27**, 933.
Browner, R. F., Dagnall, R. M., & West, T. S., 1969, *Anal. Chim. Acta*, **46**, 207.
Brzustowski, T. A., & Glassman, I., 1964, *Heterogeneous Combustion, Progress in Astronautics and Aeronautics*, **15**, p. 41. Academic Press, New York.
Bulewicz, E. M., 1967, *C. & F.*, **11**, 297.
——— James, C. G., & Sugden, T. M., 1956, *P.R.S.*, **235**, 89.
——— & Padley, P. J., 1971a, *T.F.S.*, **67**, 2337.
——— & ——— 1971b, *P.R.S.*, **323**, 377
——— & ——— 1971c, *13th Comb. Symp.*, p. 73.
——— ——— & Smith, R. E., 1970, *P.R.S.* **315**, 129.
——— ——— & ——— 1973, *14th Comb. Symp.*, p. 329.
——— & Sugden, T. M., 1956, *T.F.S.*, **52**, 1475.
Bullock, B. W., Feazle, C. E., Gloersen, P., & Silverman, S., 1952, *J.C.P.*, **20**, 1808.
——— Hornbeck, G. A., & Silverman, S., 1950, *J.C.P.*, **18**, 1114.
Burgoyne, J. H., & Hirsch, H., 1954, *P.R.S.*, **227**, 73.
——— & Neale, R. F., 1953, *Fuel*, **32**, 17.
——— & Weinberg, F. J., 1953, *4th Comb. Symp.*, p. 294.
Burrows, K. M., & Horwood, J. F., 1963, *Spectrochim. Acta*, **19**, 17.
Burt, R., Skuse, F., & Thomas, A., 1965, *C. & F.*, **9**, 159.

Calcote, H. F., 1963, *9th Comb. Symp.*, p. 622.
Callear, A. B., 1965, *Appl. Opt.*, Supplement 2, p. 145.
——— & Norrish, R. G. W., 1960, *P.R.S.*, **259**, 304.
Calloman, J. H., & Ramsay, D. A., 1957, *Canad. J. Phys.*, **35**, 129.
Carabine, M. D., & Norrish, R. G. W., 1967, *P.R.S.*, **296**, 1.
Carrington, T., 1959a, *J.C.P.*, **30**, 1087.
——— 1959b, *J.C.P.*, **31**, 1418.
Cashion, J. K., & Polanyi, J. C., 1960, *P.R.S.*, **258**, p. 529, 564, & 570.
Charton, M., & Gaydon, A. G., 1956, *Proc. Phys. Soc.*, A, **69**, 520.
——— & ——— 1958, *P.R.S.*, **245**, 84.
Chedaille, J., & Braud, Y., 1972, *Industrial Flames; I. Measurements in Flames.*
Clark, G. L., & Henne, A. L., 1927, *J. Soc. Auto. Engng*, **20**, 264.
——— & Smith, H. A., 1930, *J. Phys. Chem.*, **34**, 1924.

Clark, G. L., & Thee, W. C., 1926, *Industr. Engng Chem.*, **18**, 528.
Clements, J. H., 1935, *Phys. Rev.*, 47, 224.
Clough, P. N., Schwartz, S. E., & Thrush, B.A., 1970, *P.R.S.*, **317**, 575.
———— & Thrush, B. A., 1967, *T.F.S.*, **63**, 915.
Clusius, K., & Douglas, A. E., 1954, *Canad. J. Phys.*, **32**, 319.
———— & Huber, M., 1949, *Helv. Chim. Acta*, **32**, 2400.
Clyne, M. A. A., & Thrush, B. A., 1961, *T.F.S.*, **57**, 1305.
———— & ———— 1962, *P.R.S.*, **269**, 404.
Cole, D. J., & Minkoff, G. J., 1956, *Fuel*, **35**, 135.
Coleman, E. H., & Gaydon, A. G., 1947, *Disc. Faraday Soc.*, **2**, 166 & 176.
———— ———— & Vaidya, W. M., 1948, *Nature*, **162**, 108.
Coquart, B., Couet, C., Tuan, N., & Guénebaut, H., 1967, *J. Chim. Phys.*, **64**, 1197.
Cresser, M. S., & West, T. S., 1970a, *Anal. Chim. Acta*, **51**, 530.
———— & ———— 1970b, *Spectrochim. Acta*, **25B**, 61.
Cullis, C. F., Fish, A., Saeed, M., & Trimm, D. L., 1966, *P.R.S.*, **289**, 402.
Cummings, G. A. McD., & Hall, A. R., 1965, *10th Comb. Symp.*, p. 1365.

Dagnall, R. M., Thompson, K. C., & West, T. S., 1967, *Talanta*, **14**, 557.
———— ———— & ———— 1968, *Analyst*, **93**, 72.
Dalby, F. W., 1958, *Canad J. Phys.*, **36**, 1336.
Daly, E. F., & Sutherland, G. B. B. M., 1949, *3rd Comb. Symp.*, p. 530.
Damköhler, G. & Eggersglüss, W., 1942, *Z. phys. Chem. B*. **51**, 157.
David, W. T., Brown, J. R., & El Din, A. H., 1932, *Phil Mag.*, **14**, 764.
———— Leah, A. S., & Pugh, B., 1941, *Phil. Mag.*, **31**, 156.
———— & Mann, J., 1947, *Nature*, **160**, 229.
———— & Parkinson, R. M., 1933, *Phil. Mag.*, **15**, 177.
Davies, P. B., & Thrush, B. A., 1968, *P.R.S.*, **302**, 243.
Davis, S. P., & Phillips, J. G., 1963, *Berkeley Analyses of Molecular Spectra*, **1**, Univ. California Press.
Dean, J. A., & Rains, T. C., 1969, editors *Flame Emission and Atomic Absorption Spectrometry*, Marcel Dekker, New York.
Déchaux, J. C., Flament, J. L., & Lucquin, M., 1971, *C. & F.*, **17**, 205.
Deezsi, I., 1958, *Acta Phys. Hungar.*, **9**, 125.
Déjardin, G., Janin, J., & Peyron, M., 1952, *C. R. Acad. Sci. Paris*, **234**, 1866.
———— ———— & ———— 1953, *Cahiers de Phys.*, **46**, 3.
Derwent, R. G., & Thrush, B. A., 1971, *T.F.S.*, **67**, 2036.
Deutsch, T. F., 1967, *Appl. Phys. Letters*, **10**, 234.
Diederichsen, J., & Wolfhard, H. G., 1956, *P.R.S.*, **236**, 89.
Dieke, G. H., 1955, *Physical Measurements in Gas Dynamics and Combustion*, Oxford Univ. Press, p. 467.
———— & Crosswhite, H. M., 1962, *J.Q.S.R.T.*, **2**, 97.
Dixon, H. B., & Higgins, W. F., 1928, *Mem. Manchester Lit. Phil. Soc.*, **71**, 15.
Dixon, R. N., 1959, *Canad J. Phys.*, **37**, 1171.
———— 1963, *P.R.S.*, **275**, 431.
———— 1969, *T.F.S.*, **65**, 3141.
———— Duxbury, G., & Ramsay, D. A., 1967, *P.R.S.*, **296**, 137.

Dixon-Lewis, G., 1967, *P.R.S.*, **298**, 495.
––––– 1968, *P.R.S.*, **307**, 111.
––––– & Williams, A., 1963, *9th Comb. Symp.*, p. 576.
Donovan, R. E., 1957, *J.C.P.*, **27**, 324.
––––– & Agnew, W. G., 1955, *J.C.P.*, **23**, 1592.
Douglas, A. E., 1951, *Astrophys. J.,* **114**, 466.
––––– & Routly, P. M., 1955, *Astrophys. J.* Suppl. Ser., **1**, 295.
Downs, D., Street, J. C., & Wheeler, R. W., 1953, *Fuel,* **32**, 279.
Dressler, K., & Ramsay, D. A., 1959, *Phil. Trans. Roy. Soc.* **A**, **251**, 553.
Drowart, J., Burns, R. P., De Maria, G., & Inghram, M. G., 1959, *J.C.P.*, **31**, 1131.
Durie, R. A., 1951, *P.R.S.*, **207**, 388.
––––– 1952*a*, *Proc. Phys. Soc.* **A**, **65**, 125.
––––– 1952*b*, *P.R.S.*, **211**, 110.
––––– & Ramsay, D. A., 1958, *Canad. J. Phys.*, **36**, 35.
Dyne, P. J., & Style, D. W. G., 1947, *Disc. Faraday Soc.*, **2**, 159.

Earls, L. T., 1935, *Phys. Rev.*, **48**, 423.
Echigo, R., Nishiwaki, N., & Hirata, M., 1967, *11th Comb. Symp.*, p. 381.
Edse, R., Rao, K. N., Strauss, W. A., & Mickelson, M. E., 1963, *J. Opt. Soc. Amer.*, **53**, 436.
Egerton, Sir Alfred C., & Minkoff, G. J., 1947, *P.R.S.*, **191**, 145.
––––– & Pidgeon, L. M., 1933, *P.R.S.*, **142**, 26.
––––– & Powling, J., 1948, *P.R.S.*, **193**, 172 & 190.
––––– & Rudrakanchana, S., 1954, *P.R.S.*, **225**, 427.
––––– & Thabet, S. K., 1952, *P.R.S.*, **211**, 445.
Eiseman, B. J., & Harris, L., 1932, *J. Amer. Chem. Soc.*, **54**, 1782.
Eltenton, G. C., 1947, *J.C.P.*, **15**, 455.
Eméleus, H. J., 1925, *J. Chem. Soc.*, p. 1362.
––––– 1926, *J. Chem. Soc.*, p. 2948.
––––– 1927, *J. Chem. Soc.*, p. 783.
––––– 1929, *J. Chem. Soc.*, p. 1846.
––––– & Downey, W. E., 1924, *J. Chem. Soc.*, p. 2491.
––––– & Purcell, R. H., 1927, *J.Chem. Soc.*, p. 788.
––––– & Riley, H. L., 1933, *P.R.S.*, **140**, 378.
Erhard, K. H. L., & Norrish, R. G. W., 1956, *P.R.S.*, **234**, 178.
––––– & ––––– 1960, *P.R.S.*, **259**, 297.
Ewing, G. E., Thompson, W. E., & Pimentel, G. C., 1960, *J.C.P.*, **32**, 927.

Fair, R. W., & Thrush, B. A., 1969, *T.F.S.*, **65**, 1208.
Fairbairn, A. R., 1962*a*, *8th Comb. Symp.*, p. 304.
––––– 1962*b*, *P.R.S.*, **267**, 88.
––––– 1963, *P.R.S.*, **276**, 513.
––––– & Gaydon, A. G., 1954, *T.F.S.*, **50**, 1256.
––––– & ––––– 1955, *5th Comb. Symp.*, p. 324.
––––– & ––––– 1957, *P.R.S.*, **239**, 464.
Farber, M., & Darnell, A. J., 1954, *J.C.P.*, **22**, 1261.

Farmer, A. J. D., Hasson, V., & Nicholls, R. W., 1972, *J.Q.S.R.T.*, **12**, 627.

Feast, M. W., 1950, *Proc. Phys. Soc.*, **A**, **63**, 772.

Fenimore, C.P., 1971, *13th Comb. Symp.*, p. 373.

Ferguson, R. E., 1955, *J.C.P.*, **23**, 2085.

―――― 1957, *C. & F.*, **1**, 431.

―――― & Broida, H. P., 1955, *5th Comb. Symp.*, p. 754.

Ferriso, C. C., & Ludwig, C. B., 1964, *J.C.P.*, **41**, 1668.

Fink, E. H., Welge, K. H., 1967, *J.C.P.*, **46**, 4315.

Fischer, H., Duchane, E., & Büchler, A., 1965, *J. Opt. Soc. Amer.*, **55**, 1275.

Fish, A., 1966, *P.R.S.*, **293**, 378.

―――― Haskell, W. W., & Read, I. A., 1969, *P.R.S.*, **313**, 261.

Fontijn, A., Ellison, R., Smith, W. H., & Hesser, J. E., 1970, *J.C.P.*, **53**, 2680.

―――― Sabadell, A. J., & Ronco, R. J., 1970, *Anal. Chem.*, **42**, 575.

Fowler, A., & Badami, J. S., 1931, *P.R.S.*, **133**, 325.

―――― & Gaydon, A. G., 1933, *P.R.S.*, **142**, 362.

―――― & Vaidya, W. M., 1931, *P.R.s.*, **132**, 310.

Fox, M. D., & Weinberg, F. J., 1962, *P.R.S.*, **268**, 222.

Fraser, P. A., Jarmain, W. R., & Nicholls, R. W., 1954, *Astrophys. J.*, **119**, 286.

Fraser, R. P., 1959, *7th Comb. Symp.*, p. 783.

Friedman, R., & Cyphers, J. A., 1956, *J.C.P.*, **25**, 448.

Fristrom, R. M., & Westenberg, A. A., 1965, *Flame Structure*, McGraw-Hill, New York.

Fuwa, K., Thiers, R. E., Vallee, B. L., & Baker, M. R., 1959, *Anal. Chem.*, **31**, 2039.

Garner, W. E., & Hall, D. A., 1930, *J.Chem. Soc.*, p. 2037.

―――― & Johnson, C. H., 1927, *Phil. Mag.*, **3**, 97.

―――― & ―――― 1928, *J. Chem. Soc.*, p. 280.

―――― ―――― & Saunders, S. W., 1926, *Nature*, **117**, 790.

Garton, W. R. S., & Broida, H. P., 1953, *Fuel*, **32**, 519.

Garvin, D., 1954, *J. Amer. Chem. Soc.*, **76**, 1523.

Gausset, L., Herzberg, G., Lagerqvist, A., & Rosen, B., 1965, *Astrophys. J.*, **142**, 45.

Gay, N. R., Agnew, J. T., Witzell, O. W., & Karabell, C. E., 1961, *C. & F.*, **5**, 257.

Gaydon, A. G., 1934, *P.R.S.*, **146**, 901.

―――― 1940, *P.R.S.*, **176**, 505.

―――― 1942*a*, *P.R.S.*, **179**, 439.

―――― 1942*b*, *P.R.S.*, **181**, 197.

―――― 1943, *P.R.S.*, **182**, 199.

―――― 1944, *P.R.S.*, **183**, 111.

―――― 1946, *T.F.S.*, **42**, 292.

―――― 1955, *P.R.S.*, **231**, 437.

―――― 1968, *Dissociation Energies and Spectra of Diatomic Molecules*, 3rd ed., Chapman & Hall, London.

―――― & Guedeney, F., 1955, *T.F.S.*, **51**, 894.

―――― & Hurle, I. R., 1963, *The Shock Tube in High Temperature Chemical Physics*, Chapman & Hall, London.

Gaydon, A. G., & Kopp, I., 1971, *J. Phys.*, B, 4, 752.

—— & Moore, N. P. W., 1955, *P.R.S.*, 233, 184.

—— —— & Simonson, J. R., 1955, *P.R.S.*, 230, 1.

—— & Pearse, R. W. B., 1939, *P.R.S.*, 173, 37.

—— Spokes, G. N., & van Suchtelen, J., 1960, *P.R.S.*, 256, 323.

—— & Whittingham, G., 1947, *P.R.S.*, 189, 313.

—— & Wolfhard, H. G., 1948, *P.R.S.*, 194, 169.

—— & —— 1949a, *3rd Comb. Symp.*, p. 504.

—— & —— 1949b, *Rev. Inst. Français de Pétrole*, 4, 405.

—— & —— 1949c, *P.R.S.*, 199, 89.

—— & —— 1950, *P.R.S.*, 201, 570.

—— & —— 1951a, *P.R.S.*, 205, 118.

—— & —— 1951b, *P.R.S.*, 208, 63.

—— & —— 1952, *P.R.S.*, 213, 366.

—— & —— 1954, *Fuel*, 33, 286.

—— & —— 1970, *Flames, their Structure, Radiation and Temperature*, 3rd ed., Chapman & Hall, London.

Geib, K. H., & Vaidya, W. M., 1941, *P.R.S.*, 178, 351.

Generalov, N. A., & Losev, S. A., 1966, *J.Q.S.R.T.*, 6, 101.

Gerbach, R., 1966, *J. Appl. Spectrosc.*, 4, 247.

Gilbert, P. T., 1961, *Flame Spectra of the Elements*, Beckman Research Report No. 10, Fullerton, California.

—— 1963, *Proc. 10th Colloq. Spectrosc. Internat.*, p. 171.

—— 1964, *Proc. 10th Nat. Analysis Instrum. Symp.*, p. 193. Plenum Press, New York.

Girard, A., 1963, *Appl. Optics*, 2, 79.

Gloersen, P., 1958, *J. Opt. Soc. Amer.*, 48, 712.

Goldfinger, P., 1955, *Mem. Soc. Roy. Sci. Liège*, 15, Sp. No. 378.

Gole, J. L., & Zare, R. N., 1972, *J.C.P.*, 57, 5331.

Goodfellow, G. I., 1966, *Anal. Chim. Acta*, 36, 132.

Gordon, R. D., & King, G. W., 1961, *Canad. J. Phys.*, 39, 252.

Gray, B. F., 1970, *C. & F.*, 14, 273.

—— & Yang, C. H., 1969, *C. & F.*, 13, 20.

Gray, P., 1954, *P.R.S.*, 221, 462.

—— Hall, A. R., & Wolfhard, H. G., 1955, *P.R.S.*, 232, 389.

—— & Yoffe, A., 1949, *P.R.S.*, 200, 114.

Greaves, J. C., & Garvin, D., 1959, *J.C.P.*, 30, 348.

Griffing, V., & Laidler, K. J., 1949, *3rd Comb. Symp.*, p. 432.

Guénault, E. M., 1934–37, *Ann. Report Safety in Mines Res. Board*, 13, 66; 14, 68; 15, 59; 16, 64.

Guénebaut, H., Couet, C., & Houlon, D., 1964, *C. R. Acad. Sci. Paris*, 258, 3457.

—— & Gaydon, A. G., 1957, *6th Comb. Symp.*, p. 292.

Gurvich, L. V., *et al.*, 1962, *Thermodynamic Properties of Individual Substances*, Vol. 2, Akad. Nauk, U.S.S.R.

Guyomard, F., 1951, *C. R. Acad. Sci. Paris*, 233, 237.

Hall, A. R., & Pearson, G. S., 1969, *12th Comb. Symp.*, p. 1025.

────── & Wolfhard, H. G., 1956, *T.F.S.*, **52**, 1520.

Halmann, M., & Laulicht, I., 1966, *Astrophys. J.* Suppl. Ser. **12**, 307.

Halstead, C. J., & Jenkins, D. R., 1969, *12th Comb. Symp.*, p. 979.

────── & ────── 1970, *C. & F.*, **14**, 321.

Hancock, G., Ridley, B. A., & Smith, I. W. M., 1972, *J.C.S. Faraday II,* **68**, 2117.

Hand, C. W., 1962, *J.C.P.*, **36**, 2521.

Harned, B. W., & Ginsburg, N., 1958, *J. Opt. Soc. Amer.*, **48**, 178.

Harrison, G. R., Lord, R. C., & Loofbourow, J. R., 1949, *Practical Spectroscopy*, Blackie, London.

Harteck, P., & Kopsch, U., 1931, *Z. phys. Chem.*, **12B**, 327.

Hartley, H., 1932–33, *Inst. Gas Engrs Trans.*, p. 466.

Hartley, W. N., 1894, *Phil. Trans. Roy. Soc.*, **185A**, 161.

────── 1907, *P.R.S.*, **79**, 242.

Harvey, R., & Jessen, P. F., 1973, *Nature, Phys. Sci.*, **241**, 102.

Haslam, R. T., Lovell, W. C., & Hunneman, R. D., 1925, *Industr. Engng Chem.*, **17**, 272.

Hasson, V., & Nicholls, R. W., 1971, *J. Phys. B*, **4**, 1769.

Hébert, G. R., Innanen, S. H., & Nicholls, R. W., 1967, *Identification Atlas of Molecular Spectra*, **4**, York Univ., Toronto.

Herman, R. C., Hopfield, H. S., Hornbeck, G. A., & Silverman, S., 1949, *J.C.P.*, **17**, 220.

────── & Hornbeck, G. A., 1953, *Astrophys. J.*, **118**, 214.

Herrmann, R., & Alkemade, C. T. J., 1963, *Flame Photometry*, Interscience Publishers (translated by P. T. Gilbert).

Herzberg, G., 1932, *Z. phys. Chem. B.17*, 68.

────── 1945, *Infra-red and Raman Spectra*, Van Nostrand, New York.

────── 1950, *Spectra of Diatomic Molecules*, Van Nostrand, New York.

────── 1961, *P.R.S.*, **262**, 291.

────── 1966, *Electronic Spectra of Polyatomic Molecules*, Van Nostrand, New York.

────── 1971, *The Spectra and Structures of Simple Free Radicals*, Cornell University Press, Ithaca.

────── & Franz, K., 1931, *Z. Phys.*, **76**, 720.

────── & Herzberg, L., 1953, *J. Opt. Soc. Amer.*, **43**, 1037.

────── & Johns, J. W. C., 1969, *Astrophys. J.*, **158**, 399.

────── & Ramsay, D. A., 1952, *J.C.P.*, **20**, 347.

────── & ────── 1953, *Disc. Faraday Soc.*, **14**, 11.

────── & ────── 1955, *P.R.S.*, **233**, 34.

────── & Shoosmith, J., 1956, *Canad. J. Phys.*, **34**, 523.

────── & Travis, D. N., 1964, *Canad. J. Phys.*, **42**, 1658.

Hoare, D. E., & Walsh, A. D., 1954, *T.F.S.*, **50**, 37.

────── ────── & Li, T. M., 1971, *13th Comb. Symp.*, p. 461

Hobbs, R. S., Kirkbright, G. F., Sargent, M., & West, T. S., 1968, *Talanta*, **15**, 997.

Hollander, T., 1964, *Ph. D. Thesis*, Utrecht, p. 119.

Hornbeck, G. A., 1948, *J.C.P.*, **16**, 1005.

—— & Herman, R. C., 1949, *J.C.P.*, **17**, 842.

—— & —— 1950, *J.C.P.*, **18**, 763.

—— & —— 1951*a*, *J.C.P.*, **19**, 512.

—— & —— 1951*b*, *Industr. Engng Chem.*, **43**, 2739.

—— & —— 1954, *Nat. Bur. Stand. Circular*, **523**, 9.

—— & Hopfield, H. S., 1949, *J.C.P.*, **17**, 982.

Hottel, H. C., & Mangelsdorf, H. G., 1935, *Trans. Amer. Inst. Chem. Eng.*, **31**, 517.

—— & Sarofim. A. F., 1967, *Radiative Transfer*, McGraw-Hill, New York.

Houghton, J. T., & Smith, S. D., 1966, *Infra-red Physics*, Clarendon Press, Oxford.

Howard, J. B., 1969, *12th Comb. Symp.*, p. 877.

Hsieh, M. S., & Townend, D. T. A., 1939, *J. Chem. Soc.*, p. 332.

Huggins, W., 1880, *P.R.S.*, **30**, 576.

Hurle, I. R., Price, R. B., Sugden, T. M., & Thomas, A., 1968, *P.R.S.*, **303**, 409.

Ibiricu, M. M., & Gaydon, A. G., 1964, *C. & F.*, **8**, 51.

Jachimowski, C. J., & Houghton, W. M., 1970, *C. & F.*, **15**, 125.

—— & —— 1971, *C. & F.*, **17**, 25.

Jack, D., 1927, *P.R.S.*, **115**, 373.

—— 1928*a*, *P.R.S.*, **118**, 647.

—— 1928*b*, *P.R.S.*, **120**, 22.

Jain, D. C., & Sahni, R. C., 1966, *Proc. Phys. Soc.*, **88**, 495.

James, C. G., & Sugden, T. M., 1955a, *P.R.S.*, **227**, 312.

—— & —— 1955b, *Nature*, **175**, 252.

—— & —— 1958, *P.R.S.*, **248**, 238.

Jenkins, D. R., 1966, *P.R.S.*, **293**, 493.

Jensen, D. E., & Jones, G. A., 1972, *J. Chem. Soc. Faraday Trans. I*, **68**, 259.

—— & —— 1973, *J. Chem. Soc. Faraday Trans. I*, **69**, 1448

Jessen, P. F., 1971, *Gas Council Tech. note L.R.S.T.N. 207.*

—— & Gaydon, A. G., 1967, *C. & F.*, **11**, 11.

—— & —— 1969, *12th Comb. Symp.*, p. 481.

Jeunehomme, M., 1965, *J.C.P.*, **42**, 4086.

John, R., & Summerfield, M., 1957, *Jet Propulsion*, **27**, 169.

Johns, J. W. C., 1961, *Canad. J. Phys.*, **39**, 1738.

—— Priddle, S. H., & Ramsay, D. A., 1963, *Disc. Faraday Soc.*, **35**, 90.

Jonathan, N., Melliar-Smith, C. M., & Slater, D. H., 1970, *J.C.P.*, **53**, 4396.

Kane, W. R., & Broida, H. P., 1953, *J.C.P.*, **21**, 347.

Kaskan, W. E., 1959*a*, *C. & F.*, **3**, 39.

—— 1959*b*, *C. & F.*, **3**, 49.

—— & Hughes, D., 1973, *14th Comb. Symp.*, in press.

—— & Millikan, R. C., 1960, *J.C.P.*, **32**, 1273.

Kasper, J. V. V., & Pimentel, G. C., 1965, *Phys. Rev. Letters*, **14**, 352.

Kaufman, F., 1958, *P.R.S.*, **247**, 123.

—— & Del Greco, F. P., 1961, *J.C.P.*, **35**, 1895.

—— & Kelso, J. R., 1957, *J.C.P.*, **27**, 1209.

Kelly, R., & Padley, P. J., 1971, *T.F.S.*, 67, 740.
Kiess, N. H., & Bass, A. M., 1954, *J.C.P.*, 22, 569.
────── & Broida, H. P., 1956, *Canad. J. Phys.*, 34, 1471.
────── & ────── 1959, *7th Comb. Symp.*, p. 207.
Kinbara, T., Nakamura, J., & Ikegami, H., 1959, *7th Comb. Symp.*, p. 263.
────── & Noda, K., 1971, *13th Comb. Symp.*, p. 333.
Kirkbright, G. F., Semb, A., & West, T. S., 1967, *Talanta*, 14, 1011.
Kistiakowski, G. B., & Volpi, G. G., 1957, *J.C.P.*, 27, 1141.
Kitagawa, T., 1936, *Proc. Imp. Acad. Tokyo*, 12, 281.
────── 1937, *Rev. Phys. Chem. Japan*, 11, 61.
────── 1938, *Rev. Phys. Chem. Japan*, 12, 135.
────── 1939, *Rev. Phys. Chem. Japan*, 13, 96.
Knewstubb, P. F., & Sugden, T. M., 1958, *Nature*, 181, 474 & 1261.
Knipe, R. H., & Gordon, A. S., 1955, *J.C.P.*, 23, 2097.
Knox, J. H., & Norrish, R. G. W., 1954, *P.R.S.*, 221, 151.
Koirtyohann, S. R., & Pickett, E. E., 1965, *Anal. Chem.*, 37, 601.
Kondratenko, M. B., & Sokolov, V. A., 1966, *Optics & Spectroscopy*, 20, 94.
Kondratiev, V., 1930, *Z. Phys.*, 63, 322.
────── 1933, *J. Exp. Theoret. Phys. (Russ.)*, 3, 265.
────── 1936, *Acta Physicochim. U.S.S.R.*, 4, 556.
────── & Ziskin, M., 1937a, *Acta Physicochim. U.S.S.R.*, 7, 65.
────── & ────── 1937b, *Acta Physicochim. U.S.S.R.*, 6, 307.
Kondratieva, H., & Kondratiev, V., 1938, *Acta Physicochim. U.S.S.R.*, 8, 481.
Kovacs, I., 1969, *Rotational Structure in the Spectra of Diatomic Molecules*, A. Hilger, London.
Krakow, B., 1966, *Appl. Opt.*, 5, 201.
Krishnamachari, S. L. N. G., & Broida, H. P., 1961, *J.C.P.*, 34, 1709.
Krupenie, P. H., 1966, *The Band Spectrum of CO*, Nat. Bur. Stand., NSRDS-NBS 5.
Kydd, P. H., & Foss, W. I., 1967, *11th Comb. Symp.*, p. 1179.

Lagerqvist, A., Westerlund, H., Wright, C. V., & Barrow, R. F., 1958, *Ark. Fys.*, 14, 387.
Lam Thanh, M., & Peyron, M., 1963, *J. Chim. Phys.*, 60, 1289.
────── & ────── 1964, *J. Chim. Phys.*, 61, 1531.
Landsverk, O. G., 1939, *Phys. Rev.*, 56, 769.
Lapp, M., Goldman, L. M., & Penney, C. M., 1972, *Science*, 175, 1112.
Larkins, P. L., 1971, *Spectrochim. Acta*, 26B, 477.
Laud, B. B., & Gaydon, A. G., 1971, *C. & F.*, 16, 55.
Lauer, F. J., 1933, *Z. Phys.*, 82, 179.
Lawton, J., & Weinberg, F. J., 1969, *Electrical Aspects of Combustion*, Clarendon Press, Oxford.
Leah, A. S., Godrich, J., & Jack, H. R. S., 1950, *Nature*, 166, 868.
────── Rownthwaite, C., & Bradley, D., 1950, *Phil. Mag.*, 41, 468.
Learner, R. C. M., 1962, *P.R.S.*, 269, 311.
Leckner, B., 1971, *C. & F.*, 17, 37.

Lents, J. M., 1973, *J.Q.S.R.T.*, **13**, 297.

Lewis, B., & von Elbe, G., 1961, *Combustion, Flames and Explosions of Gases*, Academic Press, New York.

Lewis, M. N., & White, J. V., 1939, *Phys. Rev.*, **55**, 894.

Linevsky, M. J., 1967, *J.C.P.*, **47**, 3485.

Lipscomb, F. J., Norrish, R. G. W., & Thrush, B. A., 1956, *P.R.S.*, **233**, 455.

Liveing, G. D., & Dewar, J., 1880, *P.R.S.*, **30**, 580.

—— & —— 1891, *P.R.S.*, **49**, 217.

Lofthus, A., 1960, *The Molecular Spectrum of Nitrogen*, Univ. Oslo Spectroscopic Rep. 2.

Lowe, R. M., 1971, *Spectrochim. Acta*, **26B**, 201.

Ludlam, E. B., Reid, H. G., & Soutar, G. S., 1929, *Proc. Roy. Soc. Edinburgh*, **49**, 156.

Ludwig, C. B., Ferriso, C. C., Malkmus, W., & Boynton, F. P., 1965, *J.Q.S.R.T.*, **5**, 697.

Lyn, W. T., 1957, *J. Inst. Petroleum*, **43**, 25.

—— 1963, *9th Comb. Symp.*, **p. 1069.**

Lyon, R. K., & Kydd, P. H., 1961, *J.C.P.*, 34, 1069.

McGrath, W. D., & Norrish, R. G. W., 1957, *P.R.S.*, **242**, 265.

—— & —— 1960, *P.R.S.*, **254**, 317.

McKellar, A., 1940, *Publ. Astronom. Soc. Pacific*, **52**, 187 & 312.

McKellar, J. F., & Norrish, R. G. W., 1960, *P.R.S.*, **254**, 147.

McKinley, J. D., Garvin, D., & Boudart, M. J., 1955, *J.C.P.*, **23**, 784.

Maecker, H., & Peters, T., 1954, *Z. Phys.*, **139**, 448.

Mann, D. E., Broida, H. P., & Squires, B. E., 1954, *J.C.P.*, **22**, 348.

Marenin, I. R., & Johnson, H. R., 1970, *J.Q.S.R.T.*, **10**, 305.

Marmo, F. F., Padur, J. P., & Warneck, P., 1967, *J.C.P.*, **47**, 1438.

Marr, G. V., 1957, *Canad. J. Phys.*, **35**, 1265.

—— 1958, *Publ. Astronom. Soc. Pacific*, **70**, 197.

—— & Nicholls, R. W., 1955, *Canad. J. Phys.*, **33**, 394.

Marshall, G. B., & West, T. S., 1970, *Anal. Chim. Acta*, **51**, 179.

Mavrodineanu, R., & Boiteux, H., 1965, *Flame Spectroscopy*, Wiley, New York.

Meinel, H., & Krauss, L., 1966, *Z. Naturforsch.*, **21A**, 1878.

Melvin, A., 1969, *C. & F.*, **13**, 438.

Messerle, G., & Krauss, L., 1967, *Z. Naturforsch.*, **22A**, 2023.

Midgley, T., & Boyd, T. A., 1922, *Industr. Engng Chem.*, **14**, 894.

Miller, W. J., & Palmer, H. B., 1963, *9th Comb. Symp.*, p. 90.

Milligan, D. E., & Jacox, M. E., 1964, *J.C.P.*, **41**, 3032.

Minchin, L. T., Densham, A. B., & Wright, J., 1940, *J. Trans. Illum. Engng Soc.*, **5**, 75.

Minkoff, G. J., Everett, A. J., & Broida, H. P., 1955, *5th Comb. Symp.*, p. 779.

—— & Tipper, C. F. H., 1962, *Chemistry of Combustion Reactions*, Butterworths, London.

Mitchell, A. C. G., & Zemansky, M. W., 1961, *Resonance Radiation and Excited Atoms*, Cambridge Univ. Press.

Miyama, H., 1962, *C. & F.,* 6, 319.
——— & Kydd, P. H., 1961, *J.C.P.,* 34, 2038.
——— & Takeyama, T., 1964, *J.C.P.,* 41, 2287.
Monfils, A., & Rosen, B., 1949, *Nature,* 164, 713.
Moore, C. E., & Broida, H. P., 1959, *J. Res. Nat. Bur. Stand.,* 63A, 19.
Moore-Sitterley, C. E., 1949, *Atomic Energy Levels,* Nat. Bur. Stand. Circular No. 467.
Morgan, J. E., Elias, L., & Schiff, H. I., 1961, *J.C.P.,* 33, 930.
Mulliken, R. S., 1958, *Canad. J. Chem.,* 36, 10.
Mullins, B. P., 1949, *3rd Comb. Symp.,* p. 704.
Murphy, G. M., & Schoen, L., 1951, *J.C.P.,* 19, 1214.

Ndaalio, G., & Deckers, J. M., 1967, *Canad. J. Chem.,* 45, 2441.
Nederbragt, G. W., Van der Horst, A., & Van Duijn, J., 1965, *Nature,* 206, 87.
Needham, D. P., & Powling, J., 1955, *P.R.S.,* 232, 337.
Nenquin, G., Thomas, P., & Van Tiggelen, A., 1956, *Bull. Soc. Chim. Belg.,* 65, 1072.
Newitt, D. M., & Baxt, L. M., 1939, *J. Chem. Soc.,* p. 1711.
——— & Outridge, L. E., 1938, *J.C.P.,* 6, 752.
Newman, R., 1952, *J.C.P.,* 20, 749.
Nicholls, R. W., 1964, *Ann. Geophys.,* 20, 144.
——— & Jarmain, W. R., 1956, *Proc. Phys. Soc. A.,* 69, 253 and 741.
Norrish, R. G. W., 1953, *Disc. Faraday Soc.,* 14, 16.
——— & Porter, G., 1952, *P.R.S.,* 210, 439.
——— ——— & Thrush, B. A., 1953, *P.R.S.,* 216, 165.
——— ——— & ——— 1955, *P.R.S.,* 227, 423.
——— & Wayne, R. P., 1965, *P.R.S.,* 288, 200.

Oldman, R. J., Norris, W. B., & Broida, H. P., 1970, *C. & F.,* 14, 61.

Padley, P. J., 1960, *T.F.S.,* 56, 449.
——— & Sugden, T. M., 1958, *P.R.S.,* 248, 248.
——— & ——— 1959a, *T.F.S.,* 55, 2054.
——— & ——— 1959b, *7th Comb. Symp.,* p. 235.
Palmer, H. B., & Seery, D. J., 1960, *C. & F.,* 4, 213.
Pandya, T. P., & Weinberg, F. J., 1963, *9th Comb. Symp.,* p. 587.
——— & ——— 1964, *P.R.S.,* 279, 544.
Pannetier, G., 1951, *C. R. Acad. Sci. Paris,* 232, 817.
——— & Gaydon, A. G., 1947, *C. R. Acad. Sci. Paris,* 225, 1300.
——— & ——— 1948, *Nature,* 161, 242.
——— & ——— 1951, *J. Chim. Phys.,* 48, 221.
Parker, W. G., & Wolfhard, H. G., 1956, *Fuel,* 35, 323.
Parkinson, W. H., & Reeves, E. M., 1961, *P.R.S.,* 262, 409.
Pearse, R. W. B., & Gaydon, A. G., 1965, *The Identification of Molecular Spectra,* Chapman & Hall, London, 3rd ed. (with supplement).
Peeters, J., Lambert, J. F., Hertoghe, P., & Van Tiggelen, A., 1971, *13th Comb. Symp.,* p. 321.

Penner, S. S., 1952, *J.C.P.*, **20**, 1175, 1241, & 1334.
—— 1959, *Quantitative Molecular Spectroscopy and Gas Emissivities*, Addison, Wesley. Reading, Mass.
—— & Kavanagh, R. W., 1953, *J. Opt. Soc. Amer.*, **43**, 385.
Perche, A., Pérez, A., & Lucquin, M., 1970, *C. & F.*, **15**, 89.
—— —— & —— 1971, *C. & F.*, **17**, 179.
Pesic, D. S., & Gaydon, A. G., 1959, *Proc. Phys. Soc.*, **73**, 244.
Petrella, R. V., & Sellers, G. D., 1971, *C. & F.*, **16**, 83.
Phillips, J. G., 1948, *Astrophys. J.*, **107**, 389.
—— 1949, *Astrophys. J.*, **110**, 73.
—— & Davis, S. P., 1968, *Berkeley Analyses of Molec. Spectra*, 2, Univ. California Press, Berkeley.
Pilling, M. J., Bass, A. M., & Braun, W., 1971, *J.Q.S.R.T.*, **11**, 1593.
Pillow, M. E., 1955, *Proc. Phys. Soc.*, **A.**, **68**, 547.
Plyler, E. K., 1948, *J. Res. Nat. Bur. Stand.*, **40**, 113.
—— Allen, H. C., & Tidwell, E. D., 1958, *J. Res. Nat. Bur. Stand.*, **61**, 53.
—— & Ball, J. J., 1952, *J.C.P.*, **20**, 1178.
—— Benedict, W. S., & Silverman, S., 1952, *J.C.P.*, **20**, 175.
—— Blaine, L. R., & Tidwell, E. D., 1955, *J. Res. Nat. Bur. Stand.*, **55**, 183.
—— & Kostkowski, H. J., 1952, *J. Opt. Soc. Amer.*, **42**, 360.
Porter, G., 1950, *P.R.S.*, **200**, 284.
—— 1951, *J.C.P.*, **19**, 1278.
—— & Topp, M. R., 1968, *Nature*, **220**, 1228.
Porter, R. F., & Dows, D. A., 1956, *J.C.P.*, **24**, 1270.
Porter, T. L., Mann, D. E., & Acquista, N., 1965, *J. Molec. Spectr.*, **16**, 228.
Pretty, W. E., 1928, *Proc. Phys. Soc.*, **40**, 71.

Ramsay, D. A., 1952, *J.C.P.*, **20**, 1920.
Rank, D. H., Saksena, E. D., & Wiggins, T. A., 1958, *J. Opt. Soc. Amer.*, **48**, 521.
Rao, K. S., 1958, *Canad. J. Phys.*, **36**, 1526.
Rassweiler, G. M., & Withrow, L., 1932, *Industr. Engng Chem.*, **24**, 528.
Reeves, R. R., Harteck, P., & Chace, W. H., 1964, *J.C.P.*, **41**, 764.
Regener; V. H., 1964, *J. Geophys. Res.*, **69**, 3795.
Rekers, R. G., & Villars, D. S., 1954, *Rev. Sci. Instrum.*, **25**, 424.
—— & —— 1956, *J. Opt. Soc. Amer.*, **46**, 534.
Rice, O. K., 1933, *J.C.P.*, **1**, 625.
Robinson, J. W., 1961, *Anal. Chem.*, **33**, 1067 & 1226.
Rosen, B., 1970, Ed., *International Tables of Selected Constants; 17. Spectroscopic Data Relative to Diatomic Molecules*, Pergamon Press, Oxford.
Rosen, N., 1933, *J.C.P.*, **1**, 319.
Rosser, W. A., Wise, H., & Miller, J., 1959, *7th Comb. Symp.*, p. 175.
Rössler, F., 1954, *Z. angew. Phys.*, **6**, 175.
Roth, W., 1958, *J.C.P.*, **28**, 668.

Salooja, K. C., 1967, *C. & F.*, **11**, 247 & 511.

Savadatti, M. I., & Broida, H. P., 1966, *J.C.P.*, **45**, 2390.

Sawyer, R. A., 1945, *Experimental Spectroscopy*, Chapman & Hall, London.

Schott, G. L., 1960, *J.C.P.*, **32**, 710.

———— & Kinsey, J. L., 1958, *J.C.P.*, **29**, 1177.

Schou, S. A., 1929a, *J. Chim. Phys.*, **26**, 27.

———— 1929b, *J. Chim. Phys.*, **26**, 39.

Schwar, M. J. R., & Weinberg, F. J., 1969a, *C. & F.*, **13**, 335.

———— & ———— 1969b, *P.R.S.*, **311**, 469.

Semenoff, N., 1938, *Acta Physicochim. U.S.S.R.*, **9**, 453.

Shahed, S. M., & Newhall, H. K., 1971, *C. & F.*, **17**, 131.

Sharma, A., & Joshi, J. C., 1972, *J.Q.S.R.T.*, **12**, 1073.

———— Padur, J. P., & Warneck, P., 1965, *J.C.P.*, **43**, 2155.

Shelton, J. P., & Walsh, A., 1958, *XV Cong. Intern. Quim. Pura e Applicada*, Lisbon.

Shuler, K. E., 1950, *J.C.P.*, **18**, 1466.

———— 1951, *J.C.P.*, **19**, 888.

———— 1955, *5th Comb. Symp.*, p. 56.

———— & Broida, H. P., 1952, *J.C.P.*, **20**, 1383.

Silla, H., & Dougherty, T. J., 1972, *C. & F.*, **18**, 65.

Silverman, S., 1949, *3rd Comb. Symp.*, p. 498.

———— 1954, *Energy Transfer in Hot Gases*, **Nat. Bur. Stand. Circular, 523**, 51.

Singh, N. L., 1959, *Canad. J. Phys.*, **37**, 136.

Skirrow, G., & Wolfhard, H. G., 1955, *P.R.S.*, **232**, 78 & 577.

Smith, D. S., & Starkman, E. S., 1971, *13th Comb. Symp.*, p. 439.

Smith, E. C. W., 1940a, *Inst. Gas Engrs Trans.*, **237**.

———— 1940b, *P.R.S.*, **174**, 110.

Smith, W. H., 1969, *Astrophys. J.*, **156**, 791.

———— & Liszt, H. S., 1971, *J.Q.S.R.T.*, **11**, 45.

Smith, W. L., 1966, *Proc. Phys. Soc.*, **89**, 1021.

Spealman, M. L., & Rodebush, W. H., 1935, *J. Amer. Chem. Soc.*, **57**, 1474.

Spokes, G. N., 1959, *7th Comb. Symp.*, **p. 229**.

———— & Gaydon, A. G., 1959, *Proc. Phys. Soc.*, **74**, 639.

Steele, S., 1931, *Nature*, **128**, 185.

———— 1935, *Nature*, **135**, 268.

Stevens, R. F., O'Keefe, A. E., & Ortman, G. C., 1969, *Environ. Sci. Technol.*, **3**, 652.

Stollery, J. L., Gaydon, A. G., & Owen, P. R., 1971, *Shock Tube Research, being proceedings of the 8th International Shock Tube Symposium*, Chapman & Hall, London.

Storch, H. H., 1934, *J. Amer. Chem. Soc.*, **56**, 374.

Street, J. C., & Thomas, A., 1955, *Fuel*, **34**, 4.

Strickler, S. J., & Pitzer, K. S., 1964, "Energy calculations for polyatomic carbon molecules", in *Molecular Orbitals in Chemistry, Physics & Biology*, ed. P. Lowden & B. Pullman, Academic Press.

Strutt, R. J. (Lord Rayleigh), 1911, *Proc. Phys. Soc.*, **23**, 147.

Sullivan, J. V., & Walsh, A., 1966, *Spectrochim. Acta*, **22**, 1843.

Sulzmann, K. G. P., Myers, B. F., & Bartle, E. R., 1965, *J.C.P.*, **42**, 3969.

Symonds, J. L., 1966, *Symposium on Radio and Optical Studies of the Galaxy*, p. 89, Mount Stromlo Obs.

Tagirov, R. B., 1959, *Optics & Spectroscopy*, 6, 90.
Tewarson, A., Naegeli, D. W., & Palmer, H. B., 1969, *12th Comb. Symp.*, p. 415.
Thabet, S. K., 1951, *Ph.D. Thesis*, London.
Thee, W. C., 1929, *J. Soc. Auto. Engng*, 25, 388.
Thomas, J. R., & Crandall, H. W., 1951, *Industr. Engng Chem.*, 43, 2761.
Thomas, N., 1951, *T.F.S.*, 47, 958.
———— Gaydon, A. G., & Brewer, L., 1952, *J.C.P.*, 20, 369.
Thompson, D., Brown, T. D., & Beér, J. M., 1972, *C. & F.*, 19, 69.
Thring, M. W., 1962, *The Science of Flames & Furnaces*, 2nd ed., Chapman & Hall, London.
Thrush, B. A., 1958, *P.R.S.*, 243, 555.
Toishi, K., & Muira, S., 1955, *Nature*, 175, , 81.
Topps, J. E. C., & Townend, D. T. A., 1946, *T.F.S.*, 42, 345.
Tourin, R. H., & Krakow, B., 1965, *Appl. Opt.*, 4, 237.
Travers, B. E. L., & Williams, H., 1965, *10th Comb. Symp.*, p. 657.
Tyte, D. C., Innanen, S. H., & Nicholls, R. W., 1967, *York University, Identification Atlas of Molecular Spectra*, 5.
———— & Nicholls, R. W., 1964, *Univ. Western Ontario, Identification Atlas of Molecular Spectra*, 1.

Ubbelohde, A. R., 1933, *J. Chem. Soc.*, p. 972.
———— 1935a, *P.R.S.*, 152, 354.
———— 1935b, *P.R.S.*, 152, 378.

Vaidya, W. M., 1934, *P.R.S.*, 147, 513.
———— 1935, *Proc. Ind. Acad. Sci.*, 2A, 352.
———— 1937, *Proc. Ind. Acad. Sci.*, 6A, 122.
———— 1938, *Proc. Ind. Acad. Sci.*, 7A, 321.
———— 1941, *P.R.S.*, 178, 356.
———— 1951, *Proc. Phys. Soc.*, A, 64, 428.
———— 1964, *P.R.S.*, 279, 572.
van der Hurk, A., Hollander, T., & Alkemade, C. T. J., 1973, *J.Q.S.R.T.*, 13, 273.
Vanpee, M., 1956, *C. R. Acad. Sci. Paris*, 243, 804.
Vear, C. J., Hendra, P. J., & Macfarlane, J. J., 1972, *J.C.S. Chem. Comm.*, p. 381.
Verma, R. D., & Broida, H. P., 1970, *Canad. J. Phys.*, 48, 2991.
———— & Mulliken, R. S., 1961, *J. Molec. Spectrosc.*, 6, 419.
Volman, D. H., 1951, *J.C.P.*, 19, 668.
Von Engel, A., & Cozens, J. R., 1964, *Nature*, 202, 480.

Wagner, H. G., 1957, *6th Comb. Symp.*, p. 366.
———— 1961, *Fundamental Data obtained from Shock-tube Experiments*, p. 320, Pergamon Press.
Walsh, A., 1966, *J. New Zealand Inst. Chem.*, 30, 7.

Walsh, A. D., 1947, *T.F.S.*, **43**, 297.

—— 1953, *J. Chem. Soc.*, p. 2260.

—— 1963, *9th Comb. Symp.*, p. 1046.

Warren, G. J., & Babock, G., 1970, *Rev. Sci. Instrum.*, **41**, 280.

Warren, R. L., 1952, *J. Sci. Instrum.*, **29**, 284.

—— 1959, *Colloq. Spectroscopium Intern. VIII*, p. 213.

Watson, W. W., 1924, *Astrophys. J.*, **60**, 145.

Weinberg, F. J., 1963, *Optics of Flames*, Butterworths, London.

Weston, F. R., 1925, *P.R.S.*, **109**, 177 & 523.

Wheaton, J. E. G., 1964, *Appl. Opt.*, **3**, 1247.

White, J. U., 1942, *J. Opt. Soc. Amer.*, **32**, 285.

Whittingham, G., 1950, *Fuel*, **29**, 244.

Widhopf, G. F., & Lederman, S., 1971, *A.I.A.A. Jour.*, **9**, 309.

Willis, J. B., 1967, *Spectrochim. Acta*, **23A**, 811.

—— 1968, *Appl. Opt.*, **7**, 1295.

—— 1970, *Spectrochim. Acta*, **25B**, 487.

—— Fassel, V. A., & Fiorino, J. A., 1969, *Spectrochim. Acta*, **24B**, 157.

Withrow, L., & Rassweiler, G. M., 1931, *Industr. Engng Chem.*, **23**, 769.

—— & —— 1933, *Industr. Engng Chem.*, **25**, 923.

—— & —— 1934, *Industr. Engng Chem.*, **26**, 1256.

Wolfhard, H. G., 1939, *Z. Phys.*, **112**, 107.

—— & Gaydon, A. G., 1949, *Nature*, **164**, 22.

—— & Parker, W. G., 1949, *Proc. Phys. Soc.*, **A**, **62**, 722.

—— & —— 1950, *Fuel*, **29**, 235.

—— & —— 1952, *Proc. Phys. Soc.*, **A**, **65**, 2.

—— & —— 1953, *4th Comb. Symp.*, p. 420.

—— & —— 1955, *5th Comb. Symp.*, p. 718.

Wray, K. L., & Teare, J. D., 1962, *J.C.P.*, **36**, 2582.

Zeegers, P. J. T., & Winefordner, J. D., 1971, *Spectrochim. Acta*, **26B**, 161.

Zimpel, C. F., & Graiff, L. B., 1967, *11th Comb. Symp.*, p. 1015.

Author Index

Subject Index

(Including Symbols Used and
Values of Physical Constants)

Note. Radicals are indexed under their chemical symbols (e.g. OH and CH_3O rather than hydroxyl and methoxy), while stable chemical species are indexed under their usual names (e.g. oxygen, hydrogen sulphide, benzene, rather than O_2, H_2S or C_6H_6). Greek letters are given after Z. Physical constants are listed under the usual symbols, e.g. velocity of light under *c*. Page numbers of major references are printed in **bold type**. Sp. = spectrum; fl. = flame; Pl. = Plate.